BASIC PHYSICS

BASIC

PHYSICS

A LINEAR APPROACH

Joseph W. Straley
Department of Physics
The University of North Carolina
at Chapel Hill

PRENTICE-HALL, INC.
ENGLEWOOD CLIFFS,
NEW JERSEY

Library of Congress Cataloging in Publication Data

STRALEY, JOSEPH W.
  Basic physics.

  Bibliography: p. 332
  1. Physics. I. Title.
QC21.2.S73   530   73–3401
ISBN 0–13–066001–9

BASIC PHYSICS : A Linear Approach
Joseph W. Straley

PRINTED IN THE UNITED STATES OF AMERICA

10 9 8 7 6 5 4 3 2 1

PRENTICE-HALL INTERNATIONAL, INC., *London*
PRENTICE-HALL OF AUSTRALIA, PTY. LTD., *Sydney*
PRENTICE-HALL OF CANADA, LTD., *Toronto*
PRENTICE-HALL OF INDIA PRIVATE LIMITED, *New Delhi*
PRENTICE-HALL OF JAPAN, INC., *Tokyo*

Contents

THE CLASSICAL APPROACH TO THE UNSEEABLE

II

121

7

THE STRUCTURE OF MATTER   123

8

ELECTRICITY   143

9

MAGNETISM   168

13
THE STRUCTURE OF ATOMS   247

14
RELATIVITY   266

15
RADIOACTIVITY AND NUCLEAR TRANSMUTATION   288

16

NUCLEAR ENERGY   *308*

REFERENCES   *332*

APPENDICES   *334*

SOLUTIONS TO PROBLEMS   *342*

INDEX   *345*

Preface

During my years as a teacher I have become increasingly aware of the importance of presenting an articulate statement of the basic concepts of physics to that majority of our students whose major interest is in a nonscience field. *Basic Physics: A Linear Approach* is a result of the many efforts I have made to achieve that purpose.

I do not believe that the true face of physics can be presented on a completely nonmathematical plane, yet I am quite aware of the communication gap that appears when we express physical laws in mathematical language. In this book a sufficient number of the physical principles are expressed in algebraic form to preserve the "exact" nature of physics, but the entire logical development is given in verbal forms as well. To say something both in words and in mathematical language may hazard redundancy on occasion, but I believe this risk is worth taking. The alternative may be that the message will be eclipsed by mathematical manipulations; many algebraic expressions do poorly at speaking for themselves.

My agenda for a one-semester course in physics for nonscience majors consists of this text with the "A" sets of problems and a weekly laboratory session. However, if one wishes to operate at a more analytical level in some chapters, coverage of others may be abbreviated. Certain chapters, specifically Chapters 6, 7, 13, and 14, have been prepared with the thought that many users of this textbook will elect that route. I believe that it will be found that the overall continuity of the material will be preserved if attention in these four chapters is limited to reading the summary statements and studying the figures and their captions. The additional available time should permit one to deal effectively in the remaining chapters with a significant number of problems from the "B" lists; alternately, it may provide occasion for the exploration of related topics and for the presentation of term papers. Finally, the text with both the "A" and the "B" sets of problems supplemented by the *Study Guide* should provide enough material for a two-semester introductory course for students of the liberal arts.

In particular, I think, the *Study Guide* may be useful in conjunction with courses presented on a "contractual," a "systems approach," or a "self-paced" format.

A linear approach presumes that the author has regarded it of prime importance to avoid the all too common "···it can be shown ···" type of logic gap. The primary challenge in writing this book therefore has been that of ordering the material in such a way that each concept is recognizably supported by topics previously presented. Beyond that, each idea whose claim to authority rests upon specific experiments has been introduced only after the experiments themselves have been adequately described. In finding a nearly straight course through the physics story it has been necessary to either bypass or give scant attention to some rather conventional topics.

The text presents a quasi-historical account of man's developing comprehension of his physical environment. It is divided into three approximately equal parts which roughly correlate with three chronological periods in the development of physics on the one hand and with three rather well-defined modes of approach to physics on the other.

1. Part I (Chaps. 1–6) discusses the "physics of large objects." Here I deal with Newtonian physics and the Law of Conservation of Energy, extending it to include thermal energy.

2. Part II (Chaps. 7–11) describes the model-building activities of the nineteenth century. It presents the essence of classical chemistry, describes light as an electromagnetic wave, and gives an overview of electricity and magnetism as understood by Faraday, Oersted, and Maxwell. It concludes with discovery of the electron by J. J. Thomson and determination of the electronic charge by Robert Millikan.

3. Part III (Chaps. 12–16) presents some modern perspectives of physics as they emerge through relativity, quantum theory, and the discovery of nuclear energy.

A more complete overview of these three parts of the book will be found in the prefatory remarks at the beginnings of the parts themselves.

Many people have contributed in one way or another to this textbook, either by reading parts of the manuscript, by discussing certain ideas with me, or simply by being good listeners as I tested my ideas on them. My students in various sections of Physics 20 and the participants in the Project for the Revitalization of Physics Programs in Twenty Colleges surely deserve a lot of thanks for helping me in quite different ways to be realistic about what we can expect to achieve in a first-year course in physics. I am especially grateful for the advice and counsel of Wayne A. Bowers, W. E. Haisley, Eugen Merzbacher, Shirley E. Marshall, Guenter Schwarz, David L. Straley, Homer Wilkins, and Charles D. Wright.

Chapel Hill, North Carolina   JOSEPH W. STRALEY

To the Student

For better or worse, we live in a scientific age. Because of advances in the various branches of science, we are able to go to the Moon, to feed everyone in the United States through the efforts of only a few, to revolutionize life expectancy everywhere, and to nurture to hundreds of newly created appetites. At the same time, few of us expect to go to the Moon; not everyone gets fed; extended life expectancy has been translated into a population explosion; and the demand for gratification of many new "needs" has resulted in 5 percent of the Earth's population using a third of the world's resources.

Great problems and great opportunities are ahead for all of us. It is my hope that this book will lay the groundwork for defining some of the problems and for seeing the opportunities.

Prior to writing this book I had the privilege of presenting a course in introductory college physics to a total of several hundred nonscience majors. A substantial number of these students entered the course with misgivings as to their probable success in it and questioned it's relevance to their field of study. I hope that you, like a comforting fraction of the students who have taken a course like this at my university, will emerge with a more favorable image of yourself and with a better understanding of science as the culture in which we live. I hope that you will gain from this book some feeling for the extent to which science already has shaped your life and influenced scholarship in your field.

You will notice that I have divided each problem section into two parts. The "A" problems were prepared to enable you to test your articulation of the essentially qualitative aspects of the subject matter; the "B" problems should serve to examine your understanding at a more quantitative or mathematical level. If your instructor is interested in your progress in the more quantitative type of comprehension, I hope that you will examine the *Study Guide* that has been prepared to assist with self-study of topics students sometimes find difficult.

J.W.S

Physics:
A Science Based on
Measurement

The study of the physical and material makeup of our surroundings has not always been subdivided, as it is today, into branches called physics, chemistry, and astronomy. Even as recently as a century ago these now nearly separate disciplines were united under the name of "natural philosophy."

Some may lament the existence of these subdivisions, saying that they are not inherent in nature and as artificial structures may even constitute barriers to a genuine understanding of nature. However much we may lament this fragmentation of knowledge, most of us have had to admit its necessity simply because of our own limitations; it is no longer possible for one person to attain expertise in all branches of knowledge. Today, in fact, further subdivisions are taking place as nuclear physicists, for example, find less and less in common with atomic spectroscopists. Expanding knowledge generates specialists, and specialists talk mainly to other specialists in their own area of specialization.

When we think of physics, many of us will think of levers, pulley systems, motors, and mirrors. We recognize that the domain of physics is in the realm of physical objects and have tended to associate physics with devices capable of doing useful things.

Physics, however, is a research science concerned first and foremost with understanding and describing nature. As a consequence considerations of the utility of discovery play very little part in the thought processes of one who is engaged in research. In probing the structure of atoms and nuclei, the physicist does not know what he will find; much less can he evaluate in advance the use to which it might be put.

Our approach to physics in this textbook will be "quasi-historical" because a strictly historical approach has many disadvantages. Were we to present physics from a strictly historical point of view, much of our discussion would be expended on the description of efforts that did not produce lasting contributions, for physicists have spent much of their time going down trails that ultimately turned out to be blind alleys. Some of the means by which physicists arrived at our present understanding of certain phenomena, furthermore, were very circuitous when viewed with the advantages of hindsight. For that reason we will permit ourselves the luxury of taking shortcuts that were overlooked by those who made the first trip. Although the author does not wish to ignore history, for the purposes of brevity and clarity he will avoid the pitfalls that accompany a historical approach to this subject.

An examination of what we now know about the aspects of nature under study by physicists reveals that our knowledge can be subdivided into three major categories. These categories constitute the three "Parts" into which this book has been divided; they also coincide rather closely with three major periods in the development of the subject. In this way our discussion will provide an overall view of the history of physics which the reader can supplement by examining the list of references on page 332.

At this point we will not attempt to describe either the content of the three major parts of this book or the differences in the approach followed by research physicists in the three historical periods to which we have referred. The important fact is that physicists throughout history have had much in common. Physicists of all periods, being curious about their surroundings, have made accurate quantitative observations of what they have seen. They have summarized their observations and transmitted their descriptions to us through their writings. They have not been content with broad, verbal generalizations about nature but have attempted to state their generalizations in the form of succinct principles as free from ambiguity as possible and susceptible to confirmation by others.

This list of the attributes of physicists surely reads like a list of the virtues prepared by the virtuous and might serve as a starting point in an extended discourse on physics as an "exact" science. As a physicist, the author may not wish to renounce the virtues but he does feel impelled to qualify the "exactness" of physics. Physics is based entirely on measurements and most measurements simply are not exact. You can say unequivocally that a certain table has exactly four legs, but you cannot say that a clock reads *exactly* "17 minutes after 10 o'clock." Even if the internal mechanism were perfect, which it surely is not, the clock cannot be read with unlimited accuracy because of many practical limitations associated with the sharpness of the tips of the hands, the proximity of the hands to the face, and so forth.

Physical law operates in the shadow of this sort of uncertainty. Laws of physics can be no more exact than the measurements on which they are based; it might therefore be more accurate to describe physics as an "as-exact-as-possible" science. As we become more familiar with the methods and techniques by which we arrive at physical law, these limitations on the exactness of physics will be seen as part of the game. Measurements describe nature; better measurements give a more accurate description of nature.

The best claim we can make to exactness in physics is in our definition of terms. It is true that we cannot measure exactly the energy of an object, but we can at least have an exact notion of what it is we are trying to measure. It is this feature of physics more than any other that qualifies it as an exact science.

In the chapters that follow we will introduce a number of terms peculiar to physics. Some of these terms, such as "velocity," "acceleration," "work," and "energy," will seem familiar. The student will feel that he knows their meaning in advance, and he may very well be correct in this supposition. A term used in physics, however, may have a more specialized meaning than the same term would possess when employed in common practice. Hence, to study physics intelligently, the student may need to reexamine his understanding of words already in his vocabulary.

In studying physics one must accommodate to a style of definition that differs from the one usually found in a dictionary. Many definitions in physics are expressed algebraically and, in fact, constitute operational or procedural statements describing operations to be carried out in the laboratory. We will elaborate on this point as the need arises.

The terminology of physics, however, constitutes the language of physics. To be certain he "knows what he is talking about," the student must know exactly how his words are defined.

The simplest quantities that one has occasion to measure in physics are length, mass, and time. These are called *basic quantities* because other quantities, known as the *derived quantities*, are defined in terms of them. Basic quantities are measured in arbitrarily chosen units, whereas the units used to measure the derived quantities are determined by the choice of units of the basic quantities. One must therefore be certain of his understanding of basic quantities before attempting to understand the derived quantities.

Length

The word *length* designates the distance between two objects or two points. In order to specify this distance, some unit must be arbitrarily chosen and accepted. Different peoples have used different units of length throughout history; even today many different units effectively serve the local or regional needs of those involved in trade and commerce. Scientists, however, have universally adopted the meter as their unit of length. This unit represents the distance between two marks on a platinum-iridium bar kept at the International Bureau of Weights and Measures in Paris.*

Mass

Mass is frequently defined as the quantity of matter in an object, a definition that, unfortunately, does not say very much, for it serves mainly to transfer our ignorance from the word "mass" to the words "quantity of matter." In making measurements of the mass of an object,

---

* The primary standard of length, as adopted by the General Conference of the International Bureau of Weights and Measures (1960), is defined to be 1,650,763.73 times the wavelength of the orange-red component of radiation emitted by krypton. This decision demotes the meter bar in Paris to the role of secondary standard.

we measure, therefore, the effect the object may have on other objects in certain well-defined circumstances. For example, when we push against an object, exerting what we call a "force" upon it, we say that the object manifests the property of "inertia" in that the force was required to change the velocity of the object. In Sec. 2-3 we will associate the amount of "inertial" mass possessed by an object with the magnitude of the force required to cause its velocity to change in a prescribed way.

It is difficult to talk about masses until we have a name for a standard mass. The standard object will be called the *unit* in which the mass of other objects will be expressed.

The choice of the standard mass, the unit mass, is arbitrary. It is obviously desirable for as many of us as possible to agree to use the same unit. As it happened, the British chose to call the mass of a certain brass object the "pound." This name could have been associated with a smaller or a larger object if they had desired. On the other hand, scientists have universally adopted a unit of mass called the *kilogram*. By definition, a kilogram is the mass of a certain cylinder of platinum in Paris. This primary standard serves, then, as the reference object whose inertial (and gravitational) properties are to be compared to those of any other object in nature to which we wish to ascribe the property of "mass."

Time

Time refers to the interval that has elapsed between two events. The unit of time is the second and, unlike the units of length and mass, has been adopted universally, by scientists and nonscientists alike. At an early stage the second was defined as 1/86,400 part of an average solar day, and although this definition is no longer used,* it may serve with high precision to indicate the duration of a second. Alternately, one may regard the second as the time required for a 0.2477-m pendulum to make one complete swing.

The MKS system of units

At least half a dozen different systems of units are in use in science and commerce at the present time. Although some writers believe that becoming familiar with all units in current use is a vital part of learning physics, we will limit ourselves in this text to one system, the MKS system of units. This system is the metric system based on the meter as the unit of length, the kilogram as the unit of mass, and the second as the unit of time. However, because magnitudes may be expressed or known in units other than MKS units, it may be necessary on occasion to convert from centimeters or feet to meters, from grams or pounds to kilograms, or from days or hours to seconds. A table of equivalents will be found in the Appendix 1.

---

* In 1967 scientists adopted a unit of time based on atomic vibrations. This unit, the *physical second*, equals the time required for 9,192,631,770 cycles of a particular vibration in the cesium-137 isotope.

In making use of the information in a table of equivalents, one should have at his command a straightforward technique for the conversion of units. The following method is recommended.

It is common knowledge that if any number is multiplied by unity, the magnitude of that quantity is not affected. In setting up a procedure for conversion of units, we will make use of this fact after adopting a slightly amended definition of unity. We will regard unity as any fraction in which the numerator equals the denominator. Thus pure number ratios like $\frac{6}{6}$ or $\frac{234}{234}$ are equal to unity. However, we can stay within our definition of "unity" with ratios like

$$\frac{1 \text{ foot}}{30.5 \text{ centimeters}}, \qquad \frac{0.305 \text{ meters}}{1 \text{ foot}}, \quad \text{or} \quad \frac{0.454 \text{ kilograms}}{1 \text{ pound}}$$

The numerator in each of these cases equals the denominator.

In converting from one unit to another, we merely multiply by the appropriate form that unity must take to accomplish the conversion, permitting the words that appear in these products to be handled very much like numbers. If, for example, the word for one unit of measure appears in both the numerator and the denominator, it can be canceled in both places. The following examples will serve to illustrate the use of this technique.

$$1 \text{ mile} = 1 \text{ mile} \left(\frac{5280 \text{ feet}}{1 \text{ mile}}\right)\left(\frac{0.305 \text{ meters}}{1 \text{ foot}}\right) = 1610 \text{ meters}$$

$$25 \text{ pounds} = 25 \text{ pounds} \left(\frac{0.454 \text{ kilograms}}{1 \text{ pound}}\right) = 11.4 \text{ kilograms}$$

Accuracy

The reader who has checked every step in the logic to this point may have observed what seemed to be an error in arithmetic. In the first operation above, 1 mile is quoted as equaling 1610 meters, whereas 5280×0.305 yields 1610.4. This result is further complicated by the fact that a mile is listed in tables as being equal to 1609.34 meters, in disagreement with both. In the second operation we say that 25 pounds equals 11.4 kilograms, whereas 25×0.454 yields 11.35.

Whenever we quote a measurement, we transmit three kinds of information: (1) the magnitude of the quantity, (2) its accuracy, and (3) the units in which the quantity is expressed.

When we make a statement of equality, such as "1 foot equals 0.305 meters," it is quite clear that we have stated a magnitude (0.305) and a unit (meters). In what sense have we stated, also, the accuracy of the number? The quality of accuracy is expressed in the fact that we wrote 0.305 instead of 0.3048, which we might have used had we consulted a set of tables. In our expression the 5 in 0.305 is the nearest whole number in the third place and, in general, has precisely that meaning. When we write 0.305, we mean that this number is preferable to 0.304 or 0.306; stated otherwise, we might say that our accuracy has been expressed to three *significant figures* or to 1 part in 300 (or 1 part in 305 if you prefer).

The number of significant figures is not determined by the position of the decimal point. This fact is apparent if we rewrite our statement of equivalence in centimeters instead of meters to secure

$$1 \text{ foot} = 30.5 \text{ centimeters}$$

The accuracy is still 1 part in 300 or three significant figures.

When we carry out a chain of multiplications and/or divisions, the answer can be no more accurate than the least accurate number that was used in the process. For example, the area of the curved surface of a cylindrical segment of height h, where the radius of the cylinder is r, is

$$\text{Area} = 2\pi rh$$

If you were told that measurements have yielded $r = 1.214$ meters and $h = 0.82$ meter, you might write

$$\text{Area} = 2 \times 3.1416 \times 1.214 \times 0.82 \text{ square meters}$$

but the answer you should quote would be

$$\text{Area} = 6.3 \text{ square meters}$$

Think how badly you would have misled someone regarding the accuracy had you quoted the answer as 6.254799936 square meters, the number you secure upon carrying out the indicated multiplication.

The Units Rule

In an earlier section we said that length, mass, and time are the basic quantities of physics and that all other physical quantities are derived from them and thus are called derived quantities. A few derived quantities having at least secondary importance will now be defined in order to see the distinction between these two kinds of quantities and the units in which they are to be expressed. Among the derived quantities that may be understood at this time are area, volume, density, and angle. For each we note first that there exists an equation which plays the role of a *defining equation*. This equation relates the quantity being defined to other quantities that have been defined previously.

If we apply a rule known as the Units Rule, the defining equation not only enables us to compute a numerical value of a derived quantity but also serves to tell us the units in which the new quantity must be expressed. This rule may be stated as follows:

An expression that relates physical quantities also relates their units.

The application of the rule may be seen in the examples below.

The area of a surface always involves the product of two units of length. For example, the area of a rectangle is equal to its base times its height, and the area of a circle is equal to π times its radius squared. In each case it can be seen that the rule requires that area be expressed in square meters, usually written m².

The volume of an object always involves the multiplication of a length by a length by a length. To illustrate, it will be recalled that the volume of a parallelepiped equals base times depth times height. The volume of a sphere equals $\frac{4}{3}\pi$ times its radius cubed. Our rule, then, requires that volume be expressed in cubic meters (m³).

The density of an object is its mass per unit volume. This verbal statement may be written in the form of an equation, and thus it constitutes the defining equation for density. This equation is

$$D = \frac{M}{V}$$

where M is the mass of an object in kilograms and V is its volume in cubic meters. Inserting the known units of mass and volume into this equation, we see that density will always be expressed in kilograms per cubic meter (kg/m³).

As a final example, let us consider the unit in which we will express the measurement of an angle. An angle θ is defined as the ratio of the arc subtended by the angle to the radius of the circle of which the arc is a part. In equation form, this definition is

$$\theta = \frac{s}{r}$$

where s is the arc of a circle subtended by the angle and r is the radius of the circle. Applying our rule, we see that the angle would be measured in m/m. Although it would be quite correct to express angular measure in these units (meters per meter), the fact that these units can be canceled (m/m) means that the unit of angular measure is "unitless." The word *radian* is supplied to identify the method by which the angle has been measured. From the definition, we see that an angle of one radian is that angle which subtends an arc equal to its radius.

Unitless quantities are always independent of the units used in measuring them; the same numerical value of the ratio s/r will be secured when both s and r are expressed in feet as when they are expressed in meters. Because the numerical expression of the magnitude of a unitless quantity requires no unit, such numbers are called *pure numbers*.

SUMMARY

Physics is a description of our physical environment couched in terminology that, in so far as possible, is unambiguous. The exactness of the meanings of the terms used in physics provides the main basis for declaring physics to be an exact science. Conclusions in physics, however, arise largely from measurements that are rarely exact; thus physics emerges as an as-exact-as-possible science.

The basic quantities of length, mass, and time will be expressed in meters, kilograms, and seconds, the units that characterize the MKS system of units.

A derived quantity is related to the basic quantities of physics through a defining equation. The derived units are related to the basic units of length, mass, and time through the Units Rule, which states

An expression that relates physical quantities also relates their units.

In this text the number of significant figures in the number that represents the magnitude of a quantity will serve as an adequate statement of the accuracy of that quantity.

1. (a) What is the length of a football field in meters?
   (b) A certain loaded truck has a total mass of 7400 lb. What is its mass in kilograms?

2. The density of water equals 1 gram/cm³. Express this in (a) kg/m³ and (b) lb/ft³.

3. Convert each of the following to the indicated unit.
   (a) 646 m = _____ ft
   (b) 275 in. = _____ m
   (c) 4910 kg = _____ ton
   (d) 11 km = _____ mi
   (e) 0.76 lb = _____ kg

4. Assume that a certain quantity (ρ) in physics is defined by

$$\rho = \frac{4\pi L}{M^2 T}$$

   where L, M, and T are measured in meters, kilograms, and seconds, respectively, and both 4 and π are pure numbers. What are the units in which ρ will be expressed?

5. A property known as the momentum of an object can be determined by dividing an appropriate distance by an elapsed time and multiplying the results by the mass of the object. What is the MKS unit of momentum?

6. The ratio of the circumference of a circle to its diameter (π) is listed in a handbook as 3.1415926536. (a) To how many significant figures has π been quoted? (b) Reexpress π to five significant figures. (c) Sometimes people use $\frac{22}{7}$ as their value of π. To how many significant figures is this accurate?

7. Express in exponential notation.*
   (a) 314.7 =
   (b) 0.000,012,9 =
   (c) 93,164,000,000 =
   (d) 21,960,000,000 =
   (e) 0.000,000,034,8 =
   (f) 54.90 =

   * A review of the use of exponentials in the handling of very small and very large quantities will be found in Appendix 2.

8. Express in decimal notation.

(a) $3.47 \times 10^{-4} =$ (c) $7.92 \times 10^{-6} =$

(b) $80.3 \times 10^3 =$ (d) $0.0041 \times 10^4 =$

9. Referring to Prob. 7, determine the product or quotient of the indicated numbers to the appropriate number of significant figures.

$$\frac{(a)(b)}{(f)} = \quad ; \quad \frac{(d)}{(c)(e)} = \quad .$$

10. Quote each of the following products or quotients to the correct number of significant figures.

(a) $\pi \times 16.3 \times 0.012 =$

(b) $\dfrac{14\bar{0}0 \times 26.9}{0.111} =$ (the bar signifies that the 0 is significant)

(c) $\dfrac{0.140 \times 26.9}{0.111} =$ (no bar needed)

11. In Appendix 5 reference is made to a proton and a neutron, two particles whose masses are stated to be 1.67252×10^{-27} kg and 1.67482×10^{-27} kg respectively. (a) Express the difference between the mass of a proton and the mass of a neutron with an accuracy of three significant figures. (b) Express the mass of each of these particles with an accuracy of three significant figures. (c) Considering your answers to (a) and (b), can you describe the kind of situation which would necessitate using the numerical values of some of the physical quantities to their highest known accuracy?

THE PHYSICS
OF
LARGE OBJECTS

Our account of man's comprehension of the physical world will begin with a discussion of the principles that govern and describe both the motion and the interaction of objects which are large enough to be observed directly. The range of phenomena encompassed is enormous, covering phenomena as diverse as waves on the surface of a pond and the path traced by an earth satellite.

Although most of the principles governing these phenomena were understood by the year 1800, it is remarkable how the genesis of so many of them are traced to one man, Sir Isaac Newton (1642–1727). However, even Newton acknowledged his debt to others, saying that he had "stood on the shoulders of giants." Other giants followed, with the result that the ideas presented in Part I represent a composite of the work of Newton, Galileo, Huygens, Hooke, and countless others.

An account of the growth of physics need not bother the reader with dates or extended recitations presenting the individual contributions of the giants. Instead the flow of ideas will be the history; this history will be extended to some of the contemporary manifestations of the physical principles that govern objects of observable size.

I

Describing Motion

A popular image of physics would have it that physicists are involved in some sort of magic, in which "secret formulas" are conjured up by highly intelligent and enigmatic individuals. Presumably, the secret formulas they concoct become and remain the property of the person who discovers them; the formulas become a part of general knowledge only by the generosity of the discoverer or through the treachery of some individual who leaks them to the "enemy."

The truth is that physics is only a description of our physical environment. In order to describe the physical environment, the physicist must be a careful observer; but nature appears to be so uniform and consistent throughout our universe that the description of any aspect of nature made by one such observer must agree with the description of the same aspect made by another. In this sense, there are no secrets.

This chapter will be concerned with a first phase of the description of motion. We will evolve a procedure by which we can describe the motion of any object without regard to the agencies that may have produced it. We will apply this procedure to four special kinds of motion: (1) uniformly accelerated motion, (2) simple harmonic motion, (3) uniform circular motion, and (4) satellite motion.

In Chapter 2 we will move on to a second phase of our description of motion, in which we will correlate the motion an object possesses, or is in the act of acquiring, with "forces" that have acted or presently are acting on the object. In doing so we will be adding to our seventeen or more years of experience with motion, which, even without a precise definition of force, have taught us that motion and force are related.

1-1 SCALAR AND VECTOR QUANTITIES

Two types of quantities are used in physics, distinguished by whether or not it is necessary to state a direction as well as a magnitude in order to say all that we need to know about the quantity.

A scalar quantity is a quantity that is completely specified by a statement of its magnitude with the appropriate units.

Quantities like volume, density, and temperature fulfill this criterion because there is certainly no way in which any sense of direction can be associated with any of them. Scalars submit to ordinary arithmetic processes; if one adds 3 pints of water to a vessel containing 4 pints of water, he can be sure that he has then a total of 7 pints of water.

In this text, however, we will have occasion to deal with a number of situations in which a statement of direction is necessary to describe the quantity completely. For example, an automobile that leaves a given garage and travels 60 miles in a given direction will not arrive at the same location as another automobile leaving the same garage traveling in a different direction. A change of position will be called a displacement and will serve as the prototype of all vector quantities. We will utilize the following definition:

A vector quantity is a physical quantity that, like a displacement, is completely specified only if one tells both its scalar qualities and its direction; to qualify as vectors, two quantities of a given kind must combine like two displacements.

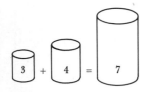

FIG. 1-1

Scalar addition. Three pints plus 4 pints equals 7 pints.

Among the vector quantities with which we will deal in this text are displacements, velocities, accelerations, forces, and electric fields.

Vector addition

The mathematical manipulation of vector quantities uses a variety of mathematics that we will call vector algebra. We will limit our attention, at this point, to vector addition and vector subtraction of displacements. Ultimately, of course, we will be far more concerned with the application of this procedure to other directed quantities. Therefore we will return to this subject as other such quantities are defined.

Figure 1-2 represents a scale diagram of the travel of an insect that, starting from the point O, crawled a distance of 3 cm north, then turned and crawled 4 cm east. It is quite clear that whereas the algebraic addition of 3 cm and 4 cm always leads to a sum of 7 cm, the vector addition of 3 cm and 4 cm has led to an answer of 5 cm. This sort of addition will be written as follows:

$$\vec{3} + \vec{4} = \vec{5}$$

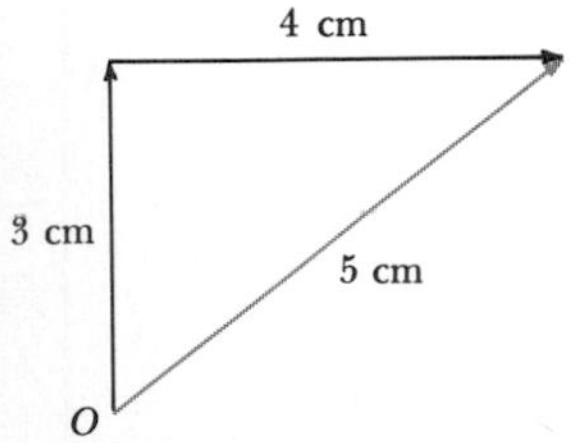

The vector addition of 3 cm and 4 cm may lead to any distance between 1 cm and 7 cm, depending on the angle between the two vectors.

FIG. 1-2

Vector addition of two displacements. Three cm plus 4 cm can equal 5 cm.

Vector subtraction

In order to devise a process by which one would subtract two vectors, it would be well to start by thinking of a circumstance in which one would require a procedure for subtracting two scalar quantities. When this has been achieved, we will extend the procedure to the subtraction of two vector quantities. Let us assume that on a given morning Mr. John L. Smith's car contains 6.8 gal of gasoline and that his bank balance is $169.50. On the following morning, at the same hour, his car contains 10.4 gal of gasoline and his bank balance is $153.02. Both the volume of gasoline and the bank balance are scalar quantities, and the difference between the magnitudes of each of these quantities represents the change of the quantity in question. In either case, the rule that one might evoke to describe the change could be expressed as follows:

The change in a given quantity is that quantity which, when added to the original value, will yield the final value.

Thus, in the case of the gasoline, the change in volume was 3.6 gal (an increase), since that number added to the original value (6.8) yields the final value (10.4). The change in Smith's bank balance was —$16.48 (a decrease) because that is the number which, when added to the original ($169.50), yields the final value ($153.02).

We will use precisely the same language in describing the change of a vector quantity. The insect to which reference was made above traveled a distance of 3 cm north and later was located 5 cm from its starting point. The change involved in the second action of the insect was the vector that when added to the first displacement yielded the net displacement. Thus the "change" represents the difference between the net displacement and the initial displacement and equals 4 cm east.

Let us apply this procedure to a similar situation, differing largely in the language we use to describe it: The radar system at a given airport determines that an airplane is 62 miles due north of the airport; later it shows that the same plane is 76 miles northeast of the airport. What change in position has taken place?

In analyzing this problem, one makes a scale diagram of the situation. First, he places a dot to mark the position of the airport. Second, letting 1 cm represent 10 miles, he draws a line 6.2 cm long directly up the page, since this, in the usual convention in map making, represents north. Third, he draws a line 7.6 cm long up and to the right, which makes a 45° angle with the first line. Finally, he asks himself this question: "What vector must I add to the original vector to secure the final vector?" Figure 1-4 shows that the magnitude of that vector is 5.4 cm and that it points at an angle of 8.4 degrees relative to the "horizontal." In fact, then, the plane must have traveled 54 miles in a direction 8.4 degrees south of east, which represents the change in the position of the plane relative to the airport.

The reader may note that this problem might have seemed far more difficult had we attempted to solve it by using trigonometry. The data, however, provided by the radar system were only accurate to two significant figures. Thus since your ruler can provide this much accuracy, the result secured graphically is fully as good as you could secure by any procedure, however sophisticated it might appear.

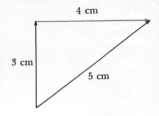

FIG. 1-3

Vector substraction. The difference between the final displacement and the initial displacement is the vector that must be added to the initial displacement to secure the final displacement.

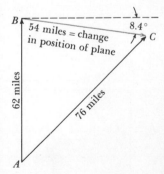

FIG. 1-4

The change in a vector. The vector that must be added to AB to yield $\overrightarrow{AC}$ is the vector $\overrightarrow{BC}$.

The *component* of a vector in a given direction is its effective value in that direction. For example, a person swimming across a river 100 ft wide may travel an actual distance much larger than 100 ft if he is carried along by the stream (Fig. 1-5).

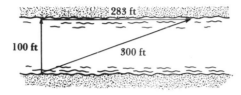

FIG. 1-5

Components of a vector. The man travels 300 ft in order to cross a stream 100 ft wide. He is carried 283 ft downstream in the process. The "effective" component of the 300-ft displacement is equal to 100 ft.

Normally we will be interested in talking about the two rectangular components of a given vector. Frequently one can state that one of these is the "effective" component. In the case of the man swimming across the stream, the effective component of his displacement was the 100 ft of travel directly across the stream, while his displacement of 283 ft downstream was "ineffective" so far as his goal was concerned.

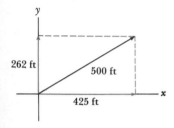

FIG. 1-6

Rectangular components of a vector. The projections of the 500-ft displacement on the x and the y axes yield the x and y components of the displacement.

The procedure by which one may evaluate the components of a vector is illustrated in Fig. 1-6. One starts with knowledge of the magnitude and direction of a given vector. An arrow of length proportional to the magnitude of the vector is drawn in the appropriate direction on rectangular graph paper. One then drops a perpendicular from the tip of the vector onto each of the coordinate axes. One measures the projection of this arrow in the two directions and translates these lengths back into magnitudes, using the same scale factor. These components will then be identified as the x and y components or as the east and north components, whichever best suits the situation.

It is useful to observe that the two rectangular components of a vector represent two vectors whose vector sum equals the original vector.

1-2 SPEED AND VELOCITY

Average velocity

The following sentence is apt to appear in any list of physics problems: "A car travels a distance of 100 miles in 2 hours along a straight road . . . ." This stereotyped phrase constitutes a partial description of the motion of an object. It is not a complete description of the motion because under most circumstances a car traveling a distance of 100 miles would stop for red lights, slow down for school buses, and increase its speed on the downslope of any hill that might be encountered.

Reading the quoted statement, one is prone to say to oneself that the car had an average speed or an average velocity of 50 miles per hour. While physicists and nonphysicists alike have mastered the basic concept involved here, both being very conscious of speed and speed laws, the former are more likely to use the word "velocity" and to write the definition in an algebraic form.* Letting the symbol Δd represent the distance traveled and Δt represent the time required to travel that distance, we may write

$$\text{Average velocity} = \frac{\text{distance traveled}}{\text{lapsed time}}$$

or
$$v_{\text{av}} = \frac{\Delta d}{\Delta t} \qquad (1\text{-}1)$$

Distinction between speed and velocity

A driver of a racing car on an oval track is very much concerned with the rate at which his car is traveling; but when one quotes a number like 163 mi/hr with reference to a car on such a track, no purpose is served by indicating a direction of motion. The scalar measure of rate of travel lacking any reference to the direction of motion will be referred to as the *speed*.

Any object that is in motion, however, must travel in some particular direction at every instant. Thus a complete description of the motion at any instant requires a specification of the direction of travel. The designation of both the speed of an object and its direction of travel is called the *velocity* of the object and carries with it all the attributes of a vector quantity.

In conformity with the practice followed by most writers of physics texts, I will use the word velocity more frequently than speed. I will certainly use the word velocity when the concept of direction is important, but, beyond that, I will usually use the word velocity in situations in which the direction of motion will make no difference one way or the other. In general, then, I will limit my use of the word speed to situations in which the introduction of a consideration of the direction of motion either would be meaningless or would complicate the situation.

In order for velocity to qualify fully as a vector quantity, we must be sure that two velocity vectors combine like two displacements. As evidence that velocity satisfies this requirement, let us consider a specific example. Assume a man to walk directly eastward across a flatcar moving north. Assume his velocity relative to the flatcar to be 8 ft/sec (east) and the velocity of the flatcar to be 24 ft/sec (north). In 1 sec he will travel 8 ft to the east and 24 ft to the north. His displacement relative to ground in 1 sec is therefore $\sqrt{8^2 + 24^2} = 25.3$ ft in a

---

* Physicists also like to use shorthand notation. The symbol Δ (delta), for example, signifies a change of some quantity; this symbol will always appear immediately to the left of the quantity whose change is under discussion. Δv may be read as "delta v" or as "the change of v" as you prefer.

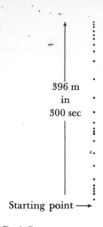

396 m
in
300 sec

Starting point ⟶

FIG. 1-7

A picture of the motion of a car. The dots represent instantaneous positions of the car at 10-sec intervals.

direction 18.4 degrees to the east of north. Velocities combine like displacements because, in fact, they are displacements (per unit time, to be sure).

A picture of motion

If one is to study motion in detail, one requires specific information regarding the location of the object under study at every instant of time. Eventually, in this chapter, we will refer to some sophisticated equipment that assists in providing such information, but we will start with a very unsophisticated example.

If a passenger should throw a handkerchief out the window of his car every 10 sec of travel, and if we assume, for the sake of this discussion, that each handkerchief settles directly to the ground, the littered highway would then present a picture of the motion to an observer overhead. Figure 1-7 is a simplified picture of the motion of such a car over a portion of a trip that it might make over a straight roadway. For simplicity, we show only a sequence of dots to represent the location of successive handkerchiefs. Figure 1-8 shows the same information re-plotted as a position versus time graph.

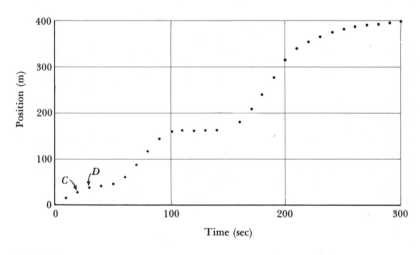

FIG. 1-8

A position versus time graph of data from Fig. 1-7. The graph presents the same information as the "picture," but details of the motion are now more apparent.

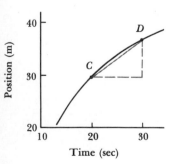

FIG. 1-9

Position versus time graph. A magnified view of a portion of Fig. 1-8.

Let us examine certain points at random in Fig. 1-8—for example, points C and D as shown in the magnified view of this region (Fig. 1-9). The position at the instant C may be read as 29.7 m from the starting point. Similarly, the position at the point D may be read to equal 36.9 m from the starting point. The distance traveled during the 10-sec interval was therefore equal to (36.9−29.7) or 7.2 m. Because the car traveled this distance of 7.2 m in 10 sec, its average velocity was equal to 0.72 m/sec during the interval CD. This procedure can be repeated to

18

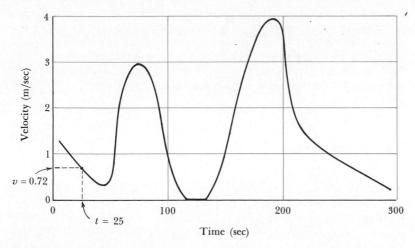

FIG. 1-10

Velocity versus time for motion shown in Fig. 1-7. An average velocity was computed for each pair of adjacent points in Fig. 1-7 or Fig. 1-8. These velocities were plotted and a smooth curve was drawn.

determine the average velocity for every 10-sec interval during the motion. Each of the average velocities secured in this way might be said to "characterize" the interval in question and could be associated perhaps most accurately with the midposition of the interval. One can then prepare an average velocity versus time graph as shown in Fig. 1-10.

We may now summarize what has been accomplished. We went from a "description" of motion in which we knew only the time and position at the start and the finish of a trip to a situation in which the time and position were known at the end of 10-sec intervals. With this data we saw that position versus time and average velocity versus time graphs could be prepared.

Clearly one can secure even more exact data. Instruments can be devised by which the position of an object can be ascertained every second, every one-thousandth of a second, or even every one-millionth of a second. In each case, we can divide distances traveled by the appropriate time intervals to secure an average velocity versus time graph. As the time intervals involved in the measurements become smaller, the average velocity in any one of the intervals approaches what we would call the *instantaneous* velocity. Furthermore, the instantaneous velocity might well be read directly from some sort of instrument. Such an instrument, a speedometer, has in fact been invented. We are unable, however, to attach a speedometer to every object we wish to discuss; most of our studies of motion will have to be made by observing the position of the object as a function of the time. Our data will be good only if the time interval between observations is small.

A procedure by which one can improve on his determination of the velocity versus time graph is suggested by Fig. 1-11. The data from the picture of the motion are plotted and a smooth curve is drawn through

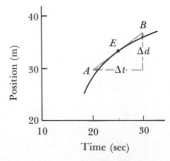

FIG. 1-11

Determining the instantaneous velocity. The ratio $\Delta d/\Delta t$ is the slope of AB and equals the instantaneous velocity at E.

the average positions of the data points. To find the instantaneous velocity at some point E, one draws a tangent to the curve at that point—that is, a line having the direction of the curve at that point. Next, one forms a triangle by drawing a horizontal and a vertical line, both of which intersect the tangent. The ratio $\Delta d/\Delta t$ is the slope of both the line AB and the curve at the point E and equals the instantaneous velocity at E.

1-3 ACCELERATION

When traveling in a car, one is constantly aware not only of one's velocity but also of changes in that velocity. Changes in the velocity may be both a source of gratification and a source of distress. A car with a powerful engine may provide passengers with the pleasure of sudden bursts of speed or of swinging gaily around sharp curves. On the other hand, a car might be accelerated by being hit from behind or by running into a tree.

Any change of velocity, whether a change in magnitude or a change in direction, will be called an *acceleration*. The average acceleration during a given time interval will be defined as the change in velocity Δv divided by the lapsed time Δt,

$$a_{\mathrm{av}} = \frac{\Delta v}{\Delta t} \tag{1-2}$$

In this section we will discuss only the case of linear motion, leaving the question of the acceleration to be associated with a change of direction for Sec. 1-6.

The nearly identical appearance of the definitions of velocity and acceleration suggests that the procedure we have described for determining the velocity may be adapted for use in the determination of the acceleration. By analogy with our discussion in Sec. 1-2, we may determine the acceleration as follows:

> The instantaneous acceleration of an object traveling in a straight line may be secured from a velocity versus time graph of its motion by evaluating the slope of a line tangent to the curve at the point in question.

Applying this procedure to the data shown in Fig. 1-10 (Fig. 1-12), one secures the curve shown in Fig. 1-13.

An acceleration has both a magnitude and a direction. That accelerations combine like displacements (the final test of their qualification to be called vector quantities) is apparent from the fact that acceleration bears the same relationship to velocity that velocity bears to displacement; hence since velocity is a vector, acceleration must also be a vector.

Acceleration will be expressed in meters/second², which may be read "meters per second squared" or, with more comprehension, as "meters per second each second." The correctness of these units may

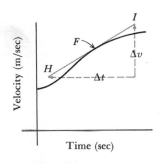

FIG. 1-12

Determining the instantaneous acceleration. The ratio $\Delta v/\Delta t$ is the slope of *HI* and equals the instantaneous acceleration at F.

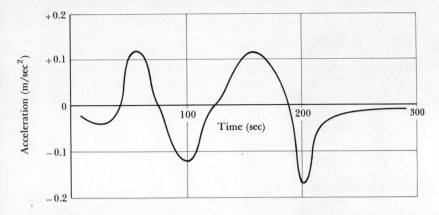

FIG. 1-13

Acceleration versus time of car depicted in Fig. 1-7. This curve was secured from Fig. 1-10 by plotting time versus the slope of the velocity curve. Notice that the acceleration curve crosses the t axis (i.e., $a = 0$) at the points at which the velocity curve has zero slope.

be verified by means of the Units Rule. Noting that acceleration is given by a unit of velocity divided by a unit of time, we have

$$a = \frac{\Delta v}{\Delta t} = \frac{\text{meters/sec}}{\text{sec}} = \frac{\text{meters}}{\text{sec sec}} = \frac{\text{meters}}{\text{sec}^2}$$

1-4 PICTURES OF MOTION

Physics would be in a primitive stage were it not for the development of devices that improve the ability of observers to see and therefore describe nature. These technological advances take the form of instruments, such as telescopes, cathode-ray oscilloscopes, watches, and meter sticks. Scientists use these devices with such regularity, and as so much a part of their day-to-day work, that scientific writers often fail to acknowledge the important role that technology has played in the progress of physics. Nonscientists watching physicists at work often tend to confuse physics with the instruments that physicists use. In order to avoid confusion between science and technology, an effort will be made to draw the attention of the reader to the goal of making precise observations and to the subservient role that instruments play in the process.

Taking a picture of motion

The instruments under consideration in the subsequent sections will consist of a camera, a darkened room, and a flashing light capable of illuminating a portion of the room at regularly spaced intervals of time. A light source possessing the appropriate characteristics is shown in Fig. 1-14.

FIG. 1-14

A strobe lamp. Several commercial forms of this lamp are available. The lamp emits a succession of bright flashes, each of very short duration. The number of flashes emitted per second can be varied at will. (*General Radio Company*)

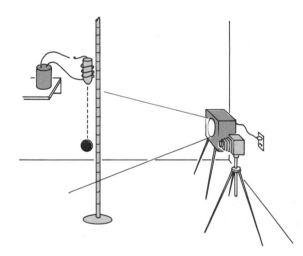

FIG. 1-15

Arrangement for taking a picture of the motion of an object. In the case shown, a freely falling object is being photographed using a strobe lamp as the light source to secure pictures like those shown in Fig. 1-17.

In order to secure a picture of the motion of an object, we will take a photograph of the object while it is being illuminated by the intermittent light source. For example, let us suppose that we are concerned with the description of the motion of a freely falling object. An arrangement like that shown in Fig. 1-15 should suffice.

We also will be concerned in this section with the motion of an object suspended by means of a spring. If one end of a spring is attached to a firm support and a massive object is attached to the other end of the spring, one will observe an elongation of the spring due to the force transmitted to it by the action of gravitation on the mass. A given object ultimately will stretch the spring to a length that is determined by the magnitude of the mass, and, at rest, the object will find a well-defined equilibrium position. If the object is displaced from its equilibrium position and released, it will undergo an oscillation about the equilibrium position. In practice, the maximum displacement during the vibration will gradually decrease until the object is once more stationary in its original position. Ideally, in the absence of friction, one would expect the oscillation to continue indefinitely without decrease in the magnitude of the maximum displacements.

Pictures of motion obtained by this technique are shown in Fig. 1-17. In the case of Fig. 1-17(*b*) and (*d*), the camera was moved steadily to the left during the exposure, with the result that the picture constitutes a position versus time graph of the motion of the object. In the case of the photograph of the mass on a spring, this technique also has the advantage of enabling one to photograph several complete oscillations of the object without the confusion that is inevitable if the positions of the object in successive oscillations should overlap.

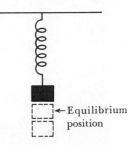

FIG. 1-16

A mass on a spring. The motion of an object subjected to a springlike force is of considerable importance in physics.

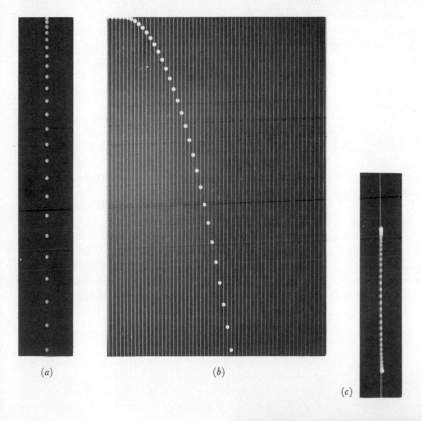

(*a*)   (*b*)

(*c*)

FIG. 1-17

Pictures of motion: (*a*) a freely falling object; (*b*) a freely falling object with camera in motion; (*c*) a mass on a spring; (*d*) mass on a spring with camera in motion.
(*Phillips Hall*)

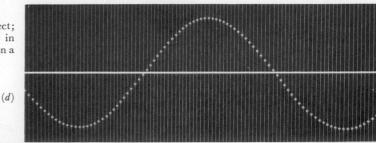

(*d*)

In the previous section we described the technique by which one can secure pictures of motion in a line. In this section we will analyze the two types of linear motion displayed in Fig. 1-17.

A freely falling object

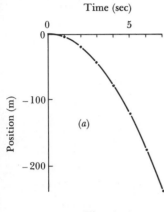

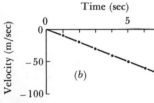

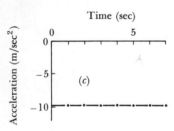

FIG. 1-18

The motion of a freely falling object:
(a) position versus time;
(b) velocity versus time;
(c) acceleration versus time.
Since the UP direction is regarded as the positive direction, the signs of all positions, velocities, and accelerations are negative.

A graph showing position versus time of a freely falling object can be secured by measuring successive positions of the object in Fig. 1-17(a). Such a graph is shown in Fig. 1-18(a). A smooth curve has been drawn through the average locations of the points because such a curve tends to average out random errors in the measurements and thus describes the motion of the object more accurately than the points themselves. Measurements of the slope of this curve yield the velocity versus time graph shown in Fig. 1-18(b). This graph shows that the velocity of a freely falling object increases linearly with time; that is, the velocity versus time curve is a straight line. The acceleration of the object is equal to the slope of this line. Since the line is straight and therefore has a constant slope, we see that the acceleration of gravity is constant. The constancy of the acceleration of a freely falling object is shown graphically in Fig. 1-18(c).

The experiment we have described has led to important observations: (1) The acceleration of gravity near the surface of the Earth is uniform (within experimental error, the acceleration of gravity is apparently the same at the ceiling as it is at the floor), and (2) the numerical value of the acceleration of gravity, from these measurements, is equal to 9.80 m/sec².

Although the acceleration of gravity is nearly uniform if one compares its value at one point in a laboratory with its value at another point in the same room, significant variations are found from position to position on and above the surface of the Earth.* Variations are found with latitude and, to a lesser extent, with longitude. The change with altitude is quite impressive, falling to half of the above value—that is to 4.90 m/sec²—at an altitude of approximately 1700 miles.

Simple harmonic motion

Photographs of the motion of a mass on a spring have been shown in Fig. 1-17(c) and (d). The latter differs from the former only in that the camera was moved steadily to the left while the camera shutter was open; it therefore presents an accurate position versus time graph. One can obtain a velocity versus time graph either directly from this photograph or from the graph obtained by replotting the data available from Fig 1-17(c), as shown in Fig. 1-19(a). The velocity versus time curve [Fig. 1-19(b)] and the acceleration versus time curve [Fig.

---

* Precise measurements show that the acceleration of gravity at sea level at the Equator, at 45° latitude, and at the North Pole equals 9.780490, 9.806294, and 9.832213 m/sec² respectively.

1-19(c)] were each obtained by making many measurements of the slope of the preceding curve.

It is important to notice certain qualitative features of the three graphs. Any of the positions on the displacement versus time graph at which the slope equals zero (i.e., at a peak and at a trough) corresponds to an instant in time at which the velocity equals zero. Similarly, any of the positions on the velocity versus time curve at which the slope equals zero corresponds to an exact instant in time at which the acceleration is equal to zero. These successive operations yield an acceleration curve that resembles the displacement curve in many respects; they have the same general shape and they each become equal to zero at the same instant. However, one important difference must be noted. A crest on the displacement curve is coincident with a trough on the acceleration curve—that is, for a mass on a spring the acceleration is negative whenever the displacement is positive, and vice versa.

If one makes measurements with adequate care, a most important quantitative feature will emerge from the data used to plot Fig. 1-19(a) and (c)—namely, that the ratio of the instantaneous numerical values of the acceleration and displacement is equal to a constant; the acceleration is proportional to the negative of the displacement.

Periodic or oscillatory motion, motion that continually repeats itself, is very common in nature; we have only to think of the waving of a flag, the swaying motion of a tree, or the intermittent lapping of the waves on the seashore to find examples of such motion. Some of these motions are more complicated than others. In physics, the simplicity of a phenomenon is largely determined by whether or not the phenomenon can be represented by a simple mathematical expression. By this criterion, the motion of a mass on a spring is the simplest harmonic motion that exists and is therefore called in fact *simple harmonic motion*. It is defined as harmonic motion in which the acceleration is always proportional to the negative of the displacement.

1-6 UNIFORM CIRCULAR MOTION

Suppose that an object on a very slick surface were to be attached by means of a string to a firm post or pin and then set into motion in a circular path about the pin. A strobe picture taken by a camera located directly above this moving object would appear as in Fig. 1-20(a).

If we could attach a speedometer to the object, the speedometer would read a constant speed. The truth of this observation is apparent from the picture of its motion, because the object clearly has traveled the same distance between any two successive flashes of light. We have a ready means of measuring this speed, for we need only measure the distance along the arc between successive flashes and know the time interval between flashes. We would find that the speed is constant, but the object does not have a constant velocity. Velocity, as we noted earlier, is a vector quantity; the fact that the direction of the motion is continually changing means that the velocity is also continually changing.

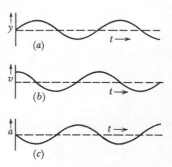

FIG. 1-19

Motion of a mass on a spring: (a) displacement or position; (b) velocity; (c) acceleration. Notice that the acceleration equals zero when the displacement equals zero; notice also that the acceleration is negative when the displacement is positive, and vice versa. Measurements show that the acceleration is proportional to the negative of the displacement.

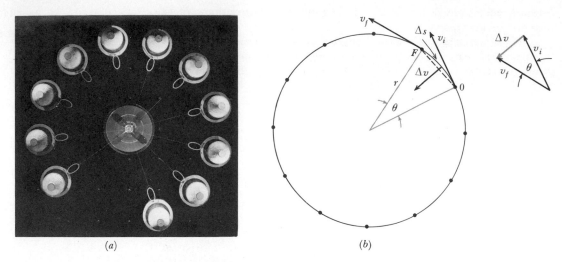

FIG. 1-20

(a) A picture of uniform circular motion. (b) Proof that the acceleration is toward the center of the circle.

(*Photograph from PSSC*, Physics, *D. C. Heath & Co., Lexington, Mass. 1965*)

Centripetal acceleration

We must now consider the problem of determining the magnitude and the direction of the acceleration that should be associated with this changing velocity vector. In Fig. 1-20(b) we identify the velocity at some arbitrary instant as v_i, the initial velocity, and the velocity at some later instant as v_f, the final velocity. The acceleration will be equal to the difference between these vectors divided by the lapse of time Δt.

In Sec. 1-1 we showed that the change in any vector can be thought of as that vector which, when added to the original vector, yields the final vector. This procedure has peen applied to the problem of finding the change in velocity Δv, in drawing the small triangle in the upper-right corner of Fig. 1-20(b). First we made use of the fact that one can move a vector about at will, provided one does not change either its magnitude or its direction, and then placed the bases of the two arrows that represent v_i and v_f together.* Second, we have drawn the small arrow Δv. This is the velocity vector that must be added to v_i to yield the final velocity v_f; it therefore represents the change in the velocity. Finally, we have moved Δv back to the circle, attaching it to the midpoint of the arc because this change in velocity "characterizes" the interval. Inasmuch as this change in velocity points toward the center of the circle, we can be sure that the object is being accelerated toward the center of the circle. The centrally directed accelera-

---

* Moving a vector doesn't change it if neither the direction nor the magnitude of the vector is changed in the process. Placing two vectors side by side is logical; if you were comparing two partially filled milk bottles, you probably would do the same.

tion of an object in uniform circular motion is called the centripetal or the central acceleration.

We may now derive an equation for the magnitude of the centripetal acceleration in terms of measurable quantities involved in the motion. Return to Fig. 1-20(b), and notice the pie-shaped figure formed by the two radii and the arc $\overset{\frown}{OF}$, and the triangular figure formed by v_i, v_f, and Δv. Because one of the radius vectors is perpendicular to v_i and the other is perpendicular to v_f, we can be sure that the angle θ in one of these figures is equal to the angle θ in the other.

If both figures were triangles, we could say that they are "similar" and therefore conclude that their sides are proportional. As it happens, we can arrive at that conclusion even though one of the figures is not, strictly speaking, a triangle. To understand how this can happen, note that the acceleration we would secure by dividing Δv by the time lapse Δt would be at best only an average for the interval described by $\overset{\frown}{OF}$; it would be an instantaneous acceleration only if that interval was vanishingly small. It can be shown (but will not be shown here) that in the limit of very small intervals the arc $\overset{\frown}{OF}$ and the chord $\overline{OF}$ become indistinguishable. Therefore we can treat the pie-shaped figure as if it were a triangle. The ratio of any two sides in one of these figures is therefore equal to the ratio of the corresponding sides in the other. Since v_i is numerically equal to v_f, we can replace them both by the symbol v and express these ratios by

$$\frac{\Delta v}{v} = \frac{\Delta s}{r}$$

Therefore

$$\Delta v = \frac{v\,\Delta s}{r}$$

From the definition of acceleration,

$$a = \frac{\Delta v}{\Delta t} = \left(\frac{v\,\Delta s}{r}\right)\frac{1}{\Delta t} = \frac{v}{r}\left(\frac{\Delta s}{\Delta t}\right)$$

Since $v = \Delta s/\Delta t$, we have

$$a_c = \frac{v^2}{r} \tag{1-3}$$

where we have introduced the subscript c simply to identify this expression as the equation for centripetal or central acceleration.

One can see the logic of a centripetal acceleration by recalling the experience of riding on a merry-go-round. You were riding on a wheel that was rotating at constant angular velocity. You traveled precisely the same distance during each second of travel. You were not, however, traveling with constant velocity because you were constantly changing your direction. If at any instant you had continued to travel with constant velocity, you would have continued straight along a tangent to the circle on which you were moving. But you did not continue in a straight line; you were constantly turning away from that direction toward the center of the circle on the circumference of which you were traveling.

1-7 SATELLITE MOTION

Let us assume that a satellite has been placed in orbit about the Earth and that we wish to secure a picture of its motion. It is not possible, of course, to prepare stop frame photographs of the satellite from a point in outer space, but electronic devices (radar beams and computers) on the surface of the Earth are capable of mapping the course of a satellite and providing such detailed information of its position from moment to moment that such a picture can be reconstructed

FIG. 1-21

A picture of satellite motion. The velocity of the satellite changes continuously both in direction and magnitude. The acceleration, found by determining the change in velocity from point to point, is always directed toward the center of the Earth.

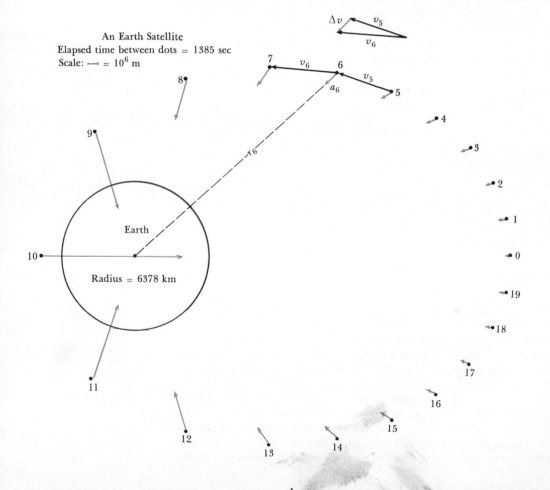

from the data (Fig. 1-21). As in our previous pictures of motion, the dots represent instantaneous locations of the satellite at uniformly spaced intervals of time.* Two general properties of the motion are apparent.

1. The dots are not uniformly spaced; thus we know that the speed of the satellite is continually changing.

2. The line of motion is a curved path, and since the direction of motion is continually changing, we know that the satellite experiences a centripetal acceleration at all times.

These two observations tell us that the satellite experiences both of the previously discussed kinds of acceleration simultaneously; stated differently, the acceleration of the satellite at any instant is directed in such a way that it possesses components both parallel and perpendicular to its motion.

The procedures we have utilized in earlier sections will enable us to determine the actual acceleration at any instant. From the locations of the dots, we can determine the velocity of the satellite at two adjacent positions along the orbit. These two velocities will differ both in magnitude and in direction. Therefore, by finding the vector difference between them, we can determine the direction and magnitude of the acceleration that characterizes that particular interval. This vector can be associated with the actual acceleration at a point halfway between the two positions. The result of several such vector substractions is shown by the arrows in the figure.

The first of two important features of satellite motion that emerges from this analysis is that a satellite in motion about the Earth or the Sun is continually being accelerated toward that object. This situation is apparent from the fact that the arrow representing the acceleration of the satellite points directly toward the center of the Earth (or Sun) at all points along the path.

A second and equally important conclusion can be learned from a mathematical analysis of the magnitude of the acceleration. The magnitude of the acceleration of the satellite varies inversely with the square of its distance from the center of the Earth (or Sun). This conclusion may be written in the form of an equation

$$a_s = \frac{k}{r^2} \qquad (1\text{-}4)$$

and is established as an experimental deduction from the picture of the satellite's motion by preparing a graph of the acceleration of the satellite versus the reciprocal of the square of the distance from the center of the Earth (Fig. 1-22). The constant k in this equation may depend on properties of the satellite that are not changing during the motion; in Chapter 4 we will find that k depends on the mass of the object about which the satellite is orbiting.

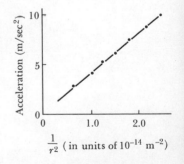

FIG. 1-22

A graph of acceleration versus $1/r^2$ for an Earth satellite. An analysis of the motion shown in Fig. 1-21 shows that the acceleration is inversely proportional to the square of the distance of the satellite from the center of the Earth.

* The "picture" was generated by a computer. The author has reason to believe that an actual satellite would assume this orbit if properly launched.

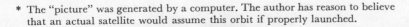

SUMMARY The quantities that are measured and expressed in our descriptions of nature are either scalar or vector quantities. Each scalar measurement when quoted should tell three things about itself: (1) its magnitude, (2) its accuracy, and (3) its units. A vector quantity requires, in addition, a statement of direction.

Vector quantities when added or subtracted combine like displacements.

The average velocity of an object during a time interval Δt is defined as

$$v_{av} = \frac{\Delta d}{\Delta t} \qquad (1\text{-}1)$$

Velocity is a vector quantity whose magnitude is called the speed.

The average acceleration of an object during a time interval Δt is defined as

$$a_{av} = \frac{\Delta v}{\Delta t} \qquad (1\text{-}2)$$

Acceleration is a vector quantity.

Motion in a line can be analyzed by preparing a position versus time graph. The slope of this curve at any point on the curve equals the instantaneous velocity at that point. The same procedure applied to a velocity versus time graph yields the instantaneous acceleration.

This procedure was applied to two cases:

1. A freely falling object was found to have a constant acceleration near sea level of 9.8 m/sec².

2. A mass on a spring moves in such a way that its acceleration is always proportional to its displacement. Motion having this property is called simple harmonic motion.

An object traveling with uniform circular motion experiences an acceleration toward the center of the circle in which it is traveling because the direction of the velocity vector is continuously changing. The centripetal acceleration is given by

$$a_c = \frac{v^2}{r} \qquad (1\text{-}3)$$

where v is the object's speed and r is the radius of the circle.

In satellite motion, both the magnitude and the direction of the velocity vector change continuously. The acceleration of an Earth satellite is always directed toward the Earth; the magnitude of the acceleration varies inversely with the square of the distance from the Earth's center to the satellite.

A ruler, a protractor, and graph paper will be useful in solving some of the following problems.

1-A1. Prepare a list of quantities, each of which is completely specified if one states only its magnitude.

1-A2. Prepare a list of quantities, each of which is completely specified only if one states a direction as well as a magnitude.

1-A3. A bus travels 20 miles north, then 30 miles northwest. Make a scale diagram and describe the final position of the bus.

1-A4. A bus travels 30 miles northwest, then 20 miles north. Make a scale diagram and describe the final location of the bus.

1-A5. The speed of light is 3.00×10^8 m/sec. Express this speed in (*a*) ft/sec, and (*b*) mi/sec.

1-A6. Can an object possess acceleration if it is at rest?

1-A7. Your roommate (assumed also to be taking this course) tells you that "centripetal acceleration" is a fiction invented by a' professor just to get the right answer. Can you find any merit to this idea? What argument would you use to convince your roommate that a centripetal acceleration is just as valid a concept as a linear acceleration?

1-B1. An object experiences three successive displacements as follows:
(*a*) 3.5 m north
(*b*) 6.1 m west
(*c*) 5.0 m southwest (45° west of south)
Find the final position of the object relative to its starting point.

1-B2. An object experiences the displacements listed in Prob. 1-B1 but in the opposite order [(*c*) then (*b*) then (*a*)]. Find the final position of the object relative to its starting point.

1-B3. An object experiences a displacement of 7.5 m north. Next it experiences a second displacement with the result that it is then 3.2 m due west of its original starting position. What was the second displacement (magnitude and direction)?

1-B4. The acceleration of an automobile is held steady at 1.8 m/sec². (*a*) How much will the velocity increase in 5 sec? (*b*) If the velocity was 16 m/sec at the start of this 5-sec interval, what will be the velocity at the end of the interval? (*c*) What was the average velocity during the interval? (*d*) How far did the automobile travel?

1-B5. An object falls from rest for 10 sec. (*a*) What velocity does it possess at the end of 10 sec? (*b*) What is the average velocity during the interval? (*c*) How far does the object travel?

1-B6. Repeat Prob. 1-B5, assuming that the object started with a velocity of 98 m/sec.

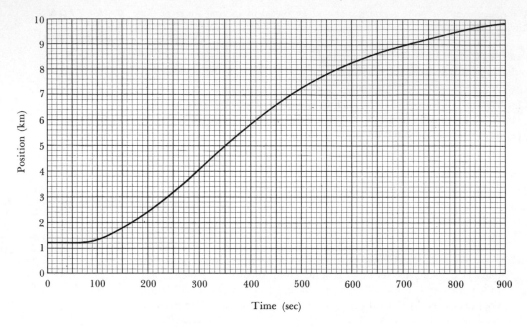

FIG. 1-23

1-B7. A position versus time graph of an object in motion along a line is shown in Fig. 1-23. (*a*) Prepare a velocity versus time graph based on this figure. (*b*) Prepare a sketch showing approximately how the acceleration varies with time.

1-B8. A child in a swing with chain supports 3 m long passes through the lowest part of its path at a velocity of 4.4 m/sec. What is the child's centripetal acceleration at this point of its motion?

1-B9. An airplane banking sharply has a centripetal acceleration at a given instant exactly equal to g ($= 9.8$ m/sec²). If its speed was 100 m/sec, what was the radius of curvature of the path of the plane?

1-B10. At one instant a racing car is traveling with a velocity of 60 m/sec directly toward the north. Three seconds later it is traveling at a velocity of 60 m/sec in a direction of 20 degrees to the west of north. (*a*) What was its change of velocity? (*b*) What was its average acceleration?

1-B11. At one instant an ice skater has a velocity of 10 m/sec directly eastward. Two seconds later his velocity equals 14 m/sec in a direction 15 degrees to the north of east. (*a*) What was the skater's change of velocity? (*b*) What was his average acceleration?

1-B12. What is the acceleration of gravity at an altitude of 1600 km (1000 miles) above the surface of the Earth, given that g (surface) equals 9.8 m/sec² and that the radius of the Earth equals 6380 km?

Forces and Motion

In the preceding chapter we described motion without regard to how motion may be produced. We discussed a general procedure by which one can utilize a picture of the motion to determine both the velocity and the acceleration of the object in question. We applied this procedure to four cases: (1) uniformly accelerated motion in a straight line, (2) simple harmonic motion, (3) uniform motion in a circular path, and (4) general motion in a plane with specific regard for satellite motion.

A mere description of motion is not, in general, a satisfactory end in itself; one wishes in addition to find a cause of motion if any cause can be identified. As we shall discover in this chapter, when the motion of an object is examined in detail, one can usually find some external agency or object that is responsibile either for the motion or for certain aspects of the motion.

Many readers encountering physics for the first time express the feeling, frequently in rather subtle ways, that physics does not deal with the real world of their own experience. The examples used to illustrate principles seem overly idealized; only in the physicist's laboratory does nature seem to conform to the mathematical equations that presumably describe nature. In the laboratory measurements are made of the acceleration while falling of a smooth, steamlined steel bar; measurements show that the acceleration of the object equals 9.801 m/sec², but just outside the window a leaf flutters quietly to the ground. In the laboratory the student suspends a mass by means of a spring. He observes its acceleration to be proportional to the negative of its displacement; meanwhile a truck rumbles by the building and rattles the windows. The student wonders, "Do these ideal situations presented in the laboratory have a significant relationship to situations we encounter outside the laboratory?"

A fundamental assumption of science is that nature conforms to relatively simple basic laws. At the same time, nature is complex, in the sense that most phenomena occurring in nature involve the simultaneous occurrence of a combination of phenomena that, taken individually, may be rather simple. One cannot hope to understand a

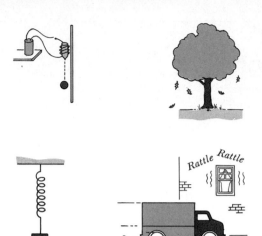

FIG. 2-1

Is physics relevant? Do the idealized situations presented in physics textbooks have a significant relationship to situations encountered in everyday life?

complex combination of phenomena without mastering the individual components. Therefore we must deal first with the individual basic phenomena.

In order to deal with these phenomena, we must design experiments that will isolate each phenomenon under investigation. Competing phenomena must be eliminated as much as possible. In most of this chapter, for example, we will discuss motion in the absence of the troublesome force known as friction. We all know that if one were to roll a basketball, a baseball, a golf ball, and a hard steel sphere across a smooth, horizontal floor, each with the same initial velocity, all the objects would eventually coast to rest. The fact that they will stop is not as important as the certain knowledge that when these objects do stop, they will be arranged in the order in which they have been listed, the steel ball being farthest away. Implicit in this observation is the assumption that the floor offers some sort of resistance to the motion of the objects—less resistance to the motion of the steel sphere than to the golf ball, less resistance to the golf ball than the baseball, and so forth. We elect to study the behavior of the one of these objects that seems to be most free from this troublesome phenomenon of friction. Even as we study the object with the least evidence of frictional resistance, we project our thinking to an ideal that we will never encounter in a physics laboratory, an object entirely free from friction. When we have fully comprehended the behavior of the ideal frictionless object, we can turn our attention to experimental investigations of friction itself. Hopefully we will then be able to blend this knowledge into an understanding of the performance of the nonideal situations.

2-1 NEWTON'S FIRST LAW OF MOTION

If one rolls a sphere across a smooth, horizontal floor, the object will travel with nearly uniform speed in a straight line for a very great distance. If the object encounters an obstruction, its speed and/or its direction of motion may change. If the sphere is initially at rest on a smooth floor, it will remain at rest until someone "pushes" it. These observations constitute special cases of the application of the experimental law known as the Law of Inertia, or Newton's First Law of Motion:

> **An object at rest will remain at rest and an object in motion will continue to travel in a straight line at constant speed unless acted upon by a net external force.**

This statement not only describes all that we have said above about the steel sphere but it also summarizes all the experience scientists have had with this aspect of motion.

FIG. 2-2

Newton's First Law. An object at rest will remain at rest and an object in motion will continue to travel in a straight line at constant speed unless acted on by an external, unbalanced force.

If the statement is accepted, we are required to determine the cause for any seeming departure from the statement. For example, a ball thrown at an angle relative to the horizontal does not travel in a straight line with uniform speed. Furthermore, even the Moon fails to travel in a straight line. Has the Law of Inertia already failed? In these cases, we can "rescue" the Law of Inertia by identifying a force, the force of gravity, which we can say is responsible for the departure of the ball in the one case and the Moon in the other from straight-line motion.

One can see from even this simple illustration that physics is, in a sense, a patchwork. The patches, however, are to be associated with phenomena that have a separate identity and that are subject to separate independent checks. The nature of these independent confirmations will emerge as we progress.

2-2 INERTIAL FRAMES OF REFERENCE

When the Law of Inertia was stated, reference was made to objects "at rest" or "traveling in a straight line at constant speed." It is not enough, however, to state that an object is simply at rest or traveling at constant velocity; one must state that the object is at rest or moving with respect to something: to the Earth, to the laboratory, or to some other material, physical device. The coordinate system relative to which the motion of

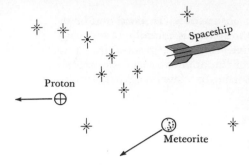

FIG. 2-3

Inertial frames of reference. The proton, the spaceship, and the meteorite constitute separate and equally valid inertial frames of reference provided that they travel with constant velocity with respect to the fixed stars.

objects is measured is called a *reference frame* or a *reference system*. We must examine the universe and locate a reference frame relative to which the Law of Inertia appears to be strictly correct.

Observations carried out by astronomers and astrologers over a period of 3000 years show that the relative positions of most of the stars (as distinguished from the planets in our solar system) have changed very little throughout this vast sweep of history. We therefore treat the fixed stars as the primary physical framework for all applications of Newton's First Law of Motion. Having defined this primary frame, and having ascertained that Newton's First Law of Motion appears to be valid in that frame, we can define an inertial frame of reference as follows:

> An inertial frame is any reference system at rest in, or moving with constant velocity with respect to, the fixed stars.

It is easy to imagine an unlimited number of inertial frames of reference; for example, platforms or perhaps even tiny particles like electrons or protons traveling through space in a straight line, at uniform speed, relative to the fixed stars would be inertial frames. In any of these frames of reference, the Law of Inertia applies. On the other hand, if one should examine any frame and find that it is experiencing an acceleration, whether linear or centripetal, with respect to the stars, one can be sure that the Law of Inertia does not apply. In such a system, a moving object subject to no force whatsoever would appear to travel in some path other than a straight line. As an example, imagine rolling a steel sphere across the floor of an accelerated subway train.

The Earth as an inertial frame

None of us have had any actual experience with an inertial frame of reference because the surface of the Earth does not quite qualify. The Earth rotates about its own axis of rotation once each 24 hours; the equation for centripetal acceleration [Eq. (1-3)] will show that any object at the Equator experiences an acceleration toward the center of the Earth equal to 33.7×10^{-3} m/sec^2. The Earth also travels in a nearly circular orbit about the Sun; an application of the equation for centripetal acceleration will show that each point on this moving Earth is accelerated toward the Sun with an acceleration equal to 4.4×10^{-3} m/sec². Furthermore, the entire solar system may well experience an acceleration due to some translational acceleration of our galaxy through space and/or rotation of the galaxy about its center of mass. Fortunately, all these accelerations are either small enough to be neglected or well enough established that appropriate corrections may be introduced in any calculations in which this departure from the Law of Inertia could affect the results.

2-3 INERTIAL MASS

Let us discuss the acceleration of an object from the vantage point of an observer in an inertial frame of reference. For convenience, we will neglect the small accelerations mentioned above and assume that the laboratory is such a frame. Accordingly, when we refer to the acceleration of an object, we will mean its acceleration with respect to the laboratory. As before, we will idealize the situation, proposing the existence of horizontal surfaces and frictionless carts, and thereby eliminating, for the moment, the need to consider forces other than those that we can apply ourselves.

The prototype of all of our experiments is shown in Fig. 2-4. A long, horizontal, frictionless track is perpendicular to a massive wall fixed to the Earth. The experimenter possesses a group of identical carts and compression springs. One or more springs can be installed between the wall and the cart or a combination of carts. The spring can be held in a compressed configuration by a string that, when burned, will release the spring whereupon it will exert a force on the cart or carts. Pictures of the motion of the object can be taken, from which one can make measurements of the acceleration.

At this point in our argument we do not have a unit in which force can be expressed. Therefore we will regard force simply as that property of the spring that produces acceleration. From our experience with springs we will accept the concept that the force exerted by a spring is not constant; rather, it is large when the spring is compressed

FIG. 2-4

Inertial mass. A cart is propelled forward by a compression spring when the string is burned. The inertial mass of a given cart is compared with that of another cart by comparing their accelerations under the action of an identical spring.

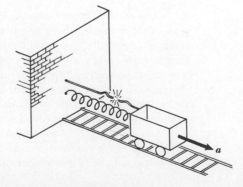

and zero when the spring reaches its equilibrium length. We will
assume that the force exerted by the spring at a specific intermediate
length will be well defined. In addition, we will assume that two identical springs arranged parallel to each other will exert a combined force
exactly twice the force of a single spring.

Experiments with a single spring

We place one spring between a cart and the wall, compress it,
and tie the string, holding the spring in the compressed configuration.
The string is then burned, and a picture of the motion of the accelerating cart is secured. From this picture we determine the acceleration of
the cart at the instant the spring has reached some previously agreed-to
length. We then attach two carts rigidly together and repeat, permitting the same spring to accelerate both carts. One continues making
observations with as many identical carts as he chooses under the
action of the same spring. Our prior experience is adequate to enable
us to predict the result of these experiments: the single cart will experience a larger acceleration than 2 carts, 2 carts will experience a larger
acceleration than 3 carts, and so on.

We can assume that the action of the spring was the same in all
cases; that is, we can assume that the "force" exerted by the spring at a
given length is the same in all cases. Thus there is some property,
related to what we might call the "quantity of matter" in the carts,
that is responsible for the difference in behavior of one cart in contrast
to 2 carts or 3 carts and so forth. This property is called the *inertia*, the
inertial mass, or simply the *mass* of the object.

The data that will be secured by measuring the accelerations
in the various pictures of motion will reveal, within experimental
error, accelerations of a, $a/2$, $a/3$, etc., when masses of m (1 cart), $2m$
(2 carts), $3m$ (3 carts), etc., respectively, were accelerated. These data
show that the ratio of the accelerations for any combination of masses
under the action of a specific force is the inverse ratio of the masses;
that is,

$$\frac{a_1}{a_2} = \frac{m_2}{m_1} \qquad (2\text{-}1a)$$

Calibrating a set of inertial masses

The experiments we have just described would enable one to cali-
brate a set of masses. Suppose that the mass of a single cart were to be
given a name; the mass of a cart might be called a "kilogram." Combinations of carts would then constitute a set of masses equal to 2 kg,
3 kg, and so on. A half-kilogram cart could be manufactured and
checked to see that it actually receives an acceleration twice that given
to the 1-kg cart. Furthermore, an arbitrary mass could be measured by
comparing the acceleration it would receive to that experienced by a
1-kg cart. Masses so determined would thereby be expressed in terms
of the arbitrarily chosen initial mass, or in the "system of units" in
which the kilogram is the arbitrarily chosen unit of mass.

Force is the factor that tends to produce an acceleration of a mass relative to an inertial frame of reference. In the experiments described in the preceding section, a spring produced an acceleration. The spring therefore exerted a force. The qualifying phrase "tends to produce" was used because we know that force cannot produce an acceleration if an equal but opposite counterforce acts on the object. A man pushing on a wall will not ordinarily cause the wall to accelerate.

The statement that force is the factor that tends to produce an acceleration carries with it some subtle requirements for the correct use of the term. Let us think of two examples.

1. A car travels down a highway at uniform velocity; that is, its acceleration is equal to zero. We may be able to identify many forces acting on the car, but since the acceleration equals zero, the net force must also equal zero. It is the net or unbalanced force that has a direct mathematical relation to the acceleration.

2. It is well known that an object at a great distance from the Earth (say 4000 miles) experiences an acceleration toward the Earth even though the object is far outside the atmosphere of the Earth and away from any identifiable physical means by which the Earth can pull on the object. Because the object experiences an acceleration, however, we know that the object experiences a force and that the Earth apparently can exert a force on an object even though no visible or physically observable strings link the Earth to the object. In other words, our definition of force requires us to accept the notion that "action at a distance" is possible.

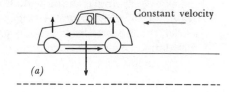

(a)

FIG. 2-5

Force. (a) If the car is traveling at constant velocity, the net forward force must equal the net backward force and the net upward force must equal the net downward force. (b) If the satellite experiences an acceleration, we know it experiences a force.

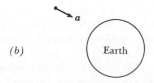

(b)   Earth

In summary, if an object travels at constant velocity relative to an inertial frame, we know that the net force on the object equals zero. If the object is being accelerated relative to an inertial frame, we know that a force is present.

Newton's Second Law concerns the manner in which the velocity of an object changes when the object is acted on by an unbalanced force. The law may be stated as follows:

If an object is acted on by an unbalanced force, the object will experience an acceleration proportional to that force.

Let us examine the experimental basis for this law. We have defined force as that which tends to produce acceleration; furthermore, Eq. (2-1a) may be rewritten as follows:

$$m_1 a_1 = m_2 a_2. \tag{2-1b}$$

That is, when a specific force (one spring) acts successively on each of several objects, the product ma has a single well-defined value for all the objects.

Experiments with two, three, etc., springs

We may now recall that in Sec. 2-3 we defined a process by which we could place two springs side by side and thereby secure a force that is twice as large as the force due to a single spring. Experiments show that when two springs are used, the value of ma common to all the objects involved in the observations is precisely twice that observed when one spring is employed. Furthermore, three springs result in a value of ma three times as large as that observed with one spring; in short, the product ma is proportional to the force. Because we have not yet defined a unit of force, we are at liberty to identify the product ma with the force and write

$$F = ma \tag{2-2}$$

We have set the force identically equal to the product ma and are therefore in a position to state the unit in which force must be expressed. Applying the Units Rule, and recalling that mass is measured in kilograms while acceleration is measured in meters per second squared, the unit of force is *kilogram meters per second squared*.

Force plays a very important part in physics, and the word must therefore be used many times in our discussions of physical phenomena. It becomes inconvenient to have to print or say "kilogram meters per second squared" every time we wish to make reference to the amount of force involved in some situation. For this reason, the unit has been "nicknamed" the *newton*(N); this nickname will be used as the MKS unit of force throughout this book, that is,

$$1 \text{ newton} \equiv 1 \frac{\text{kg m}}{\text{sec}^2}$$

where the symbol $\equiv$ means "identically equal to." The full name of the unit of force is kg m/sec², a fact that the reader should keep in mind because he will have occasion to apply the Units Rule to other equations in which force appears.

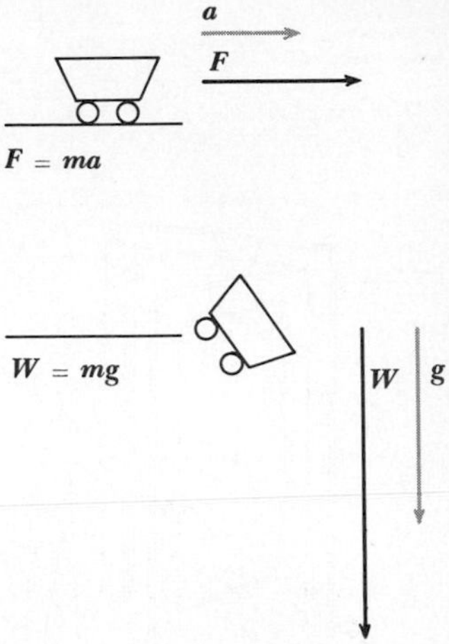

2-6 WEIGHT

We have all been aware since childhood that objects fall toward the ground whenever given the opportunity. Our more recent experiences would lead us to say this in a slightly more sophisticated fashion. Hopefully, the reader might say something like "any object released in the vicinity of the Earth will experience an acceleration toward the center of the Earth." Through Newton's Second Law this statement is tantamount to saying that the Earth exerts a force on any object in its vicinity. This force is called the force of gravitation or the *weight* of the object. Because the weight is a force, it will be measured in the same units as any other force, namely, in newtons.

A freely falling object near the surface of the Earth experiences an acceleration of 9.8 m/sec²; even a feather experiences this acceleration if permitted to fall in a vessel from which the air has been removed. However, at an altitude of about 1700 miles a freely falling object experiences an acceleration due to gravity that is half this amount, 4.9 m/sec². In either case, the acceleration is due to the force of gravity acting on the object. Letting g represent the acceleration of gravity and w represent the weight or the force that produces this acceleration, Eq. (2-2) becomes

$$w = mg. \tag{2-3}$$

Equation (2-3) says nothing new because it only expresses the relationship between a force and the acceleration produced by the force. The equation, however, should enable one to distinguish between mass, a scalar measure of the inertial property of an object, and weight, a vector quantity that equals the force exerted by some external agent on the object. According to the equation, a kilogram object near the surface of the Earth has a weight of 9.8 newtons (N); the same 1-kg object has a weight of 4.9 N when at an elevation of 1700 miles above the surface of the Earth. The mass of an object is an unchanging property of the object;* the weight of an object depends on the circumstance in which the object is placed.

---

* The statement that the mass of an object is an unchanging property of the object ignores the conclusion of the theory of relativity that the mass of an object increases with its velocity. Our statement remains valid within the precision with which measurements can be carried out, provided that the velocity of the object is less than approximately 20,000 miles per second!

$F = ma$

$W = mg$

FIG. 2-6

Weight is a specific force; g is the acceleration produced by a force equal to the weight of the object.

2-7 NEWTON'S THIRD LAW OF MOTION

Consider a person opening a door. Two objects are involved, the door and the hand of person opening the door. From the point of view of that person, his hand exerts a force against the door but the door also exerts a force against the hand. These two forces are, in fact, equal in magnitude but opposite in direction.

Whenever any object experiences a force, another object or agent is present to produce that force. Thus we cannot conceive of a force acting on an object without the presence of another object. Object *A* exerts a force on object *B*; object *B* exerts an equal and opposite force on *A*. One of these forces we may call the *action*; the other we will call the *reaction*.

These observations are summarized by Newton's Third Law as follows:

FIG. 2-7

Action and reaction. The force exerted by the hand against the door is equal but opposite to the force exerted by the door against the hand. Only the force represented by the colored arrow acts on the door.

When two objects interact with each other, the force exerted by the first object on the second is equal and opposite the force exerted by the second object on the first.

One must be careful not to confuse the situation depicted in Fig. 2-7 with the conditions described by Newton's First Law (Fig. 2-5(*a*). It is true, of course, that we show two equal but opposite forces in both figures, but the forces shown in Fig. 2-7 do not cancel each other. Only one of the forces acts on the door, and it therefore constitutes an unbalanced force as far as the door is concerned. In the absence of friction or some other retarding force, the door will be accelerated.

2-8 CENTRIPETAL FORCE

Newton's First Law states that, in the absence of an unbalanced force, any object will move at constant speed *in a straight line*. If this expression is to be taken literally, it must be assumed that an object traveling in a circular path at constant speed experiences an unbalanced force.

We have already seen in Sec. 1-6 that an object traveling at uniform speed in a circular path experiences an acceleration

$$a_c = \frac{v^2}{r} \tag{1-5}$$

The unbalanced force that must be applied to an object to constrain it to travel in a circle will therefore be given by

$$f_c = \frac{mv^2}{r} \tag{2-4}$$

Since the centripetal acceleration is directed toward the center of the circle in which the object is traveling, this force must act toward the center of the circle also. We call this force the *centripetal force*.

Writers of physics text books have yet to resolve a problem in semantics that undoubtedly has caused untold anguish to students. We speak of gravitational force, frictional force, and centripetal force. The first two terms describe a force due to an agent or an agency: a force due to gravity or a force due to friction, both of which are forces acting on the object. The word "centripetal," however, is a word that only tells something about the geometry of the situation; it is a "central" force that must be provided to produce the centripetal acceleration, perhaps provided by gravity, the pull of a string, or any one of several other possible agents or agencies.

Figure 2-8(*a*) is a strobe photograph of an object on a smooth, horizontal surface traveling in a circular path about a central pin and held in its circular path by means of a string. The string exerts a force on the object toward the center of the circle; this force, as shown in Fig. 2-8(*b*), produces an acceleration in the direction of the force. When the string force ceases to act (the string is burned when the puck reaches the "12 o'clock" location), the object proceeds to travel in a straight line.

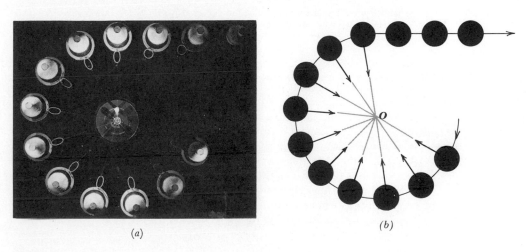

(*a*)   (*b*)

FIG. 2-8

Centripetal force. The pull of the string toward the center of the circle produces an acceleration toward the center of the circle. After the string is burned at the 12 o'clock position, the object travels in a straight line.
(*Photograph from PSSC*, Physics, *D. C. Heath & Co., Lexington, Mass. 1965*)

Fictitious forces

Some readers may feel at this point that our statement about the direction of the force is in conflict with experience. He may say, "But when I am going around a curve in an automobile, I experience a force out from the center, not toward the center!" It is true that the driver of a car making a right turn may exert an outward force against the door, but we are discussing here the force against the driver, not the force against the door. The door exerts a force against the driver toward the center of the circle. One needs only to think of the unpleas-

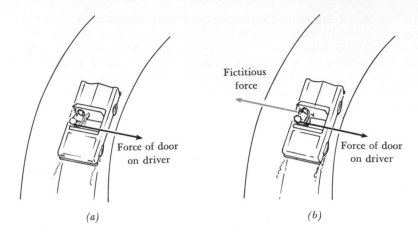

FIG. 2-9

(*a*) Driver viewed from an inertial frame. The driver experiences the unbalanced force exerted by the door against his shoulder. This force is toward the center of the circle and produces the centripetal acceleration. (*b*) Driver viewed from a non-inertial frame. The driver of the car unconsciously thinks of motion relative to his immediate environment. He tends to move relative to this environment (the car) and states that he experiences a force out from the center. This is a fictitious force.

ant consequences that could result were the door to be faulty and fail to exert a force against the driver toward the center of the circle. The car would continue to follow the curved road; the driver would obey Newton's First Law and travel in a straight line directly into the ditch!

In thinking about the preceding example, it would be well to keep in mind that the car is not an inertial frame of reference; rather, because it is being accelerated, it is a noninertial or accelerated frame. The driver in this frame may feel a very vivid tendency to travel to one side as the car goes around a curve, and he may wish to identify this tendency with a force. Having spent most of his life in an inertial frame, he tends to transpose his experience in inertial frames to the accelerated frame. However, such a "force" cannot be ascribed to an identifiable agent and therefore will be called a *fictitious force*.

2-9 FRICTIONAL FORCES

In the opening sections of this chapter it was stated that a book dealing with the basic principles of physics must start by dealing in an individual fashion with physical phenomena. In doing so we found it appropriate to talk of frictionless carts, frictionless tables, and horizontal surfaces. In the real world surfaces are never quite frictionless nor horizontal. We must now complicate the situation by introducing some thoughts about how friction may affect the motion of an object in a number of specific cases.

Assume that a workman were to attempt to push a box weighing

300 N across a horizontal but rough floor. (The mass of the object is approximately 30.6 kg or 67.3 lb.) He starts cautiously, exerting a force of only 10 N and the box does not move. Because, by Newton's First Law, the net force must equal zero, the frictional force must be 10 N but in the direction opposite the direction of the applied force. The workman increases his force to 30 N, but the box still does not move; the frictional force is now equal to 30 N. He increases the force he is exerting until, at the instant the applied force barely exceeds 59 N, the box starts to move across the floor at constant velocity. The frictional force is now a steady force of 59 N and in a direction opposite the motion.

Experience shows that the force of friction does not change appreciably with velocity—that is, for the case cited, a force of 59 N will maintain the box at a low velocity or once set in motion, at a high velocity as well.

If the workman exerts a force in excess of that required to maintain a constant velocity, the box will accelerate. For example, if the workman exerts a force of 69 N, the extra 10 N by which his applied force exceeds the force of friction will constitute an unbalanced force that will produce an acceleration. Since the acceleration is equal to the unbalanced force divided by the mass, in this case, the box would experience an acceleration equal to 10 N/30.6 kg, or 0.327 m/sec². Since the frictional force is independent of the velocity, the box will continue to gain speed at this rate indefinitely.

FIG. 2-10

Frictional forces on a horizontal surface. The frictional reaction force will be equal to the force exerted by the workman provided that the box remains at rest or in uniform motion.

FIG. 2-11

The accelerated box. If the workman exerts a force larger than the frictional force, the excess will constitute an unbalanced force and the box will accelerate.

2-10 NEWTON'S LAW OF UNIVERSAL GRAVITATION

We have had several occasions to draw attention to the fact that the acceleration of gravity is not uniform but varies from place to place in the vicinity of the Earth. But its numerical value does not differ very much from one point on the surface of the Earth to another. It is equal to 9.78243 m/sec² in the Canal Zone, 9.79609 m/sec² in Denver, Colorado, 9.82534 m/sec² in Greenland, and 9.80629 m/sec² at sea level at 45° latitude. It decreases markedly if one leaves the Earth. For

example, if one doubles his distance from the center of the Earth by
ascending to an altitude of 6380 km ($\approx$ 4000 miles), the numerical
value of g falls to one-fourth its value at the surface of the Earth,
approximately 2.5 m/sec².

The force of gravitation is responsible for the acceleration an
object experiences. Hence the force of gravitation or the weight of an
object varies in the same manner as the acceleration referred to above.
A 1-kg object will weigh 9.78243 N in the Canal Zone, since, from
$w = mg$, the weight of a 1-kg object is numerically equal to the local
value of the acceleration of gravity.

Experience shows that in calculations of the force of gravity one
may treat an extended object such as the Earth as if the entire mass
were located at a point that we will call the *center of mass*. This situation
can also be shown mathematically by regarding the object as an array
of individual mass points and determining their combined effect. In
the case of a symmetric object such as a sphere, the center of mass
corresponds to the physical center of the object itself.

The variation of gravitational force is described by Newton's Law
of Universal Gravitation, according to which each object in the universe
exerts a force of attraction on every other object in the universe. A given
object whose center of mass is located at a distance r from another
object exerts a force of attraction on it, and, by Newton's Third Law,
the second object exerts an equal but opposite force on the first. The
magnitude of the force is proportional both to the mass of the first object
and to the mass of the second object; the force is therefore proportional
to the product of the masses. The force of attraction decreases if the
objects are moved apart, falling off inversely with the square of the
distance between the objects. Letting m and m' represent the masses of
the two objects, r the distance between them, and G a constant of
proportionality, Newton's Law of Universal Gravitation may be written in the form

$$F = -\frac{Gmm'}{r^2} \tag{2-5}$$

The minus sign signifies a force of attraction.*

Had Newton been able (*a*) to put an Earth satellite into an orbit
similar to that shown in Fig. 1-21 and (*b*) to track it, he would have
secured a "picture" of its motion. Undoubtedly this picture would have
led him to the observation that the acceleration of the satellite varies
inversely with r^2, as we saw in Sec. 1-7. His equation, $F = ma$, would
then have produced Eq. (2-5).

Later (Sec. 4-4) we will present an alternate logic by which one
may arrive at this equation. For the moment, however, it will suffice to
point out that the verification of this force law is made possible by
correlations of data in a form known as Kepler's laws, which describe
the motions of planets in our solar system. As such, the Law of Universal Gravitation is clearly not "universal" in the sense of being

---

* A minus sign is used because the direction of the force is such as to decrease
the distance between m and m'.

clearly applicable to objects at remote positions in space. It is difficult to choose a date to associate with the acceptance of the assumption of the universality of this law. At the least, one can say that this assumption has endured a century of debate and further experimentation.

Determining G

The constant of proportionality G, which appears in Eq. (2-5), cannot be determined from study of the planets alone, because the masses m and m' of the interacting objects cannot be known until after G has been determined. Therefore G must be determined from measurements of the force acting between known masses within a laboratory. The gravitational force between any two objects of "laboratory" size is extremely small, but such a force can be measured by a delicate instrument known as a "Cavendish balance," first used for this purpose by Sir Henry Cavendish in 1798. With a typical model of such a balance, it is possible to measure directly the force that a sphere having a mass of 0.15 kg exerts on another sphere having a mass of 1.5 kg, when their centers are at a distance of 0.10 m apart. The observed force is equal to 1.5×10^{-9} N. (The force one exerts in lifting a postage stamp is 10,000 times larger than this!) Introducing the above data into Newton's equation [Eq. (2-5)], one may solve for G and find

$$G = 6.67 \times 10^{-11} \text{ N m}^2/\text{kg}^2$$

It is useful to express g, the acceleration of gravity, in terms of G, the universal gravitational constant. Consider a mass m at the surface of the Earth, whose mass is m_e. The mass is located at a distance r from the Earth's center of mass. The force acting on this mass from Eq. 2.3 can be expressed by

$$F = -mg$$

where we use a negative sign because the force of gravity is always down (hence negative), and, by custom or habit, physicists always express g as a positive number; alternately, from Eq. (2.5) it can be expressed by

$$F = -\frac{Gmm_e}{r^2} \tag{2-5}$$

FIG. 2-12

The Cavendish balance. Masses m and m' are at the ends of a lightweight rod suspended by a thin torsion fiber. When masses M and M' are placed as shown, the gravitational force of M' on m' and of M on m results in a twist of the fiber. The amount of twist (and therefore the force) is measured by observing the change in position of the reflected spot of light.

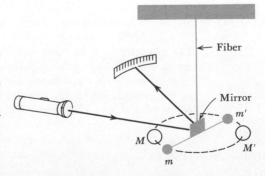

Because both equations express the weight of the object, we can equate the right-hand sides of the preceding equations and cancel m from the resulting equation to find

$$g = \frac{Gm_e}{r^2} \qquad (2\text{-}6)$$

At the Earth's surface, of course, r is equal to the radius of the Earth ($= 6.378 \times 10^6$ m).

Measurements of g, r, and G enable one to calculate m_e, the mass of the Earth, but because we have not discussed Newton's Law of Universal Gravitation in detail, we will defer this calculation until Sec. 5-6.

2-11 THE "FORCE-FIELD" INTERPRETATION OF g

The gravitational force on an object placed at a point in space at an enormous distance from any of the planets or from any of the many stars that dot our sky will be negligibly small. Such a point in space, therefore, may be regarded as neutral as far as gravitation is concerned. A point at a distance of a foot or even at a distance of 240,000 miles from the Earth's surface is not, however, "neutral" because an object will experience a force if placed there. Although we have very little knowledge of the mechanism by which the Earth exerts a force on an object in space, there is no doubt about the existence of the force. Any point in space, however, is affected by the presence of a neighboring massive object. Because all points in the vicinity of a massive object are affected by the object, we can say that a "region of influence" surrounds such an object or that a gravitational field exists in the space about the object.

A convenient way to describe the gravitational field at a point in space is to state both the magnitude and the direction of the force that a 1-kg object experiences if placed there. This situation is expressed by the *gravitational field intensity*, defined as the force that would be experienced by a 1-kg object if placed at the point under investigation. Thus from Eqs. (2-5) and (2-6) respectively

$$\text{Gravitational field intensity} = \frac{F}{m} = -\frac{Gm_e}{r^2} \qquad (2\text{-}7a)$$

hence
$$\text{Gravitational field intensity} = -g \qquad (2\text{-}7b)$$

The negative signs in these equations serve to remind us that the force on an object due to gravity is downward, the negative direction.

Two interpretations of g are therefore possible. It may be regarded as the magnitude of the acceleration that an object will experience when placed at a point in space, and is expressed in meters per second squared; or it may be regarded as the magnitude of the gravitational field intensity, in which case it is expressed in newtons per kilogram. Although it is apparent that the units employed in the two interpreta-

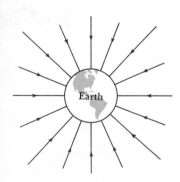

FIG. 2-13

The gravitational field of the Earth. The direction of the Earth's gravitational field at a point in space is the direction of the force on an object if placed there.

tions are fully equivalent, there is some advantage to using the units that are consistent with the interpretation one has in mind.

Lines of gravitational force

The gravitational field in the external vicinity of an object can be represented by radial lines drawn toward the center of the object as in Fig. 2-13. Any one of these lines describes the direction of the force experienced by an object placed on the line.* It would be a gross misinterpretation of the sketch, of course, to suggest that there are strings or bands connecting the Earth to the object.

The variation of the gravitational field intensity in the external vicinity of the Earth is shown in Fig. 2-14. This graph was secured by introducing into Eq. (2-7a) the now-known mass of the Earth (5.98×10^{24} kg), the known value of G (6.67×10^{-11} N m²/kg²), and a range of values of r.

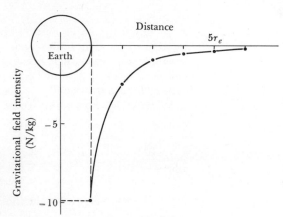

FIG. 2-14

The gravitational field of the Earth. The magnitude of the gravitational field intensity of the Earth at a point in space is equal to the force experienced by a 1 kg mass if placed there. The force varies inversely as the square of the distance from the center of the Earth.

An inertial frame of reference has been defined as a coordinate system at rest in, or moving with constant velocity with respect to, the fixed stars.

SUMMARY

Newton's Laws of Motion are experimental laws that appear to be valid for objects in an inertia frame.

1. The Law of Inertia: An object at rest will remain at rest and an object in motion will continue to travel in a straight line at constant speed unless acted on by a net external force.

2. If an object is acted on by an unbalanced force, the object will experience an acceleration proportional to that force. This may by written as

$$F = ma \qquad (2\text{-}2)$$

* We will limit our attention to the gravitational field outside an object. It can be shown, however, that inside a uniform sphere the gravitational field decreases linearly and becomes equal to zero at the center.

3. When two objects interact with each other, the force exerted by the first on the second is equal but opposite the force exerted by the second object on the first object.

The inertial property of an object is known as its inertial mass and, in MKS units, is measured in kilograms.

The weight of an object is the force the object experiences due to gravitation; thus, like force, it must be expressed in newtons (kg m/sec²). This force, acting on a freely falling object, produces the acceleration of gravity. Hence

$$w = mg \qquad (2\text{-}3)$$

The net force that acts on an object to constrain it to travel at constant speed in a circular path produces the centripetal acceleration and thus is called the centripetal force. By Newton's Second Law, then,

$$F_c = m\frac{v^2}{r} \qquad (2\text{-}4)$$

Any force acting on an object in an inertial frame must be attributable to an identifiable agent. Apparent departures from Newton's laws that one may observe from the vantage point of an accelerated or noninertial frame cannot be ascribed to an identifiable agent and are called "fictitious forces."

Newton's Law of Universal Gravitation is

$$F = -\frac{Gmm'}{r^2} \qquad (2\text{-}5)$$

where $\qquad G = 6.67 \times 10^{-11} \, \text{Nm}^2/\text{kg}^2$

The gravitational field that pervades the space around an object may be defined at any point in that space as the force that a kilogram mass would experience if placed there. This yields an alternate "force-field" interpretation of g, since from Eqs. (2-5) and (2-3) respectively

$$\text{Gravitational field intensity} = -\frac{Gm_e}{r^2} \qquad (2\text{-}7a)$$

$$\text{Gravitational field intensity} = -g \qquad (2\text{-}7b)$$

QUESTIONS
AND
PROBLEMS

2-A1. People are inclined to believe that a moving object must have a net force acting on it to "maintain the motion" of the object. Can you explain why this impression is widely accepted? How would you proceed to explain the fallacy of this notion?

2-A2. Assume that you perform some very simple physics experiment on an airplane while in flight; for example, you throw and catch a ball or time a vibrating object suspended by a spring. In what way(s), if any, would the results of your experiment be affected by the motion of the plane?

2-A3. What is the acceleration of a ball at the instant it reaches the highest point of its flight?

2-A4. A mass of 12 kg resting on a smooth, horizontal table experiences a steady horizontal force of 40 N. (*a*) What is the weight of the object? (*b*) What acceleration does it experience?

2-A5. A man places a block of ice on a long, rapidly moving conveyor belt. Because the frictional resistance is so small, the block of ice does not attain the speed of the belt at once; rather it slides as it gains speed. Make a sketch of the situation showing the block (*a*) shortly after it is placed on the belt and (*b*) after it has attained the speed of the belt. Draw all arrows* that represent forces acting *on the block of ice* at the stages mentioned.

2-A6. Consider a baseball that has been thrown at an angle of 45 degrees relative to the horizontal. Sketch the trajectory of the ball and, at representative points along the path, draw arrows* depicting forces acting *on the ball*.

2-A7. Figure 1-17(*d*) depicts position versus time for an object in vibration while suspended by a spring. Make a sketch of this figure and, on the sketch, draw arrows at a representative number of positions showing the forces acting *on the object*.

2-A8. What is the distinction between mass and weight?

2-A9. A "1-1b object" has been identified in this textbook as an object which possesses a *mass* of 1 pound. However, you undoubtedly express your weight (a force) in pounds. (*a*) Assuming that we will continue to regard a pound as a unit of mass, what meaning can we ascribe to "a force of 1 pound?" (*b*) What would "a force of 1 kilogram" mean?

2-A10. A sturdy rope is used to constrain a massive object to travel in a circular path about a fixed support. (*a*) Does the rope exert a force on the object? (*b*) Does the rope exert a force on the fixed support? (*c*) Does the object exert a force on the rope? (*d*) How do these forces compare in magnitude and direction? (*e*) Which, if any, is a centripetal force? (*f*) How does the direction of the centripetal force compare with the direction of the acceleration of the object?

2-B1. A roadway offers a constant frictional resistance force of 300 N to the motion of a 900-kg car. What driving force is needed to give this car an acceleration of 3.5 m/sec^2?

2-B2. The sketches in Fig. 2-15 depict an enthusiastic sports fan expressing his feeling by jumping up. The first sketch shows him in the process of increasing his speed in the upward direction; the second shows him a moment after he has left the ground; and the third sketch shows him at the apex of his jump. Trace the sketches and draw arrows* on your sketches showing all forces acting *on the man* at each stage. Explain the agency responsible for each force that you have identified.

---

* In Probs. 2-A5, 2-A6, 2-A7, 2-B2, and 2-B3, in which arrows are to represent forces experienced by some object, note that the arrows should have lengths in proportion to the relative magnitudes of the forces being represented. As examples, see Figs. 2-5, 2-9, 2-10, and 2-11.

FIG. 2-15

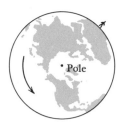

• Pole

FIG. 2-16

2-B3. Figure 2-16 depicts a man standing on the Earth at the Equator. Resketch this scene and show all forces acting *on the man*. Given that the man has a mass of 70 kg, that the centripetal acceleration at the Equator equals 0.034 m/sec², and that a free-fall experiment shows $g = 9.780$ m/sec², determine the magnitude of each of the arrows you have drawn.

2-B4. (a) What is the gravitational force acting on a 70-kg object on the surface of the Earth?

(b) What is the centripetal force acting on a 70-kg man in orbit at a distance of 600 km above the surface of the Earth? (The radius of the Earth equals 6378 km.)

2-B5. The gravitational force acting on a 180-lb man at sea level equals 802 N. Calculate the gravitational force on this man at several altitudes up to about 20,000 km. Prepare a graph of the gravitational force as a function of position using these data. In what way, if any, would this question be changed if we had used the word "weight" instead of "gravitational force"?

2-B6. An astronaut approaching a moon of one of our planets estimates its radius to be approximately 1000 km. When he lands on this moon, he finds that his weight is one-tenth that of his weight on Earth. From this information and the known properties of the Earth (see Appendix 3), compute the mass of the moon on which he has landed.

Work, Energy, and Momentum

The nonscientist is inclined to associate the word "work" with any activity that involves effort, fatigue, or discipline. A porter holding a suitcase may feel that he is doing "work"; a student reading this text may say, "This is hard work"; an executive presiding over a meeting may say that he is "at work." In scientific usage, however, work is defined in such a way as to express the visible output of the action of a force; and therefore neither the porter, the student, nor the executive is credited with doing any significant amount of work when engaged in the activities described above. Because we limit ourselves to quantities that can be objectively measured, we look only at the mechanical changes that take place as a result of an effort; we are not interested in the effort itself. Neither the porter, the student, nor the executive has produced a clear-cut mechanical change as a result of his effort. Although the student could point to the fact that he has turned several pages, and the executive could point to notes that he has prepared during the meeting, the visible effect of the activities we have described is almost nil. Hence, physically, we cannot credit any of these "workers" with having done a significant amount of work.

3-1 WORK

We will define *work* objectively and precisely through the following definition:

Work equals the product of the effective force in the direction of the motion times the distance traversed during the motion.

In equation form, letting f represent the effective force in newtons and s represent the distance traversed by the force, we may write

$$W = fs \qquad (3\text{-}1)$$

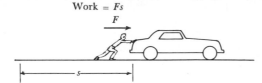

Work = Fs

F

FIG. 3-1

Work. The work done by the man in pushing the car is equal to the force he exerts multiplied by the distance through which the force acts.

The work W done by the force may then be expressed, by the Units Rule, in newton-meters, or kg m²/sec², or "joules." All three units are equivalent; the last name, honoring J. P. Joule for his contributions to our understanding of work and energy, is a nickname in the same sense that the word "newton" is a nickname for "kg m/sec²." We will express work in joules almost exclusively throughout this textbook, but it is important to remember the full name of the unit as well as its abbreviation.

In order to visualize the amount of work represented by the joule, one may note that a typical college textbook weighs approximately 5 N. One would do approximately 1 joule of work if he were to lift a textbook vertically upward a distance of 0.20 m (approximately 8 inches).

The effective force

In our definition of work we have used the term "effective force." The effective value of any vector quantity is the component of the vector in the desired direction. In this case, the desired direction is the direction of the motion. (See Sec. 1-1 to review the technique by which one may determine the component of a vector.) Most of the

FIG. 3-2

(a) Force parallel to the motion. In this case the force is fully "effective" and the work is equal to the force times the distance. (b) Force perpendicular to the motion. No work is done by a force that acts in a direction perpendicular to the motion.

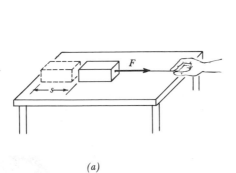

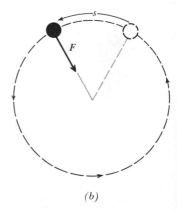

(a)

(b)

situations we will encounter, however, will involve one or another of the following extreme situations:

1. The most common situation will be that in which the force acts in the direction of the displacement. In Fig. 3-2(a) an object that is free to move in any direction is acted on by a force and caused to travel in the direction of that force. In this situation, the force is fully "effective," and one simply multiplies the magnitude of the force by the distance traversed in order to determine the amount of work done by the force.

2. A number of situations arise in which the force acts in a direction perpendicular to the motion. In Fig. 3-2(*b*) a string is depicted exerting a force on an object traveling in a circular path; this is the centripetal force discussed in Sec. 2-9. Such a force serves only to change the direction in which an object travels; under the circumstances shown here, the force neither increases nor decreases the magnitude of the velocity of the object. Such a force is completely "ineffective" in the sense in which we use the term; its "effective value" is zero and no work is done on the object.

Our unambiguous definition of work

The thoughtful reader may well reflect on the extent to which the definition of work that we have adopted constitutes a withdrawal of physics from consideration of a larger number of matters that may concern him in his social, cultural, or political life. Our definition of work makes no allowance for fatigue or effort and eliminates from further consideration a host of problems that then become the domain of psychology or physiology. Physics chooses to concentrate its attention on the object that experiences a force rather than on the agent that provides the force. Thus this branch of knowledge might well be said to be depersonalized and more involved in effects than in causes. One might even wonder if a discipline that arbitrarily chooses to limit its concern so drastically would reserve enough working space to arrive at meaningful conclusions. However, this has not been a problem; sufficient terrain has been left to permit physics to make substantial contributions to an enormous technological and philosophical revolution. It is perhaps more meaningful for us to notice that physicists had no choice about this limitation. Physics became an "exact" science not because measurements were exact (measurements are never exact) but rather because physicists were exact in their definition of terms. It is unlikely that any other definition of work could have been as free from ambiguity as the one we have given above. The definition of work is basic to all that follows; had this term been defined in an ambiguous way and its meaning been subject to debate, all conclusions that might have followed the definition would have been subject to the same debate and doubt.

Before considering the physicists' definition of energy, let us reflect on the way this term is used in everyday life. A man is referred to as "energetic"; such a man is apt to be a busy and active individual. Although such a person might be bumbling and ineffective, the term energetic is usually applied to a person capable of getting things done. A person who is tired or ill might say that he "lacks energy" and that he is not capable of doing the amount of work that he would be able to do under more favorable circumstances.

Various forms of energy

We will define energy as *the capacity of an object or system of objects to do work*. Hence the physicists' definition of energy corresponds almost identically to the normal usage of the term illustrated by the foregoing paragraph. The energy possessed by an object equals the amount of work it can do under the most favorable circumstances; in talking about the energy of an object, we may ignore all the practical details that may be involved in actually realizing the complete conversion of energy into work.

Energy may exist in many forms. A lump of coal may be burned to heat a boiler, which in turn may drive a turbine; the coal possesses *chemical* energy. Water that has fallen over Niagara Falls may, on impact, cause a turbine to do work; the water possesses energy of motion or *kinetic* energy. A weight lifted high above the ground may possess energy due to its position, and a coiled watch spring may possess energy due to its state of strain; the weight and the watch spring possess *potential* energy. The water in a boiler may possess energy associated with its elevation of temperature; we say that it possesses *thermal* energy. In later chapters we will have occasion to extend this list: *electrical* energy will be discussed in Chapter 8 and *nuclear* energy will be discussed in Chapters 15 and 16.

Since energy is measured by the work an object can do, energy, like work, will be expressed in joules.

3-3 KINETIC ENERGY

The kinetic energy of an object is the work it can do because of its motion.

Let us examine the process by which an object in motion can be made to do work. From Newton's First Law we know that an object in motion will continue to travel in a straight line unless acted on by external force. If some agency exerts a force on such an object in a direction opposite its direction of motion, the object in turn will exert a force on that agency. This force will continue to act over the distance traversed by the object in being brought to rest, and therefore the object does work on the agent. Let us examine the work done on an agent in more detail.

Consider an object of mass m traveling with a velocity v. Let an external agent exert a steady force F on the object in a direction such as to oppose the motion; the object, by Newton's Third Law, will exert an equal but opposite force F' against the agent, that is,

$$F' = -F \qquad (3\text{-}2)$$

The constant force F will produce a uniform acceleration of the object in the direction of F—hence in a direction opposite the direction of its motion: this "deceleration" may be expressed by Newton's Second Law:

$$F = ma$$

where a is the acceleration experienced by the object and F is the force acting on it.

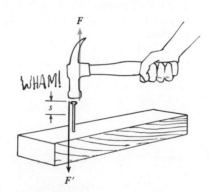

FIG. 3-3

Kinetic energy of a moving object. The moving hammer possesses the capacity to do work. When it engages the nail, it exerts a force F' on the nail. The reaction force F acts on the hammer and ultimately stops it.

We must now digress briefly to secure a relationship between velocity and acceleration in the case of uniformly accelerated motion. Letting v_0 and v represent the initial and final velocities, respectively, we may write

Change of velocity = acceleration × elapsed time

or
$$v - v_0 = at \qquad (3\text{-}3)$$

and   Distance traveled = average velocity × elapsed time

or
$$s = \left(\frac{v + v_0}{2}\right) \times t \qquad (3\text{-}4)$$

Dividing Eq. (3-3) by Eq. (3-4), we can eliminate the elapsed time to obtain

$$\frac{v - v_0}{s} = \frac{a\cancel{t}}{\left(\dfrac{v + v_0}{2}\right)\cancel{t}}$$

Cross-multiplying, one finds

$$v^2 - v_0^2 = 2as \qquad (3\text{-}5)$$

For the case in which an object is brought to rest, the final velocity v is equal to zero.

We are now in a position to obtain an expression for the kinetic energy of the moving object because its kinetic energy is defined as the work it can do in being brought to rest. Therefore we have

$$\text{Kinetic energy} = F's = -Fs$$
$$= -mas$$
$$= -m\frac{-v_0^2}{2}$$

or
$$\text{Kinetic energy} = \tfrac{1}{2}mv_0^2 \qquad (3\text{-}6)$$

In most applications, we will wish to express the kinetic energy in terms of the velocity possessed by the object; the subscript will usually be omitted.

It is instructive to repeat the preceding argument computing the work done by an object in being uniformly decelerated from an initial velocity v_0 to a final velocity v, where v is not equal to zero.

$$\text{Work done by object} = F's = -Fs$$
$$= -mas$$
$$= -m\left(\frac{v^2 - v_0^2}{2}\right)$$
$$\text{Work done by object} = \tfrac{1}{2}mv_0^2 - \tfrac{1}{2}mv^2 \qquad (3\text{-}7)$$

Although no new information is secured from this more general treatment of the problem, the resulting equation underscores the fact that the work done by the object is equal to the decrease in the quantity $\tfrac{1}{2}mv^2$; that is, it is equal to the decrease in the kinetic energy. The composite quality of an object expressed by $\tfrac{1}{2}mv^2$ is therefore an important property of any object in motion.

3-4 POTENTIAL ENERGY

The potential energy of an object is the work it is capable of doing because of its position or state of strain.

An object at a height h above the ground is capable of doing work in descending to the ground; an elongated spring is capable of doing work in the process of returning to its equilibrium length. Each of these objects therefore possesses potential energy.

It is always necessary to identify a *reference position* at which an object possesses zero potential energy. It is rather natural to say that a spring in its most relaxed condition possesses no potential energy; it is also quite natural to associate zero potential energy with a mass that is completely removed from the gravitational field of the Earth. Other choices of "zero" potential energy are possible, however. For example, the low side of Niagara Falls is 123.0 m above sea level while the high side is 50.8 m higher, 173.8 m above sea level. An engineer interested in making use of the energy possessed by the water above the Falls would be inclined to think only of the amount of work the water would be capable of doing in descending to the low side of the Falls. Another observer might well want to make some point of the fact that the water below the Falls still possesses energy relative to water at sea level. As we shall see in the discussion below, one may choose whatever "reference level" he wishes. In practice, one is concerned only with the change in potential energy, a property that is independent of the arbitrary choice of the reference position.

Three specific situations will be discussed to illustrate the evaluation of potential energy and to lay the groundwork for later applications of the definition of potential energy.

The energy of an object in a uniform gravitational field

If one limits his experience to observations accessible to him by foot or automobile, he will be able to detect only small variations in the acceleration of gravity with changes of altitude. If, for example, he should measure the acceleration of gravity at the first floor of a skyscraper and find it to be equal to exactly 9.80000 m/sec², he will find that g decreases by only 0.00031 m/sec², to 9.79969 m/sec², when he ascends to the top floor, assumed to be 300 ft higher. Even if he were to ascend to an elevation of 32 km (20 mile), where 95 percent of the atmosphere lies below him, the acceleration of gravity would have decreased by only 0.06 m/sec², to 9.74 m/sec². It is therefore common practice in dealing with activities near the surface of the Earth to regard the acceleration of gravity, and hence the weight of objects, as constant.

If a reference level is chosen relative to which the potential of an object is to be expressed, an object descending from a height h above that level will experience a steady force equal to mg. Making use of suitable strings and pulleys, the object could be constrained to descend at constant velocity and do work on some other object, such as a machine that might pump water or generate electricity.

Were the object to descend to the reference position, it would exert a force mg through a distance h and therefore do an amount of work equal to mgh. Since this is the amount of work the object is capable of doing in descending from the height h, it represents the potential energy of the object at the height h.

$$\text{Potential energy} = mgh \qquad (3\text{-}8)$$

It must be emphasized that h is the vertical distance of the object

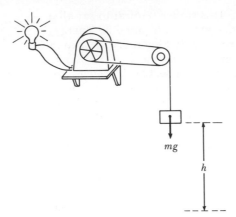

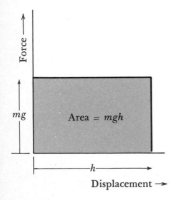

FIG. 3-4

Potential energy of an object near the surface of the Earth. As the object descends, it experiences a steady force equal to mg. If it descends at uniform speed, the string exerts a steady force equal to mg on the machine. In descending a distance h, the work done on the machine equals mgh.

FIG. 3-5

Force versus displacement in a uniform gravitational field. The work done by an object in descending a distance h in a uniform gravitational field equals the area under the curve.

from the reference level. If the object is above the reference level, h will be a positive number, and the potential energy of the object will be positive; if the object is below the reference level, h will be negative, and the potential energy will be negative. An object that descends from a position at which its potential energy is "zero" to a position at which its potential energy is negative will experience a decrease in potential energy and therefore can do work in the process.

Figure 3-5 presents a graph showing force (y axis) versus displacement (x axis), in the case of a uniform gravitational field. Note that the work the object is capable of doing, and hence the energy possessed by the object, is equal to the area under the curve.

The energy of a stretched spring

In Fig. 3-6(a) we show a relaxed spring lying on a horizontal table. Figure 3-6(b) shows the same spring after being stretched. For most springs, the force that must be applied to hold the spring at a given elongation x is proportional to the amount of elongation. This sort of behavior can be represented by the equation

$$f = -kx \qquad (3\text{-}9)$$

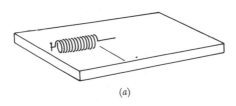

(a)

$f = -kx$

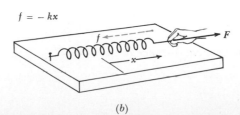

(b)

FIG. 3-6

(a) A relaxed spring. (b) The same spring elongated. The hand exerts a force F against the spring; the spring exerts a force f against the hand. The force exerted by the spring is clearly in a direction opposite to the elongation.

where f is the force exerted by the spring, x is its elongation, and k is a constant of proportionality known as the *force constant* of the spring. The negative sign is necessary because the equation relates two vector quantities that are certain to have opposite directions. For example, if one displaces the free end of the spring to the *right*, the force exerted by the spring will be directed to the *left*.

We may now derive an expression for the energy stored in a stretched spring—that is, an expression for the work that a spring can do in returning to its equilibrium length. However, the force exerted by the spring is not constant; thus we cannot secure the desired result simply by multiplying a force times a distance. The force varies in a linear fashion, so we can compute the work by multiplying the average force times the change in elongation. The average force in returning to the equilibrium elongation equals $-kx/2$ and the change in elongation is $-x$ (a decrease in length). The work done by the spring, that is, its original potential energy, is

$$\text{Potential energy of a spring} = \tfrac{1}{2}kx^2 \qquad (3\text{-}10)$$

This result agrees with the answer we would get if we write an expression for the area under the curve shown in Fig. 3-7; the area under the curve is half the area of a rectangle with one side of length kx, the other side of length x. A graph of displacement versus potential energy is shown in Fig. 3-8.

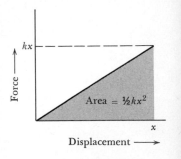

FIG. 3-7

Graphical representation of the force exerted by a spring. When a stretched spring does work on some other object, that object is displaced in the same direction as the force f. The total work that a stretched spring can do is equal to the area under the curve.

The energy of an object in a nonuniform gravitational field

We have already seen that the gravitational field intensity of the Earth (the force experienced at a particular point by a kilogram mass if placed there) varies inversely with the square of the distance of the point from the center of the Earth. In equation form,

$$\text{Gravitational field intensity} = -\frac{Gm_e}{r^2} \qquad (2\text{-}7a)$$

For most of us, limited to moving only a few kilometers relative to the center of the Earth 6380 km away, the gravitational field intensity is a constant. With the advent of the "space age," however, we have become more conscious of the possibilities of traveling significant distances from the Earth. Then the force experienced by an object will not remain constant, and Eq. (3-8) will not apply since it relates to the case of a uniform gravitational field.

When one considers the relationship of an object to the Earth's gravitational field on the grand scale depicted in Fig. 3-9, the surface of the Earth is not as significant as it seemed in our earlier discussion. Both Eq. (2-5) and Eq. (2-7a) show that the force of gravity at large distances ultimately will become so small as to be negligible. Although we are at liberty to choose the reference position wherever we like, it is

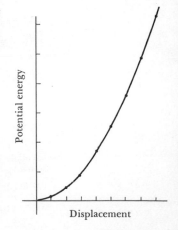

FIG. 3-8

Potential energy versus displacement for a linear spring. The energy of a stretched spring varies as the square of its elongation.

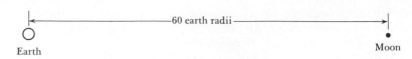

FIG. 3-9

An object at a significant distance from the Earth. The object is shown at a distance of 60 earth radii from the Earth, approximately the distance from the Earth to the Moon. The acceleration of gravity due to the Earth is equal to 0.0027 m/sec² at this distance.

rather natural to associate the condition of zero potential energy with the situation in which an object experiences *no* gravitational force—that is, at infinity. Although the actual attainment of zero potential energy is impossible, a very close approximation to this condition has already been attained by astronauts.

An object initially at infinity experiences a small but significant force when moved to a point at a finite but large distance from the Earth. The force law may not be the same as the law that would be obeyed were the object to be attached to a spring, but qualitatively the situations are analogous. Whether the object is pulled by a spring or by gravitation, work is done upon it. The potential energy originally present would have been transformed, in part at least, into work. If brought to rest at some new location, closer to the center of attraction, less energy remains to be transformed into work. In both cases, the potential energy decreases.

We have decided to associate zero of potential energy with the object at infinity. If the object is to have less potential energy at all finite points, it must follow that the gravitational potential energy of an object at any finite position must be *negative*.

A graph displaying the variation of the gravitational field inten-

FIG. 3-10

Gravitational field intensity of the Earth. The force on an object is a downward force, hence the negative signs. Note that the scale used on the *y* axis is nonlinear; the force varies much more rapidly than indicated by this graph. For comparison, the force due to the Moon is shown to the right (dashed curve).

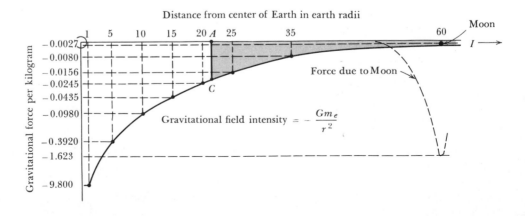

sity with distance from the center of the Earth was given in Fig. 2-18. According to that graph, the field intensity decreases from its value of −9.8 N/kg at the surface of the Earth to the very low value of −0.39 N/kg at a distance of 5 earth radii. This graph has been redrawn in Fig. 3-10, but because we must deal with an even larger range of r values, a nonlinear scale has been used on the vertical axis. Only in this way can one see the small but still significant forces present at distances as great as 60 earth radii from the Earth.

The force on a 1-kg object due to gravitation at an infinite distance from the Earth will be equal to zero. At all finite positions, however, it will experience a force, and this force will steadily increase as the object approaches the Earth.

An object that is dropped at a great distance from the Earth will accelerate, thereby gaining kinetic energy. On the other hand, by using suitable strings and pulleys, the object could transmit this energy to some other object or device and thus translate its loss of potential energy into work on that other object or device. The amount of work that the 1-kg object would do because of this force is given by the area between the curve and the x axis; for example, in traveling from infinity to the point A in the figure, the work done by a 1-kg object on some other object would be equal to the area represented by $IAC.*$

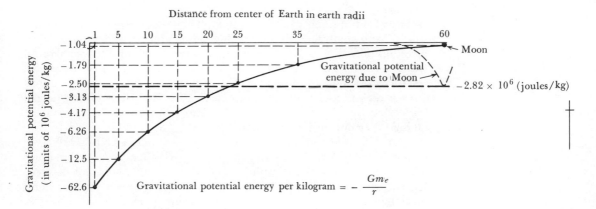

FIG. 3-11

Gravitational potential energy per kilogram due to the Earth. The potential energy of an object will be assigned a value of zero when the object is at infinity. The energy of such an object at any finite point will be less, thus will be negative, as shown on the y axis above. For comparison, the potential energy due to the Moon is also shown (dashed curve).

An algebraic expression for the area lying between the curve and the x axis cannot be secured by elementary mathematical techniques. By employing the methods of integral calculus, however, one can show

* It must be recalled that we used a nonlinear scale on the y axis. One would have to employ a linear scale on both axes in order to make serious use of measured areas.

that this area, and hence the potential energy at an arbitary value of r, is given by

$$\text{Gravitational potential energy per kilogram} = -\frac{Gm_e}{r} \quad (3\text{-}11a)$$

The gravitational potential energy of an object having a mass m is of course proportional to m. Therefore

$$\text{Gravitational potential energy of a mass } m = -\frac{Gm_e m}{r} \quad (3\text{-}11b)$$

A graphic representation of Eq. (3-11a) is shown in Fig. 3-11, in which once more we have used a nonlinear scale along the y axis because the numerical values become so small at large values of r.

Some writers refer to the gravitational potential energy per kilogram simply as the "gravitational potential." In this text we will continue to use the more descriptive, full name of the quantity.

3-5 CONSERVATIVE
SYSTEMS AND CONSERVATIVE FORCES

A *system* is an assembly of interacting objects and devices that can be studied and analyzed without taking other objects and devices into consideration. For example, we can discuss the Sun and its planets without regard for other stars in our galaxy; thus we call it the solar system. A nucleus with its cloud of electrons is treated as if other atoms are not present and is called an atomic system. In the laboratory a mass suspended by a spring from a rigid support can be analyzed without taking other objects into account.

Any system can be classified as being either a conservative or a nonconservative system, depending on whether or not the total mechanical energy (kinetic plus potential energy) is constant or conserved. For a system to qualify as conservative, all objects in the system must first and foremost be totally free from nonconservative forces, the most important of which is friction.

Strictly speaking, it is impossible to devise an arrangement of laboratory objects that qualifies as a conservative system because friction is sure to be present to a small extent even in the best of circumstances. A mass on a spring will eventually stop vibrating. A wheel on excellent bearings will ultimately come to rest. Even the solar system fails to satisfy the requirements exactly.

In physics, however, we start by idealizing the behavior of the systems that we wish to understand by dealing with them as if nonconservative forces, such as friction, were not present. We discuss the mass on the spring as if it would oscillate at constant amplitude indefinitely. We treat the solar system as if it would continue in its present motion forever. Only after we have analyzed a given system as an ideal conservative system do we start to make corrections for its nonconservative defects. All the situations discussed to this point have been treated in this manner.

Eventually the nonideality of objects in nature will have to be considered. Although several references to nonconservative systems will occur in the following pages (inelastic collisions in the next section, for example) we will defer further treatment of nonconservative systems until Chapter 6, where the concept will lead to a definition of thermal energy and to the establishment of an important principle known as the Law of Conservation of Energy.

3-6 INTERACTIONS

A sports car that collides with a fruit stand while traveling with a velocity of 60 miles per hour might well inflict the same amount of damage as a larger car traveling at a lower velocity. A man driving a nail with a tack hammer will involuntarily cause the tack hammer to hit the nail with a much larger speed than that he would utilize if he had a hammer of larger size. A football coach encourages his players to "block hard," usually meaning that they should engage their opponent at high speed. Each of these examples illustrates the fact that the force an object can exert in an impact depends on both the mass and the velocity of the object. These two properties of an object contribute to the "quantity of motion" of the object, a composite property known as its "momentum":

> The momentum of an object is a vector quantity whose magnitude is equal to the product of the mass and the velocity of the object and whose direction is in the direction of the velocity.

When we say that two objects interact with each other, we mean that they each exert a force on the other as described by Newton's Third Law. The effect of these forces, at least in all the situations we will discuss, is to change the velocity of both objects. It is our objective in the next few sections to describe how these forces affect the velocity of each of the interacting objects. We will idealize the situations to be described by assuming that no other forces are present during the interaction or, if present, that they are negligibly small in comparison with the forces of interaction. The changes in the velocities of two colliding billiard balls at the instant of impact, for example, is determined by the force of interaction, regardless of the amount of "reverse English" the cue ball may have possessed. The force of friction between a car and the roadway is so small in comparison with the force of interaction between the two cars in a collision that friction may be neglected in accounting for the changes in velocity that occur.

We can identify several distinct types of interactions in nature.

1. An *elastic* interaction is an interaction in which the total amount of kinetic energy possessed by the two objects is not changed by the interaction. Thus it takes place in a system in

the collision of two hardened steel pucks on the surface of a smooth, frozen pond. Measurements will show that the total kinetic energy after the collision will differ by no more than 1 percent from the total kinetic energy of the two objects before the collision. Little if any of the initial kinetic energy will have been used up in deforming permanently the surface of either of the two pucks. An equation for the velocity of an object after an elastic encounter with another object will be derived in Sec. 15-4 in a discussion of nuclear collisions.

2. An *inelastic* interaction is one in which there is a change in the total kinetic energy. We will consider two categories of inelastic interactions of two objects.

   a. The colliding objects may each do work in deforming the other; this loss of mechanical energy will appear as thermal energy. One needs only examine a wrecked car to become aware of the enormous amount of work that was required to bend the metal of the car into such a shape. This work was done at the expense of the kinetic energy of the colliding objects. A "completely inelastic collision" represents the extreme limiting situation in which the two objects fuse together upon impact and leave the point of interaction with a common velocity. We will see in Chapters 15 and 16 that this seemingly idealized inelastic collision is rather common in nuclear physics, where, in many cases, the nucleus of an atom may "capture" an impinging particle.

   b. An inelastic collision can occur in which kinetic energy is gained. This type of interaction includes most explosions; a rifle and the bullet it projects may serve as a simple example. When the rifle is fired, the gunpowder changes quickly

FIG. 3-12

Types of interactions. (*a*) An elastic collision—the kinetic energy remains unchanged. (*b*) Inelastic interactions: A collision—kinetic energy is lost. An explosion—kinetic energy is gained.

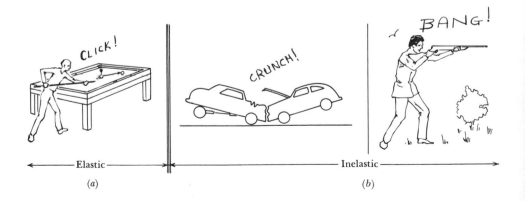

Elastic

(*a*)

Inelastic

(*b*)

into a gas that expands, propelling the bullet forward and the gun backward. In this case, there will be a considerable gain in kinetic energy; in fact, this system starts with zero kinetic energy. Inelastic collisions are very common in nuclear events; many situations will be cited in Chapters 15 and 16 in which the nucleus of an atom can be induced literally to explode on being bombarded by some tiny, subnuclear particle.

These categories of interactions cover a wide range of possibilities, varying from those that occur between objects in a conservative system to those occurring in extreme cases of objects in nonconservative systems. Considering the lack of "ideality" of phenomena as violent as collisions and explosions, one must marvel that any simple principle can describe them all. However, by making use of Newton's Laws of Motion, we will show that the total momentum of any two interacting objects remains unchanged in all the varieties of interactions just described.

3-7 THE LAW OF CONSERVATION OF MOMENTUM

Derivation from Newton's laws

In Fig. 3-13 we depict an object of mass M traveling to the right on a smooth, horizontal table with a velocity V. This object overtakes and collides with an object of mass m traveling with a velocity v in the same direction. Assuming M to be greater than m, both objects will still be traveling to the right after the collision but, as one might suspect, M will have a velocity V' less than V, while m will have a velocity v' greater than v.

To simplify matters, we will assume that the force of interaction is constant, with a steady magnitude F during the time interval Δt that the interaction occurs, after which the force of interaction drops

FIG. 3-13

A collision. (*a*) Before. (*b*) After. Because the objects exert equal forces on each other for the same lengths of time, the momentum before the interaction is the same as the momentum after the interaction. Momentum is conserved in all interactions of two objects, whether elastic or inelastic.

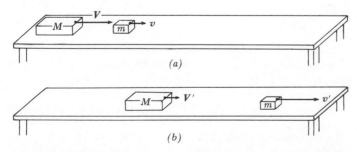

suddenly to zero. The force exerted by M on m equals $+F$, whereas, by Newton's Third Law, the force exerted by m on M equals $-F$. It is important to note that the time interval during which m experiences the force $+F$ is exactly equal to the time interval during which M experiences the force $-F$; we can represent the time interval for both objects by the same symbol Δt. Applying Newton's Second Law to the objects of mass m and M, respectively, we have

$$F = \frac{m(v' - v)}{\Delta t} \tag{3-12a}$$

and
$$-F = \frac{M(V' - V)}{\Delta t} \tag{3-12b}$$

Solving each equation for the quantity $F \, \Delta t$, and equating, yields

$$F \, \Delta t = m(v' - v) = -M(V' - V)$$

From this we see that

$$mv' + MV' = mv + MV \tag{3-13a}$$

That is, the total momentum after the interaction is the same as the total momentum before the interaction. Equation (4-12a) may be written in the alternate form

$$(mv' - mv) = -(MV' - MV) \tag{3-13b}$$

which shows that the change of momentum of one of the objects is equal to the negative of the change of momentum of the other.

These conclusions from this simple example constitute a special case of the *Law of Conservation of Momentum*, which may be stated in the following alternate but equivalent forms:

In the interaction of any two objects, tne total momentum before the interaction equals the total momentum after the interaction,

or **In the interaction of any two objects, the change in the momentum of one of the objects is equal to the negative of the change in the momentum of the other.**

The universal nature of the conservation of momentum

Let us note that in our discussion of the interaction of two objects we made no assumption about the nature of the interaction except to say that the force was a steady force, an assumption that was made for simplicity. There are many other possibilities; for example, we could have attached a hardened steel spring, a ball of modeling clay, or a small dynamite cap to the tip of one of the objects with the result that the interaction could have been either elastic, inelastic with a loss of kinetic energy, or inelastic with a gain of kinetic energy. According to

Newton's Third Law, however, the force exerted by M on m is exactly equal but opposite to the force exerted by m on M, regardless of the nature of the interaction. Therefore the conclusion we have reached applies to all of the interactions that have been mentioned.

In our example we assumed the force of interaction to be constant for the duration of the interaction Δt. The force one would encounter in general, however, would not be constant but would vary considerably during the period of the interaction. For a more general proof, one rewrites Eqs. (3-12a) and (3-12b) as

$$F\,\Delta t = m(v' - v) \qquad (3\text{-}12a')$$

and

$$F\,\Delta t = -m(V' - V) \qquad (3\text{-}12b')$$

If, then, one regards the entire process as a large number of small interactions in succession and makes use of the fact that each of these subinteractions involves equal forces for the same length of time, one sees that momentum is conserved in each of the subinteractions individually. Momentum, therefore, must be conserved for the whole operation irrespective of whether the force is or is not constant during the interaction.

Basic principles of rocketry

Suppose that a man in outer space is at rest with respect to the fixed stars and possesses no equipment aside from a rifle and a pocketful of bullets. If he wishes to propel himself in some direction, toward some particular star, for example, he could point his rifle away from the star, and, in accordance with Newton's Third Law, on firing the rifle he would attain a small velocity with respect to the fixed stars

FIG. 3-14

Rocket propulsion. (*a*) Before firing. (*b*) After firing. The man and the bullet are now both moving relative to the original inertial frame of reference. The coordinate system of the man, rifle, and remaining bullets becomes the new inertial frame of reference. By repeated firing of the rifle, the man may acquire a large velocity relative to the original frame.

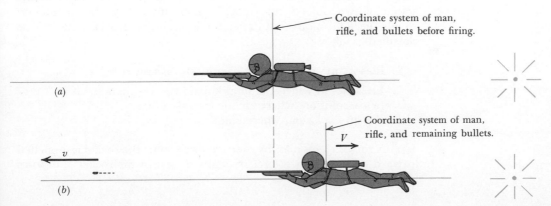

and toward the star previously designated. By Newton's First Law, we know that he will maintain whatever velocity he attains for an indefinite period of time. If he were to fire another shot, he would acquire another increase in velocity. Were he to continue to fire in the proper direction, the velocity he would ultimately attain after any number of shots would be equal to the algebraic sum of all the individual increases in velocity.

SUMMARY   Work equals the effective force in the direction of motion multiplied by the distance traversed during the motion. The effective force is the component of the force in the direction of the motion.

Energy is defined as the capacity of an object to do work. Both work and energy are measured in the same units (kg m²/sec² or joules).

Kinetic energy is the work an object can do by virtue of its motion.

$$KE = \tfrac{1}{2}mv^2 \tag{3-6}$$

Potential energy is the work an object can do by virtue of its position or state of strain. Three specific cases have been discussed:

1. PE of an object in a uniform gravitational field $= mgh$.
2. PE of a stretched spring $= \tfrac{1}{2}kx^2$.
3. PE of an object in the inverse-with-r² field of the Earth $= -Gm_e m/r$.

A conservative field may be said to exist in a given region if the total mechanical energy, kinetic plus potential, of any object located in the region is constant, or "conserved." Such an object may be said to be under the influence of a conservative force.

A conservative system is an object or a combination of objects, each of which is subject only to conservative forces.

The momentum of an object is the product of its mass and its velocity.

The Law of Conservation of Momentum states that in the interaction of two objects, the total momentum before the interaction equals the total momentum after the interaction. Interactions can be classified as follows:

1. Elastic, in which kinetic energy also is conserved.
2. Inelastic, in which the total kinetic energy changes, as in
   (a) a collision, where kinetic energy is lost.
   (b) an explosion, where kinetic energy is gained.

A rocket ejects mass by exerting force on it; the rocket is propelled forward by the reaction force of the mass against the rocket (Newton's Third Law).

3-A1. Distinguish between work and energy. Why is it correct to say that an object or system possesses energy but incorrect to say that it possesses work?

3-A2. Apply the Units Rule to determine the units in which kinetic energy is expressed.

3-A3. The following quantities have been introduced to this point in this text: mass, length, time, velocity, acceleration, force, work, energy, kinetic energy, potential energy, and momentum. (*a*) What is the MKS unit in which each of these quantities is to be expressed? (*b*) Which are scalar and which are vector quantities?

3-A4. Explain the general situation in which you would determine the energy of an object by measuring the area under a curve. Describe the procedure as applied to a particular case.

3-A5. A man finds that it requires a force of 50 N to slide a 30-kg box horizontally across a warehouse floor at constant velocity. (*a*) How much work does the man do on the box in moving it horizontally a distance of 10 m? (*b*) How much energy is gained by the box?

3-A6. What minimum force is needed to accelerate a 600-kg automobile from rest to a speed of 30 m/sec in 6 sec?

3-A7. What is the least amount of work that would suffice to accelerate a 600-kg car from rest to a speed of 30 m/sec? Does your answer depend on the magnitude of the acceleration of the car?

3-A8. What is the initial upward speed of a particle that reaches a height of 20 m above its point of departure? Does your answer depend on the angle at which the object is projected?

3-A9. (*a*) If an object or system of objects possesses energy, must it also possess momentum?

(*b*) If an object or system of objects possesses momentum, must it also possess energy?

3-A10. You squeeze a ball of putty, then compress a spring, doing the same amount of work in both cases. Discuss the prospects of recovering the energy expended in these two cases.

3-B1. A 400-kg racer is traveling at a constant speed of 40 m/sec on a horizontal, circular racetrack whose circumference equals 2510 m. The forward driving force on the car is steady at 1000 N. (*a*) What centripetal force acts on the car? (*b*) Make a sketch and a vector diagram showing the direction of the net force relative to the direction of the car. (*c*) How much work is done on the car per revolution? (*d*) What is the gain in energy per revolution?

3-B2. An automobile having a mass of 900 kg is acted on by a force of 1500 N for 12 sec starting from rest. (*a*) What acceleration does the automobile experience? (*b*) What velocity is attained? What is the average velocity of the automobile during the 12 sec? (*c*) How far does it travel? (*d*) How much work is done on the automobile?

3-B3. An electron ($m = 9.11 \times 10^{-31}$ kg) experiences a force of 1.2×10^{-14} N while moving a distance of 0.003 m. (*a*) How much work was done on this electron? (*b*) How much kinetic energy did it acquire? (*c*) What velocity did it acquire, assuming it started from rest?

3-B4. A force of 12 N elongates a certain spring by 0.02 m. (*a*) What force is required to produce a total elongation of 0.03 m? of 0.06 m? (*b*) What is the average force exerted by one stretching the spring from an elongation of 0.03 m to an elongation of 0.06 m? (*c*) How much work will be required to stretch the spring through this range? (*d*) Make a force versus elongation graph and show that the answer you found is the area under that portion of the curve.

3-B5. (*a*) Prepare a graph similar to Fig. 3-10 showing gravitational force (in N/kg) versus distance (in *meters*) in the range 6.38×10^6 m to 128×10^6 m but differing from Fig. 3-10 in that you use a linear scale on the y-axis. To improve the accuracy of the graph, plot additional points (e.g., at $r = 12.76 \times 10^6$ m, force = 9.8/4, etc.).
(*b*) From your graph determine the work in joules required to lift a 1-kg object from the surface of the earth to a position 128×10^6 m (= 20 earth radii) from the center of the Earth.
(*c*) Compare your answer to that you find using Fig. 3-11.

3-B6. (*a*) Using Fig. 3-11, compute the kinetic energy gained by a 50-kg object in falling to the Earth from an infinite distance, assuming it started from rest.
(*b*) What velocity would it acquire?

3-B7. (*a*) Using Fig. 3-11, determine the kinetic energy of a 10-kg object that has fallen from outer space to the surface of the Moon, neglecting energy due to the Earth, other planets, and the Sun. (*b*) What would be its velocity?

3-B8. An 80-kg man in outer space fires a 4-kg bullet with a speed relative to him of 800 m/sec. Find his final speed relative to his original inertial frame.

3-B9. A common method of measuring the velocity of a bullet from a given rifle is to fire the bullet into a block of wood, calculating the original velocity of the bullet from the recoil velocity of the block with the embedded bullet. Given that a 0.5-kg block recoils with a velocity of 3.8 m/sec when struck by a 0.010-kg bullet, what was the original velocity of the bullet?

3-B10. A 2-kg object on a smooth surface moving north with a velocity of 6 m/sec collides with a 5-kg object at rest. After the impact, the 5-kg object is traveling north with a velocity of 2.74 m/sec. (*a*) What are the speed and direction of the 2-kg object after the collision? (*b*) What was the total kinetic energy before and after the collision? (*c*) Comment on the inequality you found in (*b*).

Gravitation

The laws of physics, for the most part, are experimental laws. Their claim to validity, therefore, is based on the painstaking effort of careful observers. One cannot make a serious claim to understanding physics unless he knows the experimental and logical basis for these laws. While we were rather careful to establish the basis for Newton's Laws of Motion, his Law of Universal Gravitation was written without proof. In this chapter we will examine the experimental data that led to this law and to our present understanding of the place of our Earth and solar system in the larger framework of stars and galaxies.

4-1 GRAVITATIONAL FORCE AND GRAVITATIONAL ENERGY

In Sec. 2-11 it was stated that each object in the universe exerts a force of attraction on every other object and that the force acting between any two objects is proportional to the product of their masses and inversely proportional to the square of the distance between their centers. That is,

$$F = -\frac{Gmm'}{r^2} \qquad (2\text{-}5)$$

where G is a universal constant equal to $6.670 \times 10^{-11}\,\text{N m}^2/\text{kg}^2$. After noting that the act of separating two objects bears some similarity to the stretching of a spring, an expression for the gravitational potential energy was introduced in Sec. 3-4. This expression took the form

$$\text{Gravitational potential energy of two masses} = -\frac{Gmm'}{r} \qquad (3\text{-}11b)$$

The quantity on the right approaches zero as r approaches infinity. Thus the arbitrary reference position at which the gravitational po-

73

tential energy assumes a value of zero is a position at infinity. Because the gravitational potential energy decreases as two objects approach each other, the potential energy of two objects at a finite distance apart must be negative—hence the minus sign in the foregoing equation.

4-2 DESCRIBING THE SKY

Modern man spends such a large fraction of his time indoors that he has some difficulty reconciling his lack of consciousness of the appearance of the night sky with the vivid perception that was manifested by the people of Greece, Asia Minor, and North Africa, who 2000 years ago slept out of doors. Even so, most of us can identify two or more constellations, and we recognize that the group of stars that make up a given constellation maintain a fixed location with respect to each other. On a grander scale, except for the Sun, the Moon, and the nine planets, most visible celestial objects appear to maintain a fixed arrangement with respect to each other. The planets appear to wander, sometimes even changing the direction of their motion relative to the background of fixed stars.

Scientific investigations start from adequate descriptions of nature. Early descriptions told how the sky looked and how objects in the sky appeared to behave when viewed by a person standing on the Earth. These descriptions carried the kind of information one can

FIG. 4-1

Time exposure of the night sky. Any camera may be employed to make a photograph like this time exposure. The camera should be placed on a firm support in a region free from disturbing city lights.
(*Lick Observatory Photograph*)

obtain by taking a photograph of the night sky in the form of a time exposure of an hour or more. The photograph in Fig. 4-1 shows how the sky in the region of the North Star appears as seen by a camera at approximately 36° north latitude.

A description couched in the language of such an observer relates all motion to his frame of reference. All the stars appear to be at the same distance from the Earth, as if they were embedded in some vast celestial dome that rotates about the Earth. Even if the observer had no preconceptions, the description is almost certain to generate the assumption that the Earth is the center of the universe and that the surface of the Earth is a fixed frame relative to which the motion of the celestial bodies takes place.

Most of the ancients started with some preconceptions about the kinds of motion that were acceptable in a sky fashioned by a perfect Creator. Nature had to be perfect. Since circles and spheres were regarded as perfect geometric shapes, acceptable motion for the celestial objects had to be a circular motion or some combination of circular motions. Their efforts met with considerable success; a model that placed the planets and stars on an ingenious arrangement of rotating concentric spheres was devised by Claudius Ptolemy in the second century A.D. This model has continued to be useful to navigators and astronomers ever since.

4-3 KEPLER'S LAWS

Although Aristarchus of Samos in the third century B.C. had proposed a celestial model with the Sun at the center of the universe, it remained for Nicolaus Copernicus (1473–1543) to propose a heliocentric model of the universe and to provide supporting argument for it, sufficient to attract the serious consideration of other thinkers. Even his views were not accepted by a large number of his peers, and had it not been for the painstaking experimental work of Tycho Brahe (1546–1601) and the genius of Johanne Kepler (1571–1630), the Sun-centered model might well have been delayed another hundred years.

As an assistant in Tycho's laboratory, Kepler became the recipient of a wealth of experimental data recording planetary motion with a precision that stands as a monument to the patience and skill of an astronomer who worked without telescopes. The data of Tycho led Kepler to the following three experimental laws:

1. Each planet travels in a planar elliptical orbit,* with the Sun at one of the foci.

* An ellipse can be defined as the locus of a point in a plane such that the sum of its distances from two fixed points is constant. One can draw an ellipse by carrying out the following procedure: (1) Insert two tacks into a sheet of drawing paper at a distance L apart. (2) Tie the two ends of a piece of thread whose length is greater than L to the two tacks. (3) Place a pencil inside the loop of thread, pull the thread taut, and move the pencil sidewise, keeping the thread taut. The pencil will trace out an ellipse.

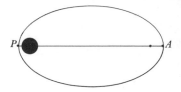

FIG. 4-2
Kepler's First Law. The orbit of any planet or satellite is an ellipse, a limiting form of which is a circle. Halley's and Encke's comets, most artificial satellites, and the planets Mars and Pluto depart significantly from circular paths. The Sun (for sun-satellites) is at one of the foci and is located in the plane of the orbit. In our equations, R equals half of the major axis PA.

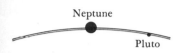

Neptune

Pluto

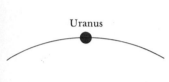

Uranus

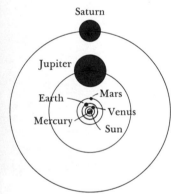

Saturn

Jupiter

Earth — Mars

Mercury — Venus

Sun

FIG. 4-4

Keplerian planetary
orbits. The planetary
orbits are all located in
essentially the same
plane except the orbit of
Pluto, whose plane is
tilted relative to that of
the other planets. (Pluto
is not in danger of
colliding with Neptune
as one might believe
from the sketch!) The
planets have been drawn
to a scale vastly
different from that of
their separations so that
they can be compared
to each other.

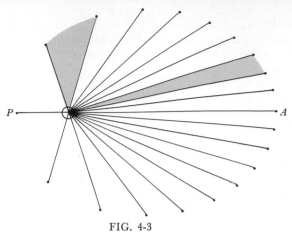

FIG. 4-3

Kepler's Second Law. The dots represent
the position of a sun-satellite at equal intervals of time. All the pie-shaped areas are
equal.

2. The radius vector one may draw from the center of the Sun to
a given planet sweeps out equal areas in equal times.

3. The ratio of the square of the period of revolution to the cube
of the semimajor axis (R) is the same for all of the planets;
that is,

$$\frac{T^2}{R^3} = K \qquad (4\text{-}1)$$

The orbit of a planet or satellite may be circular; in fact, such is
the case for most of the planets of the Sun. In the case of a circular
orbit the semimajor axis (half of the distance PA in Figs. 4-2 and 4-3)
equals half of the diameter. Thus R for a circular orbit equals the
radius.

These three laws do more than simply summarize the data of
Tycho Brahe. As generalizations of the behavior of satellites, they
predict the relationship between orbital properties of any planets that
may be discovered in the future as well as of all Sun, Earth, or Moon
satellites that may ever be launched.

Kepler's view of the solar system, enhanced by data secured
since his time, is presented in Fig. 4-4. As stated above, the orbits of the
planets are very nearly circular. Tycho's data, however, had revealed a
discrepancy of 8 minutes of arc in the position of Mars, adequate for
Kepler to conclude that its orbit is an ellipse. Had Tycho been able to
see and track the planet Pluto, whose departure from a circular orbit is
very pronounced, the need for Kepler to invoke elliptical orbits would
have been far more apparent.

The experimental basis for
Kepler's Third Law

Table 4-1 summarizes modern data in support of Kepler's Third
Law. When one considers the wide range of values of R and M represented by these planets(Jupiter has a mass 6000 times that of Mercury),

TABLE 4-1
Kepler's Third Law

SUN SATELLITES (PLANETS)

DESIGNATION	SEMIMAJOR AXIS R (m)	PERIOD T (sec)	KEPLER CONSTANT $K_s = T^2/R^3$ (sec^2/m^3)
Mercury	0.579×10^{11}	0.07601×10^8	2.976×10^{-19}
Venus	1.081	0.19414	2.984
Earth	1.495	0.31559	2.981
Mars	2.278	0.60219	3.068
(Ceres)	4.14	1.453	2.975
Jupiter	7.78	3.743	2.976
Saturn	14.26	9.296	2.980
Uranus	28.68	26.512	2.980
Neptune	44.94	52.004	2.980
Pluto	58.96	78.163	2.981

EARTH SATELLITES

DESIGNATION	SEMIMAJOR AXIS R (m)	PERIOD T (sec)	KEPLER CONSTANT $K_e = T^2/R^3$ (sec^2/m^3)
Moon	38.4×10^7	2,360,000	9.83×10^{-14}
Pacific 1	4.268	86,166	9.550
OSO 3	0.6926	5,754	9.965
Atlantic 2	4.2179	86,166	9.894
ATS 2	1.2104	13,134	9.729
Explorer 34	11.205	373,860	9.936

the constance of the value of K for Sun satellites can hardly be regarded as accidental.

Equally convincing evidence is now available from data on Earth satellites. The Moon moves in a nearly circular orbit about the Earth with a period T of 27.322 days ($= 2.36 \times 10^6$ sec) and at a distance R of 38.4×10^7 m. These data show that K_e equals 9.83×10^{-14} sec^2/m^3. A typical Earth satellite like Explorer 34 swept out an elliptical path whose semimajor axis was 11.205×10^7 m with a period of 3.7386×10^5 sec, indicating a value of K_e of 9.936×10^{-14} sec^2/m^3. Data on several Earth satellites, chosen nearly at random from a long list of satellites that have been launched in recent years, are shown in Table 4-1. Further data regarding the solar system will be found in Appendix 3.

4-4 IMPLICATIONS OF KEPLER'S THIRD LAW

*Physics is more
than a description of nature*

If the only purpose of physics was to describe motion, one might well have difficulty in making a choice between an extended Ptolemaic

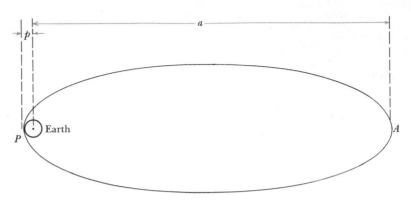

FIG. 4-5

The orbit of Explorer 34. At perigee this satellite was at an altitude of 154 miles; at apogee, 131,187 miles. Other data regarding the flight of Explorer 34 are:

$a = 2.175 \times 10^8$ m — Launched 5/24/67
$p = 0.0662 \times 10^8$ m — Weight 163 lb
$R = 1.121 \times 10^8$ m — Period 3.74×10^5 sec

geocentric (Earth-centered) model and Kepler's heliocentric (Sun-centered) system. Given sufficient patience, there is every reason to believe that one could continue to combine circular motions of various radii and periods and ultimately match experimental data as closely as one chooses. Indeed, with high-speed computers to replace the tedium of hand calculations, which have been required until very recently, a Ptolemaic type of description would now be relatively easy to carry out.

Physics, however, wishes to go farther than simply describe motion; its broader goal is to relate motion to external agencies and to lay the groundwork for the prediction of events in advance of providing the specific circumstances in which the events are possible.

It will be recalled that the general procedure for describing motion presented in Chapter 1 was only preliminary to the discussion in the next chapter, in which acceleration, and therefore all motion, of objects was ascribed to external agencies. It will also be recalled that some emphasis has been placed on the fact that one can be sure of a clear identification of each force with an external agency only if the problem has been formulated relative to some inertial system. Various examples were mentioned to show that measurements in an accelerated (noninertial) frame of reference would require the introduction of fictitious forces that could not be ascribed to any external agent.

At this stage in history it seems elementary to most of us that celestial objects like the Sun and the various planets should conform to the same laws as the smaller objects one can handle in the laboratory. This mode of thinking has not always prevailed. In fact, until comparatively recent times it was widely believed that celestial objects were subject to sundry extraterrestrial rules and whims.

The Law of Universal Gravitation is obtained by combining Newton's equation for centripetal force [Eq. (1-3)] with Kepler's Third Law [Eq. (4-1)]. We will simplify the derivation by assuming that the orbits of all the planets are circular and therefore that the gravitational force provides the centripetal force. Letting F represent the gravitational force, we have

$$F = \frac{mv^2}{r} \tag{2-4}$$

where m is the mass of the planet, v is its velocity, and r is the radius of its orbit. The velocity, however, equals the circumference of the orbit ($= 2\pi r$) divided by the time to make one complete revolution T. We now have

$$F = \frac{4\pi^2 mr}{T^2} \tag{4-2}$$

One can eliminate T^2 from this equation by introducing its equivalent from Kepler's Third Law,

$$F = \frac{4\pi^2 mr}{Kr^3} = \left(\frac{4\pi^2}{K}\right)\frac{m}{r^2} \tag{4-3}$$

This equation shows that the force of attraction exerted by the Sun on a planet is proportional to the mass of the planet and inversely proportional to the square of the distance of the Sun from the planet. It is not yet clear, from the equation at least, how the mass of the Sun influences the magnitude of the force.

In order to justify the changes in this equation that are needed to yield the Law of Universal Gravitation, let us consider a "thought" experiment involving moving mass on a very large scale (Fig. 4-6). The first sketch (a) indicates the Sun and one of the planets. The Sun exerts a force on the planet, and the planet exerts an equal and opposite force on the Sun. The second (b) suggests that some great hand has moved additional mass to the planet, doubling its mass. Equation (4-3) states that the force on the planet will thereby be doubled; Newton's Third Law requires that the force exerted by the planet on the Sun also will have doubled. The third sketch (c) suggests that the same great hand, starting from the situation in (a), has now doubled the mass of the Sun. The planet now experiences a force of attraction to the original Sun mass and to the added mass. Thus once more both the force exerted by the Sun on the planet and the force exerted by the planet on the Sun have doubled. This thought experiment demonstrates that the force experienced by either the Sun or the planet is proportional both to the mass of the planet and to the mass of the Sun. One can only conclude that the quantity in parenthesis in Eq. (4-3) must itself be proportional to the mass of the Sun—that is, that

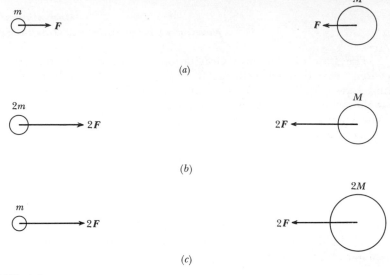

(a)

(b)

(c)

FIG. 4-6

A "thought" experiment. (*a*) The Sun M exerts a force F on a planet m, which, by Newton's Third Law, exerts an equal but opposite force on the Sun. (*b*) If one doubles the mass of the planet, the Sun experiences a force equal to $2F$. (*c*) If one doubles the mass of the Sun (keeping the mass of the planet equal to m), the force on the planet is doubled.

$$\frac{4\pi^2}{K} = Gm' \tag{4-4}$$

where m' is the mass of the Sun and G is a constant of proportionality. This means that

$$F = -\frac{Gm'm}{r^2} \tag{4-5}$$

The negative sign has been introduced to designate that the force is a force of attraction. Measurements of the force between laboratory-sized objects have been carried out using a Cavendish balance (Sec. 2-10) and show that

$$G = 6.670 \times 10^{-11} \, \text{N m}^2/\text{kg}^2$$

4-5 DETERMINING THE MASS OF THE SUN AND THE EARTH

The mass of the Sun

The value of $K_s (\approx 3.0 \times 10^{-19} \, \text{sec}^2/\text{m}^3)$ combined with the preceding measurement of G enables us to use Eq. (4-4) to determine the mass of the Sun. Solving for m', the mass of the Sun, we have

$$\text{Mass of Sun} = \frac{4\pi^2}{K_s G} = \frac{4\pi^2}{3.0 \times 10^{-19} \times 6.67 \times 10^{-11}}$$

$$\approx 2 \times 10^{30} \text{ kg}$$

where the presently accepted value is 1.987×10^{30} kg.

The mass of the Earth

The procedure we just employed may also be used to determine the mass of the Earth. Taking an average of the values of the Kepler constant listed in Table 4-1, we find that

$$K_e = 9.82 \times 10^{-14} \text{ sec}^2/\text{m}^3$$

This value of K_e when introduced into Eq. (4-4) yields a value of 6.03×10^{24} kg for the mass of the Earth.

One can also compute the mass of the Earth by a procedure that involves only measurements of phenomena on the surface of the Earth. This procedure is interesting because of the insight that it yields regarding the relationship between g, the acceleration of gravity, and G, the gravitational force constant.

Let us recall that the force experienced by any object of mass m (its weight W) at a distance r from the center of the Earth of mass M_e can be expressed alternately by

$$W = -\frac{GM_e m}{r^2}$$

and by

$$W = -mg$$

From these we find

$$g = \frac{GM_e}{r^2}$$

or

$$M_e = \frac{gr^2}{G} \qquad (4\text{-}6)$$

Thus we can compute the mass of the Earth provided that we have a means of measuring G, g, and r. Let us examine each of these quantities separately.

1. We have already made use of the assumption that G, the gravitational constant, is a universal constant. This constant has been found to equal 6.67×10^{-11} N m²/kg².

2. g, the acceleration of gravity, is determined by measurement of a picture of the motion of a freely falling object (Sec. 1-5); at the surface of the Earth $g = 9.80$ m/sec².

3. r, the radius of the Earth, is measurable by triangulation methods. The procedure requires that two observers at a considerable distance apart simultaneously measure the location of

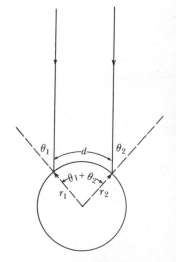

FIG. 4-7

Determining the radius of the Earth. Two observers at a known distance (*d*) apart simultaneously direct telescopes at a distant star. The sum of their measured values of θ_1 and θ_2 equals the angle between r_1 and r_2. The radius equals $d/(\theta_1 + \theta_2)$

a distant star. Originating at such an enormous distance from the Earth, the light reaching the two observers may be assumed to travel in parallel rays. Measurements show that $r = 6.38 \times 10^6$ m.

We are now in a position to determine M_e. Substituting the measured values into Eq. (4-6), we find

$$M_e = \frac{gr^2}{G} = \frac{9.80 \times (6.38 \times 10^6)^2}{6.67 \times 10^{-11}}$$

$$= 5.98 \times 10^{24} \text{ kg}$$

4-6 ANGULAR MOMENTUM

Let us imagine a puck on a smooth, horizontal surface to be tied to a rigid post by a string of length r. (Recall Fig. 2-8.) If, while holding the string taut, the puck is given a transverse push, it will travel in a circular path of radius r for a length of time determined by the extent to which friction has been minimized (Fig. 4-8). In principle, in the absence of friction, the puck will continue to travel in this circular path indefinitely.

The puck clearly possesses some variety of inertia, but since it continues to travel in a circle rather than in a straight line, we will ascribe an inertial property to the puck, which we will call "rotational inertia," and a quantity-of-motion property, which we will call the "angular momentum." These two properties play the role in rotational motion that has been played by mass and momentum, respectively, in linear motion.

The angular momentum of a mass point in motion about a fixed point is defined as

$$\text{Angular momentum} = mv_\perp r$$

where m is the mass of the object, r is the distance from the point to the object, and $v_\perp$ is the component of the object's velocity perpendicular to the radius vector. (For circular motion, $v_\perp$ is just the velocity v.) From the Units Rule, we see that angular momentum is expressed in kg m²/sec.

An isolated object or system of objects will maintain a constant total angular momentum; for example, if two systems of objects, each possessing angular momentum, interact with each other, the total (vector sum) angular momentum remains unchanged. This is the Law of Conservation of Angular Momentum

Figure 4-9 shows a portion of the elliptical path of a satellite. The satellite at one instant is located at P but an instant later will have traveled a short distance $v \, \Delta t$ tangent to the path at P. If Δt is very small, the straight line $v \, \Delta t$ and the actual path of the satellite may be regarded as indistinguishable. The area swept out is ΔA, a triangle whose base is $v_\perp \, \Delta t$ and whose altitude is r. Therefore

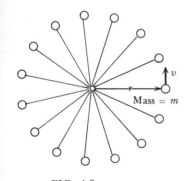

FIG. 4-8

A puck in uniform circular motion. The puck is constrained to travel in a circular path by a string. If the string provides the only force acting on the puck, the puck will travel in a circular path indefinitely. We say it possesses "rotational inertia."

Mass = m

$$\Delta A = \tfrac{1}{2}(v_\perp \Delta t)r$$

where $v_\perp$ is the component of v perpendicular to the radius vector r. The rate at which area is being swept out is then

$$\frac{\Delta A}{\Delta t} = \frac{1}{2}v_\perp r$$

$$= \left(\frac{1}{2m}\right)(mv_\perp r)$$

However, we have stated already that the angular momentum of an object equals $mv_\perp r$, so Kepler's Second Law is equivalent to saying that an object under the action of a central force maintains constant angular momentum.

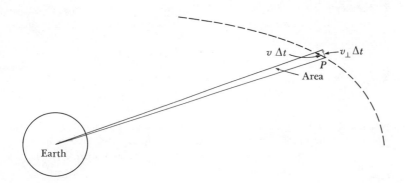

FIG. 4-9

Conservation of angular momentum. Our observation that a satellite sweeps out equal areas in equal times (Kepler's Second Law) is equivalent to saying that the satellite maintains constant angular momentum ($mv_\perp r$).

Kepler's laws are experimental laws describing the behavior of planets in our solar system, but they have equal validity to any planetary system.

SUMMARY

1. Each planet travels in a planar elliptical orbit with the Sun at one of the foci.

2. The radius vector one may draw from the center of the Sun to a given planet sweeps out equal areas in equal times.

3. The ratio of the square of the period of revolution to the cube of the semimajor axis is the same for all of the planets of a given planetary system; that is,

$$\frac{T^2}{R^3} = K \qquad \text{(a constant)} \qquad (4\text{-}1)$$

The angular momentum of a point object relative to some point in space is equal to $mv_\perp r$, where $v_\perp$ is the component of the object's velocity perpendicular to the radius vector. The total angular momentum of an isolated body or system of bodies is constant (Law of Conservation of Angular Momentum).

The Kepler constant (K) is determined by the mass of the central predominant mass—for example, $K_s = 2.98 \times 10^{-19} \sec^2/m^3$ for Sun satellites and $K_e = 9.82 \times 10^{-14} \sec^2/m^3$ for Earth satellites. This enables one to determine the mass of the Sun or the Earth from K, since

$$\frac{4\pi^2}{K} = Gm' \qquad (4\text{-}4)$$

We have shown that

1. An object that obeys Kepler's Second Law is under the action of a central force and maintains constant angular momentum.

2. Kepler's Third Law, in combination with Newton's equation for centripetal force, yields the Law of Universal Gravitation.

**QUESTIONS
AND
PROBLEMS**

4-A1. The mass of the Moon is $\frac{1}{83}$ that of the Earth. How does the gravitational force exerted by the Earth on the Moon compare in magnitude to the gravitational force exerted by the Moon on the Earth?

4-A2. Using data from Appendix 3, make a large scale diagram of the solar system modeled on Fig. 4-4. It is suggested that you scale up the size of the planets relative to the scale you use for interplanetary distances by a factor of 1000.

4-A3. Suppose that you were to take a space flight and that you encounter a planetary system consisting of a sun and two planets. Suppose that one planet is in a circular orbit of radius 3×10^{10} m and that the other is in an elliptical orbit whose semimajor axis equals 27×10^{10} m. Make a schematic diagram of this planetary system, label the important dimensions on the diagram, and state Kepler's laws as they would apply to this planetary system.

4-A4. Assume that in the case of the planetary system described in Prob. 4-A3, the inner planet makes one complete revolution about its sun in 400 days. What should be the period of the other planet?

4-A5. In attempting to make use of Kepler's Third Law, a student compares the radius of the Earth's orbit about the Sun to the radius of the Moon's orbit about the Earth. He found that the T^2/R^3 law didn't work. What was wrong with his logic?

4-A6. Forces between objects are described by the Law of *Universal* Gravitation. What evidence do you think scientists possess that this law is in fact universal, meaning that it applies to objects at remote positions in our galaxy and to objects in other galaxies as well?

4-B1. Two astronauts at a distance of 50 km apart on the Moon make simultaneous observations of the location of a particular distant star. One astronaut reports the star to be directly overhead; the other reports that the star is located at an angle of 1°39.4′ (= 0.0289 rad) relative to the vertical. What is the radius of the Moon according to this data?

4-B2. A satellite is put into a circular orbit about the Earth with a radius just half of the radius of the Moon's orbit. Apply Kepler's Third Law to determine the period of this satellite.

4-B3. The astronauts described in Prob. 4-B1 measure the acceleration of gravity on the Moon and find it equal to 1.62 m/sec². Combine this with their other data and compute the mass of the Moon.

4-B4. An object with sufficient kinetic energy to escape the Earth's gravitational field is said to possess "escape kinetic energy." Its velocity is called the "escape velocity." (*a*) Compute the escape kinetic energy of a 1-kg object launched from the Earth. (*b*) Compute its escape velocity. (*c*) How would the escape velocity of a 1000-kg object differ from that of a 1-kg object?

4-B5. With pins and string draw an ellipse on $8\frac{1}{2}$ in. × 11 in. plain paper such that its major axis is approximately twice its minor axis. Choose two points on this ellipse whose distances from one of the foci differ by a factor of exactly 2. (*a*) Draw an arrow at each of these points in proper relative lengths to represent the accelerations of a satellite when passing through them, (*b*) Using a different color, draw arrows that represent $v_\perp$ at each of these points in proper relative magnitudes, and from these find the relative magnitudes of the velocities v at the two points you have chosen. (*c*) Recalling Fig. 1-21, show how the change of velocity is related to the average acceleration during the interval.

4-B6. Many years ago (1772) Bode and Titius predicted that a planet (or planets) should be found in a circular orbit at a distance of 4.2×10^{11} m from the Sun. Much later many objects (asteroids) were found in orbits of this average radius. (*a*) What should be the period of an asteroid? (*b*) Relative to the other planets (Fig. 4-4), where should the asteroids be found?

4-B7. A bowling ball rolls in a straight line *ABCD* across a level floor with a velocity of 4 m/sec. Its closest distance of approach to some point *P* equals 10 m. At the instant the ball reaches *D*, a man with a croquet mallet gives the ball a sharp blow along the line *DP*, imparting an additional velocity of 6 m/sec to the ball in that direction. Make a scaled-up diagram of this situation and, either graphically or analytically as you prefer, (*a*) determine the area swept out in one second at two different places along *AD*, (*b*) determine the direction of the ball after being struck, and (*c*) show that the area swept out in one second after being struck is the same as the areas computed in (*a*).

FIG. 4-10

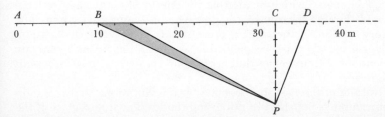

Vibration and Wave Motion

The motion of a mass on a spring has been presented as a model of simple harmonic motion—that is, as linear motion in which the acceleration of the object is proportional at all times to the negative of its displacement. Furthermore, in the ideal situation an object suspended by a spring constitutes a conservative system because friction, a nonconservative force, can be ignored. Hence the sum of the kinetic and potential energy of the system, as revealed by the constancy of amplitude of the oscillations, will remain constant.

One can identify in his environment a substantial number of objects and devices that execute a vibratory motion which approaches the ideality of a mass on a spring. A pendant tied to a rigid support by means of a thread will vibrate as a simple pendulum for a very long time and will execute motion that approximates simple harmonic motion. A piano spring, if plucked near its center, will execute simple harmonic motion; in fact, most musical tones originate from an object that is undergoing either simple harmonic motion or a simultaneous mixture of simple harmonic motions of various frequencies.

Although most sounds originate from objects that are vibrating, many sounds are not immediately recognizable as a vibratory motion. The reason is mainly because only in the very low frequency range (below 14 vibrations/sec) does the ear-brain system perceive a vibration as a sequence of separate impulses. Throughout the audible range of musical sounds (14 to 18,000 vibrations/sec), our interpretation of a musical tone is a far more subtle phenomenon, in which each frequency is interpreted by the hearer simply as a different "pitch." In addition, most of the articles that are responsible for the sounds we hear are heavily "damped," which means that the energy of vibration is quickly dissipated through friction or some other nonconservative force. The vibratory motion of the object, therefore, is of very short duration. Finally, objects that are recognizable as "noise producers" generate combinations of frequencies that are not interpreted as pleasing sounds by the hearer.

In spite of the fact that so many of the vibrations that make up our surroundings fall short of possessing the ideal characteristics of the

mass on the spring, the analysis of this idealized object constitutes
the proper place to start. As in the discussion of any topic in physics,
the phenomenon must be first observed in its purest form, stripped of
as many complexities as possible. After achieving an understanding of
the motion executed by the ideal oscillator, we can turn our attention
to less ideal oscillating devices.

5-1 A MASS ON A SPRING

It will be recalled that Chapter 1 dealt only with the description of
motion and that the discussion of the forces responsible for motion
(Newton's laws) was deferred until Chapter 2. We can now demonstrate
that because of the nature of the net force acting on an object suspended
by a spring, the object *must* execute the kind of motion we have called
simple harmonic motion.

We have noted that the force exerted by a spring on an object
is given by

$$F = -ky \tag{3-9}$$

where the negative sign appears because the net force on the object is
certain to be *down* when the object is *above* the equilibrium position,

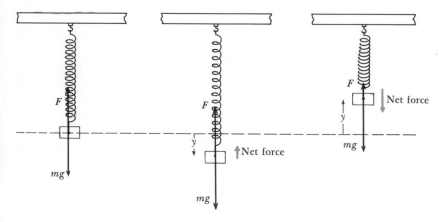

FIG. 5-1

Forces on a vibrating object. Gravitation exerts a steady force downward. At the
equilibrium position, the force due to the spring exactly equals the force of gravity.
When the object is below the equilibrium position, the net force is UP; when the object
is above the equilibrium position, the net force is DOWN.

and vice versa. The force produces an acceleration described by
Newton's Second Law of Motion; that is,

$$F = ma \tag{2-2}$$

Equating these expressions for the force, we have

$$ma = -ky$$

or

$$a = -\left(\frac{k}{m}\right)y \qquad (5\text{-}1)$$

We have done more than prove that a massive object on a spring executes simple harmonic motion; in addition to showing that the acceleration is proportional to the negative of the displacement, we have determined that the constant of proportionality is equal to k/m.

5-2 A GEOMETRIC SIMPLE HARMONIC MOTION

Our description of simple harmonic motion will be improved if we can identify an alternate example of simple harmonic motion in which the motion can be described without reference to a force. A relatively elementary mechanism provides the purely geometric model we need at this point.

Consider a wheel with a Ping-Pong ball attached to its rim, hence with the center of the ball at a distance r from the axis of rotation (Fig. 5-2). If the wheel is driven by a constant-speed motor, the ball will travel with constant speed in a circular path. The radius of the path is equal to r and the acceleration of the ball is equal to v^2/r. Letting T represent the time required for the Ping-Pong ball to make one revolution (or for its shadow to execute one complete vibration), one may express the speed of the Ping-Pong ball as

$$v = \frac{2\pi r}{T}$$

If a source of parallel light is placed in the plane of rotation of the Ping-Pong ball, one will observe that the shadow of the uniform circular motion of the ball resembles the vibratory motion of an object on

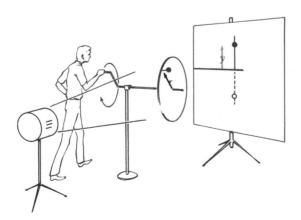

FIG. 5-2

The shadow of a uniform rotor. A Ping-Pong ball is attached to a wheel that is rotated at constant angular velocity. The source of light is in the plane of rotation. The shadow of the ball vibrates up and down; its motion resembles a mass on a spring.

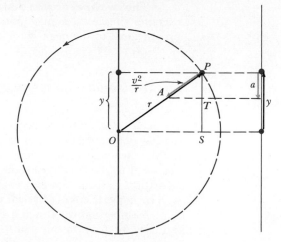

FIG. 5-3

Geometric analysis of rotor shadow. Since the acceleration of the Ping-Pong ball is along the radius vector, its shadow will be proportional to the shadow of the radius vector but in the opposite direction.

a spring. Our goal in this section is to present a geometric proof that this shadow projected by uniform circular motion is, in fact, simple harmonic motion.

The proof is remarkably simple. Because the acceleration vector of the Ping-Pong ball lies along the radius vector, the shadows of the acceleration vector and the radius vector must be in the same ratio as the vectors themselves. Furthermore, because the acceleration vector is toward the center of the circle while the radius vector is away from the center, the shadow of the acceleration vector must point down when the shadow of the radius vector points up, and vice versa.

Let us repeat this argument, expressing the ideas algebraically. The position of the Ping-Pong ball at an arbitrary instant in time is shown in Fig. 5-3. The shadow of the radius vector is the vector y. The shadow of the acceleration vector is the vector a; that is, the acceleration of the shadow of the ball has a magnitude equal to the projection of v^2/r on the screen (or on the diameter of the circle). The triangles ATP and OSP are similar and therefore their sides are proportional. Thus

$$\frac{a}{y} = -\frac{v^2/r}{r}$$

or

$$a = -\frac{v^2}{r^2}y$$

where we have introduced the minus sign because we known that the vector a is directed opposite the vector y.

The maximum displacement of the shadow from the "equilibrium position"—that is, from the center of its motion—is called the amplitude of the vibration; clearly, the amplitude is equal to r. Similarly, we see that the maximum velocity of the shadow equals $2\pi r/T$ and its maximum acceleration equals v^2/r. Introducing our expression for v into the preceding equation and simplifying, we have

$$a = -\left(\frac{4\pi^2}{T^2}\right)y \tag{5-2}$$

Thus we have proven that the shadow of uniform circular motion executes simple harmonic motion; the acceleration of the shadow is proportional to the negative of the displacement of the shadow.

5-3 THE PERIOD IN SIMPLE HARMONIC MOTION

We have seen that a massive object on a spring (Sec. 5-1) and the shadow of an object in uniform circular motion (Sec. 5-2) both execute simple harmonic motion because in each case the acceleration is proportional to the negative of the displacement. We will utilize this information to secure an expression for the period of a massive object on a spring in terms of physical properties of the object and the spring.

Assume that we attach a Ping-Pong ball to the rim of a wheel of radius r. Adjacent to this wheel, we suspend an object by a spring and set it into vibration with an amplitude of r (Fig. 5-4), illuminating them with a beam of parallel light. Rotating the wheel by hand or by a variable-speed motor, we adjust the speed of rotation until the two shadows move up and down together. Since they both execute simple harmonic motion of the same amplitude and the same period T, we can be sure that their motions will be indistinguishable.

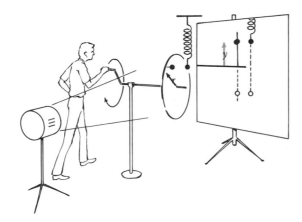

FIG. 5-4

The vibrator and the rotor shadow compared. If the operator turns the rotor at the proper constant angular velocity, the shadows of the vibrator and of the rotor become indistinguishable.

Since either Eq. (5-1) or Eq. (5-2) describes the motion of the object on the spring, we can equate the two constants of proportionality that appear in these equations, that is,

$$\frac{k}{m} = \frac{4\pi^2}{T^2} \qquad (5\text{-}3a)$$

Upon solving for T, we have

$$T = 2\pi \sqrt{\frac{m}{k}} \qquad (5\text{-}3b)$$

Propagation of a pulse

Assume that you are holding one end of a horizontal rope and that the opposite end of the rope is attached to a firm wall support. If you move your hand quickly up and back to place, a pulse will travel down the rope. This pulse will be reflected when it reaches the wall but with a "change of phase," meaning that the pulse which arrives

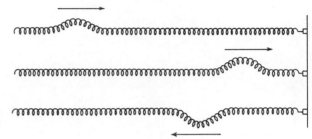

FIG. 5-5

Reflection of a pulse from a rigid support. The pulse traveling along a rope or coiled spring suffers a "change of phase" upon reflection; a crest becomes a trough.

as a "crest" will be reflected as a "trough"; furthermore, a pulse that arrives as a trough will be reflected as a crest. The speed of the pulses traveling along the rope will depend on the tension in the rope and on the mass of the rope per unit of length.

Propagation of waves

Let one end of the rope be attached to a device that is oscillating with simple harmonic motion along a line perpendicular to the rope. In this case, the "pulses" generated by the source will form a uniform traveling pattern, which is called a wave; furthermore, provided the rope is very long (in principle, if the rope is infinitely long), we need not be concerned with the complications that might be produced by the reflected waves. Focusing one's attention on a particular point along the rope, one will find that the rope at that point executes simple harmonic motion having the same period as the source.

Figure 5-6 shows a series of sketches of a substantial length of a rope along which a wave is traveling with a velocity v. In order to discuss the velocity of the wave, one must identify some specific point in the waveform whose position at a given instant can be determined with some accuracy; a convenient position will be the crest of one of the waves. These crests are separated by a distance we will call the wavelength λ (pronounced "lamb-da"). The frequency f of the source in vibrations per second is the same as the frequency of vibration at any point in the medium through which the wave is traveling. This frequency is equal to the reciprocal of the period of the vibration. For example, if the source or some point in the medium makes 10 vibrations/sec, the time required for one complete vibration to occur will equal 0.1 sec.

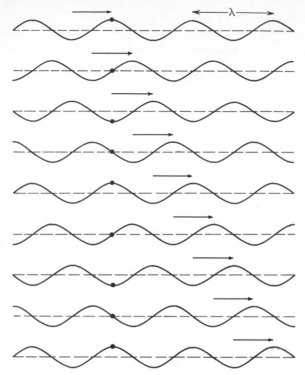

FIG. 5-6

Progress of a wave. The wave travels to the right at constant velocity; any specific point in the medium oscillates with simple harmonic motion. In a simple wave the medium is not transported with the wave; only the waveform travels.

We have identified three properties of the wave: the velocity v, the frequency f, and the wavelength λ. These properties are not independent of each other. Our immediate goal will be to determine how they are related.

Assume that at a given instant a crest leaves the source. One second later this crest will have traveled a distance numerically equal to the velocity and thus will be located at a distance equal to the velocity from the source. The source will have emitted a total of f crests; therefore in the length v there will be f crests, each having a wavelength λ (Fig. 5-7). This situation can be stated algebraically as

$$v = f\lambda \qquad (5\text{-}4)$$

Wave motion constitutes an important means by which energy is transmitted from one place to another; for this reason, the subject of wave motion will become of increasing importance in this text. Rope waves and water waves are already familiar to us because each is transmitted by a familiar and visible medium. Radio and television signals,

FIG. 5-7

Derivation of $v = f\lambda$. The sketch depicts the waves initiated by the source in one second, a number equal to the frequency f. In one second the initial wave travels a distance equal to the velocity v. Each wave has a wavelength λ. Therefore $v = f\lambda$.

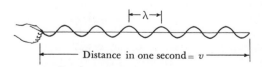

Distance in one second $= v$

as well as sound and light, are also transmitted by means of a wave motion and are familiar to us not because of their wave character but because of the energy they transport.

Transverse and longitudinal waves

Each type of wave can impose a range of frequencies on the medium through which it travels. Furthermore, different media transmit waves at velocities ranging from a few feet per second to thousands of miles per second. In addition, as we will see, waves may be *transverse* (T) or *longitudinal* (L). Typical waves manifesting the variety of characteristics to which we have referred are listed in Table 5-1.

TABLE 5-1
Characteristics of typical waves

	WAVE-TYPE	VELOCITY (m/s)	WAVELENGTH (m)	FREQUENCY (vib/sec)
Water waves in a shallow pan	L&T	0.3	0.03	10
Ocean waves	L&T	4	20	0.2
Waves traveling along ¼ in. rope	T	6	3	2
Waves in violin string—"A" string	T	304	0.70	435
Sound wave in air at 0°C	L	331.36	0.76	435
Earthquake wave (transverse component)	T	4.5×10^3	2.6×10^3	1.7
Earthquake wave (longitudinal component)	L	8.0×10^3	2.4×10^3	3.3
X rays	T	3×10^8	0.4×10^{-10}	7.0×10^{18}
Light wave (green light)	T	3×10^8	0.54×10^{-6}	5.5×10^{14}
Radio wave (WCHL)	T	3×10^8	200	1.5×10^6

A *transverse wave* is one in which each particle in the medium vibrates along a line perpendicular to the direction in which the wave is traveling. The rope waves, which served as our introduction to the subject of waves, clearly involve transverse vibration of any particular part of the rope. In the case of an earthquake, the vibrations that travel through the Earth's crust have one component which is a transverse oscillation. More important for our future discussion is light, which consists of a transverse oscillation of an electric and a magnetic field in the medium through which the light travels.

A *longitudinal wave* is one in which each particle in the medium vibrates along a line parallel to the direction in which the wave is traveling. A sound wave is an example of a longitudinal wave. When sound is propagated through the air, individual air molecules vibrate back and forth along a line parallel to the direction of motion of the wave as the wave goes by. As a result, the air experiences alternate compressions and rarefactions rather than the alternate crests and troughs of transverse waves. The compressions are regions in which the pressure, and hence the density, of the air is greater than that of the undisturbed air, whereas the rarefactions are regions in which the pressure and density are less than that of the undisturbed air.

Not all waves that occur in nature, however, can be classified as strictly longitudinal or transverse. In the case of water waves, listed in our table as "L&T," the actual motion of an individual particle of water takes place along an oval path. The particle does not simply move up and down as the wave goes by but, in addition, undergoes a small amount of back and forth motion having the same frequency. Thus a water wave combines longitudinal and transverse character.

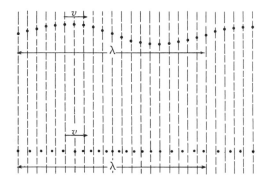

FIG. 5-8

Transverse and longitudinal waves compared. In a transverse wave, particles execute simple harmonic motion along a line perpendicular to the direction of propagation of the wave; in a longitudinal wave, particles oscillate along a line parallel to the direction of propagation of the wave.

5-5 THE BEHAVIOR OF WAVES

When a stone is dropped into a pool of water, circular wave fronts emanate from the disturbance. The wave train may consist of only a dozen or more complete wavelengths, but most of the qualitative features of wave motion are observable. The wave has a velocity and a wavelength, and it imparts a definite frequency to the water on the surface of the pool as the wave goes by.

In this section we will devote most of our attention to the behavior of water waves under varied circumstances, not so much because of any special importance of water waves themselves but because the surface waves on a pool of water are easily observed and manifest most of the characteristics of waves in general. Familiarity with the behavior of water waves can enable us to become aware of the kind of information that would lead us to ascribe wave behavior to other phenomena. We will never be able to see the waves that we will associate with light, for example, but we can observe the extent to which light, under certain controlled circumstances, behaves as waves should behave under those circumstances.

The studies described in this section have been carried out with a device called a ripple tank, which consists of a large, horizontal glass tray mounted 3 ft above approximately a white surface. One pours water to a depth of approximately $\frac{3}{8}$ in. into this tray. A small, variable-speed motor can be made to vibrate a probe that will serve as a point source of waves. The motor can also be used to vibrate a horizontal straight stick on the surface of the water, thus generating straight wave fronts. A glass plate laid in the tray will produce a shallow region; and because the velocity of a water wave depends on the depth of the water, we will be able to observe what happens to the waves when they experience a change in velocity.

By mounting a point source of light approximately 4 ft above the ripple tank, one can observe and study the waves on the white surface of the table because the light transmitted by the tray is focused on the table by the lenslike nature of the wave crests. Finally, by "chopping" the light with a rotating sectored wheel, the transmitted light can be made to flicker at the same frequency as the wave source, in which case the waves will appear to be stationary.

Ripple tank photographs

FIG. 5-9

A ripple tank. Note that in all the photographs and sketches of ripple tank patterns that follow, the waves emitted by the source travel from left to right (the "positive" direction).

Certain characteristics of the behavior of water waves are of particular importance:

1. When a straight wave front encounters an obstruction, secondary waves of the same wavelength and frequency travel away from the obstruction as if the obstruction were itself a source (Fig. 5-10).

FIG. 5-10

Effect of an obstruction.
(*From PSSC* Physics, *D. C. Heath & Co., Lexington, Mass., 1965*)

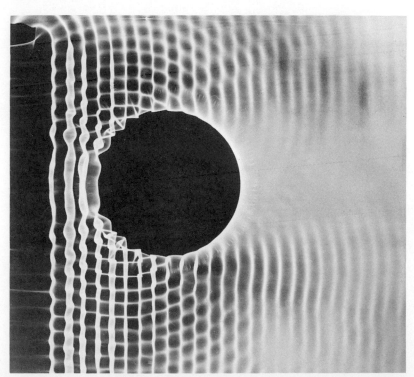

2. When a straight wave front encounters an aperture that is small in comparison with the wavelength of the incident waves, a circular wave front travels into the medium beyond, as if the aperture were a source (Fig. 5-11).

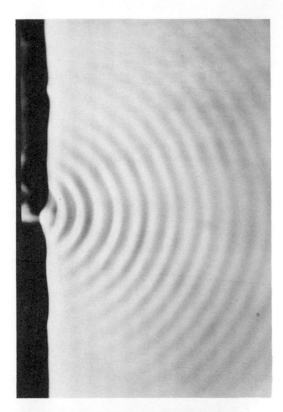

FIG. 5-11

Effect of an aperture
(*Phillips Hall*)

FIG. 5-12

Effect of a wide aperture.
(*Phillips Hall*)

FIG. 5-13
Reflection.
(*Phillips Hall*)

3. When a straight wave front encounters an aperture whose width is large compared to the wavelength, most of the wave energy that passes through the aperture continues to travel straight ahead. However, even in this case one will see evidence of a circular wave front along the edge of the advancing straight wave front (Fig. 5-12).

4. When a straight wave front encounters a plane obstruction, the wave front will be reflected in a symmetric fashion; that is, the angle between the incident wave front and the obstruction will be equal to the angle between the reflected wave front and the obstruction (Fig. 5-13).

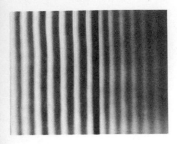

FIG. 5-14

Effect of change of
medium.
(*Phillips Hall*)

5. When a straight wave front encounters an interface between two media at which the wave undergoes a change in velocity, the wavelength of the wave will increase or decrease, depending on whether the velocity has increased or decreased. The frequency, however, will remain constant, because every wave front that encounters the interface travels into the new medium (Fig. 5-14).

6. When a straight wave front encounters the interface between two media at an oblique angle, there will be a change in the direction of propagation of the wave (Fig. 5-15).

7. If two circular wave fronts originate at points separated by only a few wavelengths, the two waves will interact, interfering or reinforcing each other, depending on certain factors we will discuss in the next section (Fig. 5-16).

5-6 INTERFERENCE AND REINFORCEMENT OF WAVES FROM TWO POINT SOURCES

The most convincing evidence that light consists of waves is found in the experimental observation that, under certain conditions, light from

FIG. 5-15

Refraction.
(*Phillips Hall*)

FIG. 5-16

Wave pattern produced by two point sources. (*a*) Two vibrating probes. (*b*) Circular wavefronts emerging from two adjacent apertures in a partition.
(*Phillips Hall*)

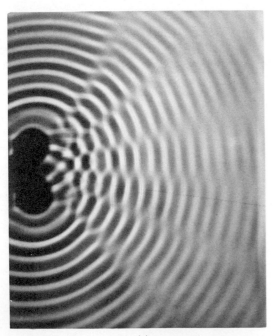

(*a*)

(*b*)

two sources may produce alternate regions of reinforcement and interference. The prototype of this experiment with light is one in which
interference and reinforcement are observed between waves from two
point sources in a ripple tank.

The two point sources in the experiments carried out in a ripple
tank can be two probes driven at the same vibrational frequency [Fig.
5-16(*a*)], or they can be two apertures upon which plane wave fronts are
incident [Fig. 5-16(*b*)]. It is the latter experimental arrangement that
most nearly corresponds to certain optical experiments that will be
described in Chapter 11. Both Figs. 5-16(*a*) and 5-16(*b*), however, show
that in the interaction of the curved wave fronts that emerge from the
probes or from the apertures, alternate regions appear in which the
medium (water in this case) is set into vibration or remains undisturbed.

The formation of nodal lines

To understand how a portion of the medium can remain undisturbed by the two sources, let us consider a point *P* in an undisturbed
zone of Fig. 5-17. This point is characterized by the fact that its distance
from one of the apertures is precisely one-half wavelength greater than
its distance from the other. Assume that at a given instant a crest has
reached *P* from aperture *a*; simultaneously, a trough will have arrived
at *P* from aperture *b*. A crest constitutes an upward displacement of the
water, whereas a trough is a downward displacement. The actual dis-

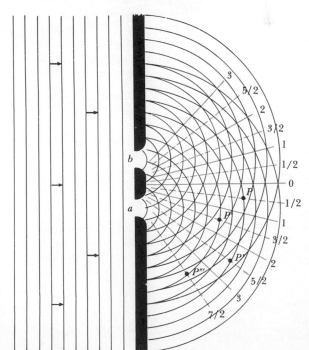

FIG. 5-17

Interference and reinforcement of waves
from two apertures (a "double slit"). Reinforcement lines are shown by solid lines,
nodal lines by dashed lines. Waves from the
two apertures are in phase on reinforcement
lines, out of phase on nodal lines.

placement of the water at P will be equal to the sum of the displacements produced by waves from the two apertures, but since these displacements are equal but opposite, the net displacement will be zero. As time passes, the two waves continue to make equal but opposite contributions to the displacement at P, with the result that the water at P will never experience a displacement. Thus at P the wave from one of the apertures cancels the wave from the other at all times. The sequence of positions along which "destructive interference" occurs is called a *nodal line*.

The same situation will take place at positions characterized by the point P', where the distance to the two apertures differs by $\frac{3}{2}$ wavelengths, or positions like P'', where the difference in the distances equals $\frac{5}{2}$ wavelengths. A *nodal line* can therefore be defined as the locus of a point whose distances from two point sources differ by $\frac{1}{2}$, or $\frac{3}{2}$, or some other odd-numbered multiple of half wavelengths.

Reinforcement of waves

In the space between nodal lines, the effect of the two waves is to produce reinforcement. For example, consider a point on a line perpendicular at its midpoint to the line connecting the two sources. The waves from the two apertures arrive at this point precisely in phase; that is, a crest arrives at this point from one aperture at exactly the same instant a crest reached this point from the other aperture. The combined effect of these two crests is twice that of either crest alone. Extending our argument as above, we can see that maximum reinforcement should occur at any point at which the distances to the two apertures differ by 0, 1, 2—that is by any whole number of wavelengths.

The phenomenon we have described is enhanced and sharpened if one uses 3, or 4, or more equally spaced apertures. A diffraction grating, which utilizes this principle to study the wave nature of light, may have as many as 100,000 lines ruled on a small glass plate, each acting as an aperture to the incident light.

SUMMARY Combining the equation for force exerted by a spring with Newton's Second Law, we have shown that for a mass on a spring,

$$a = -\left(\frac{k}{m}\right)y \qquad (5\text{-}1)$$

and thus constitutes simple harmonic motion.

A geometric simple harmonic motion was identified with the projection of uniform circular motion on a diameter for which

$$a = -\left(\frac{4\pi^2}{T^2}\right)y \qquad (5\text{-}2)$$

Combining (5-1) and (5-2) and solving for T, we found that the period of an object on a spring is given by

$$T = 2\pi \sqrt{\frac{m}{k}} \qquad (5\text{-}3)$$

Waves may be classified as transverse (vibration perpendicular to direction of travel of wave) and longitudinal (vibration parallel to direction of travel of wave).

Waves possess three properties—wavelength (λ), frequency (f), and velocity (v)—interrelated by the equation

$$v = f\lambda \qquad (5\text{-}4)$$

When a plane wave encounters a narrow aperture, the wave that emerges will be a circular wave having the aperture as its center.

Alternate regions of reinforcement and interference occur if waves are generated by two nearby point sources having the same frequency. Reinforcement occurs at a given point if the distances of that point from the two sources differ by 0, λ, 2λ, or any other whole number of wavelengths; interference (nodal lines) occur where these distances differ by $\lambda/2$, $3\lambda/2$, $5\lambda/2$, or any other odd-numbered multiple of half wavelengths.

5-A1. Ask your roommate to stand near one wall of your room and opposite an unshielded lamp. Have him move his clenched fist in a circular path at constant speed while you examine the shadow cast by the lamp. Notice the similarity of the motion of the shadow to the motion of an object suspended by a rubber band. Relate the period, amplitude, and displacement of the shadow to the period and radius of the motion of your roommate's fist. Explain to your roommate what connection you believe this experiment may have to physics.

5-A2. Make a sequence of sketches of an object vibrating on a spring (see Fig. 1-17d), depicting it in approximately eight stages of its motion in making one complete vibration. On top of these sketches, draw arrows depicting the forces acting on the object in proper relative lengths. Show that the net force is properly related to the direction and magnitude of the acceleration.

5-A3. Write Newton's Second Law as it would apply for an object suspended by a spring, given that the spring exerts a force F given by

$$F = -ky$$

where y is the elongation of the spring. How does this expression serve to prove that such an object executes simple harmonic motion?

5-A4. Without writing any algebraic equations, tell how we can be sure that the shadow of uniform circular motion must be simple harmonic motion. When certain that you fully understand the verbal logic of the situation, introduce the appropriate algebraic equations to secure a mathematical proof that the shadow of uniform circular motion executes simple harmonic motion.

5-A5. Two synchronous periodic events of the same frequency are said to be *in phase*; thus we would say that two objects moving up and down together are in phase. What then would *out of phase* mean? Examine Fig. 5-17 and reread the last line of the caption. Translate what that line is attempting to say into language more familiar to you.

5-A6. A wave traveling across the surface of a ripple tank has a velocity of 0.25 m/sec and a wavelength of 0.04 m. What is the frequency of vibration of the water as the wave goes by?

5-B1. (*a*) Consider a ball traveling at constant speed in a circular path with a radius of 0.40 m and a period of 2 sec. What are the velocity and the acceleration of this ball? (*b*) Let a lamp at a great distance from this ball form a shadow as in Fig. 5-2. What are the maximum velocity and the maximum acceleration of the shadow? (*c*) At what points or point in the motion of the shadow will the maximum velocity occur? At what points or point will it have the maximum acceleration?

5-B2. A force of 5 newtons elongates a certain spring by 0.05 m. (*a*) What is the force constant of the spring? (*b*) What elongation will occur if one end of the spring is attached to a rigid support and a mass of 1.5 kg is suspended by attaching it to the other?

5-B3. Let a 1.5 kg object be suspended by a spring whose force constant equals 250 N/m. Let the object be set into vibration with an amplitude of 0.04 m. (*a*) What net force will the object experience at its turning points? (*b*) What will be its acceleration at the turning points? (*c*) What acceleration will the object have as it passes through the equilibrium position?

5-B4. The relaxed length of a certain spring equals 2.50 m. A 5-kg object attached to this spring executes 0.5 vib/sec—that is, $T = 2$ sec. (*a*) What is the force constant of the spring? (*b*) What is the equilibrium length with the 5-kg object attached? (*c*) What acceleration will the object possess when passing through a point 0.30 m below the equilibrium position?

5-B5. Identical springs are attached to a 6-kg cart as shown in Fig. 5-18. A force of 12 N is required to displace this cart 0.1 m to the right of the equilibrium position. (*a*) How much additional energy is stored in the springs by this displacement? (*b*) What maximum velocity will be acquired by the cart when released? (*c*) Will this cart execute simple harmonic motion when released? Explain.

5-B6. In the situation described in Prob. 5-B5, (*a*) What is the effective force constant of the springs in this arrangement? (*b*) What accelera-

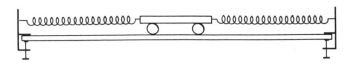

tion will the cart have at the instant it is released? (*c*) What will be the period of the vibration of the cart?

5-B7. A certain radio station operates on an assigned frequency of 1.36×10^6 vib/sec (hertz). What is the wavelength of the waves emitted by the radio station?

5-B8. Plane waves traveling with a velocity of 0.1 m/sec and having a wavelength of 0.01 m encounter an interface with which the incident waves make an angle of 30 degrees. Beyond the interface, the waves travel with a velocity of 0.07 m/sec. Make a scale diagram showing what happens to the waves at the interface.

5-B9. Using Fig. 5-17 as a model, prepare a scale diagram showing the diffraction of waves having a wavelength of 0.8 cm when they encounter two apertures (slits) whose centers are 2.0 cm apart. Do your work carefully enough to permit locating and measuring the angular orientation of the nodal lines.

Thermal Energy

At this point it will be helpful to reexamine the procedure that we have followed in defining terms used in physics. This step is desirable, not only because it will throw some light on the logic of physics but also because it will facilitate our understanding of the very important definition of temperature.

Let us start by considering the very familiar "is"-type of definition. Statements such as "A table is a piece of furniture with four legs" and "A horse is a quadruped with mane and uncloven hoof" fit this category. A definition taking this form identifies an object with a class of objects and proceeds to tell how this particular object differs from other objects of that class. Dictionaries, to be sure, tend to use "is" type of definitions, saying that "Mass is the inertial property of matter" and that "Time is the period between two events," but such statements do not suffice to establish numerical values to associate with the properties in question.

Definitions in physics

In contrast, definitions used in physics may be called *procedural* or *operational definitions*. A definition of this variety is constructed very much like the procedural instructions in a laboratory manual; each definition gives a set of directions telling the reader that if he carries out certain measurements and calculations, he will arrive at a numerical value for the quantity in question.

As an example, suppose that we were to attempt to construct at this point a definition of the word "newton." We might write "A newton is that force which when acting on a mass of 1 kilogram will produce an acceleration of 1 meter per second squared." Even this definition, which contains the word "is," constitutes a procedural instruction; one is literally instructed to go to the laboratory and measure the acceleration experienced by a 1-kg object under the action of various forces if he wishes to identify a force that has a magnitude of 1 newton.

This is not to say that dictionary-type definitions are never used

6

or usable in physics. Such statements are often used to transmit the intuitive basis for a definition. A prominent dictionary, for example, says that "temperature is the degree of hotness or coldness of anything." Far from maintaining that this definition is in error, we will use the essential idea contained in this statement as the starting point for our discussion of temperature. It is not, however, a definition of temperature because we would never be able to establish a science involving numerical relationships between properties of objects with such a definition.

6-1   TEMPERATURE AND TEMPERATURE SCALES

The subjective aspect of temperature

A statement of the temperature of an object or of a region is, for most of us, a means of transmitting information about one aspect of that object or region. We have achieved some familiarity with the Fahrenheit scale, and it has become natural to us to say that an object or region that is "hot" has a higher temperature than an object or region that is "cold." If the temperature outside is 0° Fahrenheit, we recognize that it is a cold day; if the temperature outside is 100° Fahrenheit, we regard it as a hot day. The early decision to associate larger numerical values with "hot" than with "cold" may well have been fortuitous but was, as will be seen, very fortunate.

Our discussion of temperature should not only provide us with an understanding of the procedure by which a thermometer is calibrated but also should enable us to answer certain questions regarding the role of temperature and temperature measurements in physical processes. For example, if two objects have the same temperature, what is it that is really the same about the condition of the two objects? To what extent does the statement of the numerical value of the temperature say something specific about the object itself?

FIG. 6-1

Temperature observation. "Hotness" and "coldness" can be expressed numerically.

*the physics of
large objects*

Many physical properties of materials change when the object is moved from a region that is "cold" to a region that is "hot." In some cases, as in the frying of an egg or in the melting of a lead sphere, the original condition of the object will not be fully reestablished when the object is taken back to the colder environment. On the other hand, many situations exist in which the physical property under observation depends only on the temperature, at least if one stays within a limited range of the initial temperature.

As an example, let us suppose that we have devised a precise means by which we may measure the length of a small steel bar. If one measures the length of this bar when it is in physical contact with a block of ice, then measures its length when it is exposed to the vapor immediately above a vessel of boiling water, and finally measures it after it is brought back into contact with the ice, several important facts can be observed from the data. The object will have a greater length when in contact with steam than when in contact with ice, but its length will be precisely the same when in contact with ice after being exposed to the steam as it was before being exposed to the steam. The length of the object thus serves as an indicator of the relative hotness or coldness of the ice and the steam. If the same steel bar were to be placed in contact with some third object, the length then assumed by the bar could be used as a measure of the hotness of that third object. In fact, the length of the steel bar may serve to establish a numerical value of the temperature, provided one adopts a procedure by which the length is to be interpreted.

Any object possessing a physical property that is solely temperature dependent may be used to measure hotness or coldness. If we utilize the numerical values that are assumed by any temperature-dependent physical property to describe the degree of hotness or coldness, we have established a *temperature scale*. The object may be called a *thermometer*.

A thermometer measures its own temperature

When one uses a thermometer to measure the temperature of an object, he places the thermometer in contact with the object whose temperature is being measured. The relevant physical property of the thermometer undergoes a rapid initial change but finally establishes a constant equilibrium value. Implicit in these manipulations is the assumption that when one establishes *thermal equilibrium* between the thermometer and the object under observation (i.e., when the relevant physical property of the thermometer settles down to some ultimate constant value), the two objects have the same temperature. We see, then, that in making a measurement of the temperature of an object, two objects are always involved: the medium whose temperature is being measured and the thermometer. The thermometer measures its own temperature, and we infer the temperature of the medium to be that of the thermometer which is in thermal equilibrium with it.

Experience shows that a given metal bar assumes a well-defined

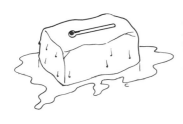

FIG. 6-2

A thermometer measures its own temperature. We assume that the immediate surroundings have the same temperature as the thermometer.

length when in contact with melting ice; placed in a drinking glass filled with ice cubes, it will have the same length as it would have in contact with the ice on a melting freshwater pond. We therefore can say that the temperature of melting ice is well defined. Experience also shows that the temperature of steam is well defined: the length of a metal bar suspended immediately above the surface of a vessel of boiling water is independent of the size of the vessel or how vigorously the water is caused to boil.

The choice of fixed points in thermometry

Since the temperatures of melting ice and of the vapor above boiling water are both well defined, one could arbitrarily associate one numerical value with the temperature of melting ice and another numerical value with the temperature of the vapor above boiling water. There are a few precautions that must be taken, of course. The water must be pure and the pressure of the atmosphere must be "standard." (Pressure will be defined and discussed in more detail in the next section).

In establishing the centigrade (or Celsius*) temperature scale, these "fixed points" are given numerical values of 0° centigrade and 100° centigrade respectively. These fixed points will be referred to as the "ice point" (0°C) and the "steam point" (100°C).

On the Fahrenheit scale, the "fixed points" are 32°F and 212°F respectively. The difference in temperature between the ice point and the steam point is therefore 180°F; hence a change in temperature of one Fahrenheit degree (1°F) is $\frac{5}{9}$ of the change in temperature of one centigrade degree (1°C). To convert from one scale to the other, one establishes the ratio

$$\frac{t(°C)}{100} = \frac{t(°F) - 32}{180} \qquad (6-1)$$

Establishing a temperature scale

One is now in a position to establish a temperature scale. A reasonable approach would be to prepare a graph showing the temperature along the x axis and the physical property of the thermometer along the y axis. Any line drawn between the two fixed points will constitute a definition of all the temperatures in the intermediate region and therefore will establish a temperature scale. The most likely line for one to elect to draw would be a straight line.

Suppose that a steel bar has a length of 0.050000 m when in contact with melting ice and a length of 0.050060 m when in contact with steam. The data are shown graphically in Fig. 6-3. A linear temperature scale has been established along the x axis, and the change in length relative to the length of the bar when in contact with ice is shown

---

* For all practical purposes, the centigrade and Celsius scales are identical. A distinction between them, however, will be noted in Sec. 6-4.

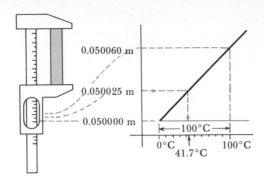

FIG. 6-3

Calibration of a steel-bar thermometer. Numerical values of 0°C and 100°C are assigned to the lengths of the bar at the ice and steam points respectively. Other temperatures are inferred by interpolation or extrapolation.

along the y axis. Intermediate temperatures are defined by drawing a straight line through the two fixed points.

Let us assume that when this steel bar is placed in contact with the contents of a vessel, it acquires a length of 0.050025 m. The temperature t of the vessel is found by interpolation; that is,

$$\frac{t}{100} = \frac{25}{60} \quad \text{and} \quad t = 41.7° \text{ centigrade (steel bar)}$$

It must be recognized that we could establish literally dozens of different thermometers were we to make use of different materials as well as different physical properties of materials. Suppose, for example, that one were to construct a mercury-in-glass thermometer as shown in Fig. 6-4. One would note the position of the mercury when the thermometer was submerged in ice and its position when enveloped in steam. One would subdivide the region between these two marks into 100 equally spaced divisions. He would submerge the thermometer in the medium under observation and read a temperature from the linear scale that had been inscrib~d on the thermometer stem.

Unfortunately, the temperature scale based on the expansion of a steel bar and the scale based on the expansion of mercury do not agree precisely: We have established two separate temperature scales, one a "steel" temperature scale, the other a "mercury-in-glass" temperature scale. In fact, each of the dozens of thermometers one might devise will yield a different temperature scale; each will there-

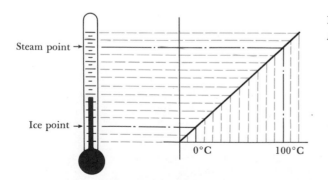

FIG. 6-4

A mercury-in-glass thermometer.

fore yield a different numerical value for the temperature of any medium, agreeing only at the fixed points of 0° and 100° centigrade. It is clear that we must select one of these thermometers as the basis for a *standard* temperature scale, and all thermometers must be calibrated by comparison with this standard. Although, in principle, one could simply make an arbitrary choice, it is found that, for physical reasons that will emerge in our later discussion, one of these choices is preferable over any of the others.

6-2 PRESSURE AND BOYLE'S LAW

The definition of pressure

Pressure is defined as force per unit area; that is,

$$\text{Pressure} = \frac{\text{force against a surface}}{\text{area of the surface}}$$

or
$$p = \frac{F}{A} \tag{6-2}$$

The pressure of the atmosphere is regarded as "standard" when it equals $1.013 \times 10^5 \, \text{N/m}^2$ ($=14.7 \, \text{lb/in}^2$.) but varies with location and depends on local weather conditions. Such a pressure on the surface of a picture window 1 m square results in a force against the surface of $10^5 \, \text{N}$ (approximately 11 tons); the window, of course, would collapse were it not for an equal force on the opposite surface.

In most of the situations in which we will deal with the pressure exerted by a gas, the gas will be confined in a vessel. In all such situations, the gas will exert equal pressure at all points on the surface of the confining vessel.

FIG. 6-5

Boyle's law. (*a*) Increasing the pressure on a confined gas at constant temperature is accompanied by a decrease in volume. (*b*) A graph of pressure versus volume for a gas at constant temperature shows that the product of pressure and volume yields a constant. The curve is known as a hyperbola.

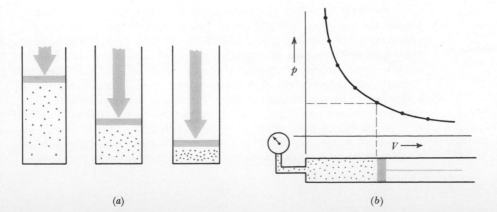

(*a*)   (*b*)

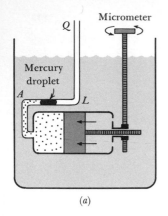

(a)

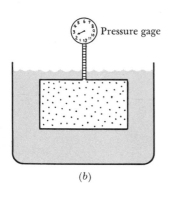

(b)

FIG. 6-6

Gas thermometers.
(a) A constant-pressure gas thermometer; the temperature is inferred from the volume of gas at a prearranged constant pressure.
(b) A constant-volume gas thermometer; the temperature is inferred from the pressure of the gas at a prearranged constant volume.
The temperature scales established by these thermometers are identical; furthermore, the temperature scale is independent of the gas used.

Let a gas be confined in a cylinder by means of a piston and assume that the cylinder and its contents can be maintained at constant temperature. If the piston is moved in such a direction as to decrease the volume of the gas, the pressure of the gas will increase but will return to its original value if the piston is restored to its prior position. Measurements show that the pressure is inversely proportional to the volume. Letting p_1 and V_1 represent the initial values of the pressure and the volume and p_2 and V_2 represent the final values, we have

$$p_1 V_1 = p_2 V_2 \qquad \text{(temperature constant)} \qquad (6\text{--}3a)$$

or since this relationship is valid for a given specimen of gas regardless of the "final" volume, we may write

$$pV = K \qquad (6\text{--}3b)$$

This experimental observation is known as Boyle's law.

The numerical value of K depends on the temperature of the gas and on the amount of gas enclosed by the vessel. As we shall see in the next section, the fact that pV for a given amount of gas is determined by the temperature enables us to establish a thermometer and a temperature scale based on a gas thermometer. This "gas thermometer" temperature scale, in fact, will serve as the basis for all measurements of temperature.

6-3 THE GAS THERMOMETER TEMPERATURE SCALE

pV as a thermometric property

We noted in the previous section that, for a given specimen of gas, the product pV has a numerical value that is determined entirely by the temperature of the gas. A thermometer that makes use of this property of a gas as an indicator of temperature is called a gas thermometer.

In practice, either the pressure or the volume is held constant. Hence the property actually under observation is the volume or the pressure of the enclosed gas. A thermometer that detects a change of temperature through a change in volume of a gas consists of a gas which is maintained at constant pressure and is called a "constant pressure gas thermometer." A thermometer that detects a change in temperature by a change in pressure consists of a gas maintained at a constant volume and is called a "constant volume gas thermometer."

A sketch of a constant pressure gas thermometer is shown in Fig. 6-6(a). It consists of a glass cylinder and a long capillary tube of known, uniform internal diameter. The gas that serves as the thermometric material is enclosed by a small drop of mercury in the capillary and by a piston in the cylindrical tube. A calibrated micrometer screw enables one to make a precise measurement of the volume available to the gas. The mercury droplet is located in the horizontal section of the capillary *AL*. Because the capillary is open at *Q*, one can be sure that the gas is under a constant pressure of 1 atm (atmosphere). The entire cylinder, and the horizontal portion of the capillary tube, must be submerged in the medium whose temperature is being measured.

Calibrating a gas thermometer

In calibrating a constant pressure gas thermometer, one follows the procedure described in Sec. 6-1. The thermometer is submerged in an ice bath (0°C) and the volume determined from the position of the mercury droplet and the position of the piston. The thermometer is then uniformly exposed to steam emerging from water boiling at atmospheric pressure (100°C) and the volume noted once more. The data are plotted as in Fig. 6-7(*a*), and a gas thermometer temperature scale is established by drawing a straight line through the two fixed points.

A constant volume gas thermometer is calibrated in a similiar manner. The volume of the gas is held constant and the pressure is read from a suitable pressure gauge when the thermometer is immersed in melting ice and in steam. Figure 6-7(*a*) indicates that the temperature scale is established by drawing a straight line through the two data points on a pressure versus temperature graph.

It may not be apparent that we are better off than we were before. One can design literally dozens of different gas thermometers using, for example, hydrogen, helium, nitrogen, or any of a number of other gases either at constant pressure or at constant volume and, in principle, each will yield a different temperature scale. In practice, however,

FIG. 6-7

Calibration of a gas thermometer. (*a*) A straight line is drawn through the steam point (100°C) and the ice point (0°C). (*b*) This line intersects the horizontal axis at a temperature of −273.15°C, and because negative pressures and negative volumes are unthinkable, we have defined a temperature scale in which a temperature lower than −273.15°C has no meaning.

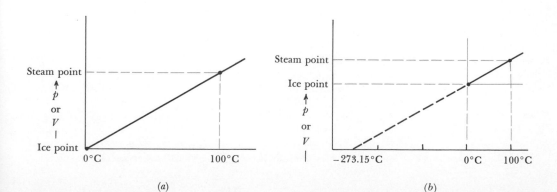

gases behave remarkably alike, and one will detect only very small differences between the scales that are obtained with the various types of gas thermometers. The behavior of gases has been studied so carefully that one can ascribe the discrepancies between gas thermometers to physical factors, such as the size of the molecules and the forces that they exert on each other. With this kind of information, one can make small corrections on the data at the two fixed points. This enables one to say that *all corrected gas thermometers yield the same temperature scale*. This fact alone undoubtedly would have impelled scientists to select the corrected or *ideal* gas thermometer temperature scale as the basis for all temperature readings. In Chapter 7 we will encounter another more compelling reason for this choice.

The "absolute zero" of temperature

An interesting consequence of our choice of temperature scale can be observed if we enlarge our graph so that the quantity p (or V) plotted on the vertical axis initiates at zero [Fig. 6-7(b)]. In this case the straight line is found to intersect the horizontal axis at a temperature of $-273.15°C$, which in turn represents the temperature of a gas were it possible to reduce its pressure (or volume) to zero. This lowest possible temperature is called the *absolute zero* temperature. No object has ever been cooled to this temperature; like a frictionless surface, it is regarded as unattainable.*

6-4 THE KELVIN TEMPERATURE SCALE

In summary of the logic to this point, recall that since the product pV for a given specimen of gas is determined by the temperature of the gas, the product pV may serve as the physical property one may use to measure temperature. Letting the ice point (0° centigrade) and the steam point (100° centigrade) serve as reference points, one may establish a centigrade (meaning 100 divisions) scale by assuming a linear relationship between the product pV and the temperature. Extension of this linear relationship to $pV = 0$ shows that a centigrade temperature lower than $-273.15°$ has no meaning.

Defining the Kelvin scale

With these facts in mind, one sees that a temperature scale can be established by assigning a value of zero to this lowest possible temperature; *one* measurable fixed point then serves to define all other temperatures. The Kelvin temperature scale is such a scale.

The single measurable fixed point on the Kelvin temperature is

---

* The statement that no object can be cooled to the absolute zero temperature is contained in a principle known as the Third Law of Thermodynamics.

the "triple point" of water, defined to have a temperature of 273.16°K.*
On this scale, melting ice has a temperature of 273.15° K. The steam
point is no longer a fixed point because as more accurate methods of
measuring pressure and volume are developed, the temperature of
steam will surely be found to differ, at least slightly, from 373.15°K.
Since the scale we have been calling the centigrade scale no longer can
be said to have 100 degrees separating the temperature of ice and the
temperature of steam, we will adopt a different name for that scale. The
name chosen to replace "centigrade" is "Celsius." Habit and malprac-
tice are very strong, however; it may be a long time before many of us
will read 37°C as "thirty-seven degrees Celsius."

The relationship between Kelvin and Celsius temperatures is,
then,

$$T = t + 273.15 \qquad (6\text{--}4)$$

where T represents the temperature in Kelvin degrees and t represents
the temperature in Celsius degrees.

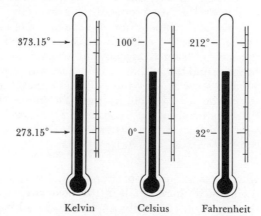

FIG. 6-8
Temperature scales compared.

Kelvin   Celsius   Fahrenheit

In many applications, one is concerned only with a change of
temperature rather than the temperature per se. It should be noted
that a *change* of temperature is exactly the same whether expressed in
Kelvin or in Celsius degrees; a change in temperature by one Kelvin
degree (1°K) is precisely a change in temperature of one Celsius degree
(1°C).

The volume-temperature and pressure-temperature equations

The equations that relate volume to the Kelvin temperature on a
constant-pressure gas thermometer and that relate pressure to Kelvin

* The "triple point" of water is defined as the situation in which the solid,
liquid, and vapor states are in equilibrium. The melting point of ice at atmospheric pressure equals 0°C, while its triple-point temperature equals 0.01°C;
clearly, the distinction between the melting point of ice and the triple point
of water is a nuance that we should acknowledge but not worry about.

temperature on a constant-volume gas thermometer are as follows:

$$\frac{V_1}{T_1} = \frac{V_2}{T_2} \qquad \text{(pressure constant)} \qquad (6\text{-}5)$$

and
$$\frac{p_1}{T_1} = \frac{p_2}{T_2} \qquad \text{(volume constant)} \qquad (6\text{-}6)$$

A distinctive difference exists between the logic by which these equations were secured and that employed in arriving at Boyle's law [Eq. (6-3)]. Boyle's law, it will be recalled, is an experimental law and thus owes its validity to painstaking observations in the laboratory. The latter two equations, however, are definitions; they are valid because of the way we define temperature.

6-5 THE GENERAL GAS LAW

Describing an experiment in which one might establish the validity of Boyle's law, we proposed an arrangement in which the pressure, volume, and temperature of a specimen of gas could be controlled or changed at will. In that discussion we noted that when the volume of the gas is changed at constant temperature, the product pV of the pressure and the volume remains constant also. It was then shown that pV may serve as a measure of the temperature of the gas, enabling us to use the arrangement as a gas thermometer. The temperature scale that resulted from these considerations established the Kelvin temperature of the gas as proportional to pV; that is,

$$pV = \alpha T \qquad (6\text{--}7)$$

where α is a constant of proportionality whose numerical value will be discussed in more detail in the next chapter. If interpreted properly, this equation is equally applicable to all gases; for this reason, it is called the *general gas law*.

In our discussion of the behavior of gases, we ignored one eventuality that should be acknowledged at this point. If one decreases both the volume and the temperature, a situation will eventually be reached at which the "gas" will begin to liquify, and, in fact, if the process is continued, all the material will change to a liquid phase. Clearly Eq. (6-7) is applicable only to values of pressure, volume, and temperature at which the material is a gas.

If no condensation occurs, however, the pressure p, the volume V, and the temperature T of a gas are related through Eq. (6-7). This means that, for a given amount of gas, if one chooses numerical values for any two of the variables p, V, and T, the third is uniquely determined.

It may be useful to note that the general gas law can be regarded as a generalization of previously stated principles and definitions. If T is held constant, Eq. (6-7) reduces to Boyle's law, whereas if p or V is held constant, one obtains Eq. (6-5) or Eq. (6-6) respectively.

If one rubs his hands vigorously together, he will be conscious not only of the fact that he is doing work but also of the fact that the temperature of his hands increases. If one pulls a nail from a piece of wood, the nail may be quite warm immediately after being withdrawn. A hand tire pump gets hot as one inflates a tire. In fact, in every mechanical situation in which the total mechanical energy (kinetic plus potential) of an isolated system decreases or in which work is done without the appearance of an equal amount of mechanical energy, some portion of the system experiences a rise in temperature.

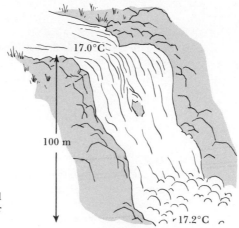

FIG. 6-9

Conversion of potential energy into thermal energy. The water below the falls is warmer than the water above the falls.

The thermal equivalent of mechanical energy

The quantitative aspects of these observations can be determined only by experimentation. The measurements are carried out in an insulated vessel (a calorimeter) containing water and a thermometer to measure the change in temperature. Four such calorimeters are shown in Fig. 6-10, in which several pieces of auxilliary apparatus have been installed as follows: (*a*) a group of paddle wheels, (*b*) a pair of disks that may rub against each other, (*c*) a pump that forces water through an orifice and back into the vessel, and (*d*) an electric heater. Each device receives energy from an identical source, symbolized in the figure by an object that descends a well-defined distance. The devices do identical amounts of work in the presence of the water that fills the calorimeter.

Careful experimentation with such calorimeters have shown that exactly 4186 joules of work must be done in order to raise the temperature of 1 kg of water by one Celsius degree (1°C) or 1°K. Furthermore, it makes no difference which of the various devices is employed; a definite amount of work produces a definite rise of temperature.

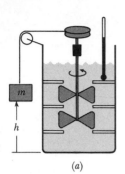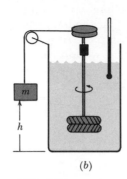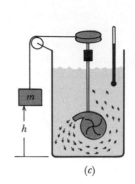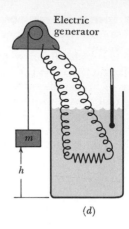

(a) (b) (c) (d)

FIG. 6-10

Measuring the thermal equivalent of mechanical energy. Regardless of the mechanism employed, 4186 joules of work will always raise the temperature of a kilogram of water by 1°K.

Thermal energy and the Law of Conservation of Energy

It will be recalled that a conservative system was defined in Sec. 3-5 as a combination of objects or devices in which the sum of the kinetic and the potential energy is conserved. It was also stated that in such a system all objects and devices are subject only to conservative forces. Friction was stated to be a nonconservative force.

When friction is present in a system, the work that objects do against frictional forces is done at the expense of kinetic and/or potential energy. Our observation that work done against friction always results in a definite rise in temperature has led physicists to the hypothesis that energy has merely been changed into another form, called *thermal energy*.

When we measure thermal energy, we observe its effect on the system to which it has been added, which, as we have seen, is a rise in temperature. The unit of thermal energy is called a kilogram calorie and is defined as the amount of thermal energy required to raise the temperature of 1 kg of water by 1°C; therefore it is the equivalent of 4186 joules. Thus the kilogram calorie is the thermal equivalent of 4186 joules or, if one prefers, 4186 joules is the mechanical equivalent of 1 kilogram calorie. The "Calorie" to which reference is frequently made at mealtime, is a kilogram calorie as defined above.

The hypothesis that a definite amount of work is exactly the equivalent of a precise amount of thermal energy has the effect of saying that the total energy of the universe remains constant in any process, whether the process occurs in a conservative or a nonconservative system. This hypothesis has endured nearly 100 years of test and discovery and forms a cornerstone of physics known as the Law of Conservation of Energy. This law may be stated as follows:

Energy can neither be created nor destroyed; it can only be transmitted from one object to another or changed into some other form.

At the time the Law of Conservation of Energy was announced, scientists were familiar with energy in only a few forms: potential energy, kinetic energy, and, finally, thermal energy. As experimental scientists have become more familiar with physical phenomena, energy in other forms has been discovered. To date, no phenomenon in nature appears to be immune from the dictates of this law; the total energy of the universe is apparently constant. Unfortunately, however, although the total energy in the universe may remain constant, energy may not remain available; energy can be "lost" even if it cannot be destroyed, or it can be changed into a form in which it cannot be fully reconverted into useful work.

Boyle's law is an experimental law that states that for any gas at constant temperature,

$$pV = K \qquad (6\text{-}3)$$

where K is a constant whose value depends on the temperature and the amount of gas, V is the volume, and p is the pressure, the force per unit area.

The Kelvin temperature scale is based on the absolute zero point and a single measurable fixed point, the triple point of water defined to equal 273.16° Kelvin; on this scale, the ice point equals 273.15° Kelvin. The Celsius temperature scale differs from the Kelvin scale only in the choice of zero, which, for the Celsius scale, is the ice point. Therefore

$$T(°\text{K}) = t(°\text{C}) + 273.15 \qquad (6\text{--}4)$$

By definition, the Kelvin scale is established by setting the temperature proportional to pV in a gas thermometer. It follows that

$$pV = \alpha T \qquad (6\text{--}7)$$

where α is proportional to the amount of gas present.

Because of the way the temperature scale is established, a temperature lower than 0°K or −273.15°C has no meaning. This temperature is called the absolute zero.

The Law of Conservation of Energy states that energy can neither be created nor destroyed; it can only be transmitted from one object to another or changed into some other form.

In nonconservative systems mechanical energy is changed into thermal energy. Experiments show that 4186 joules of mechanical energy will raise the temperature of 1 kg of water by 1°C; this amount of energy is called a kilogram calorie.

6-A1. State what is meant by an operational definition, then define mass and kinetic energy, explaining what operations or procedures are implied by your definitions.

6-A2. Compute entries (*a*), (*b*), (*c*), and (*d*) for the small table of equivalent temperatures shown.

CELSIUS	KELVIN	FAHRENHEIT
100°C (Steam point)	(*a*) ———	212°F
0°C (Ice point)	273°K	32°F
−196°C (Nitrogen point)	(*b*) ———	(*c*) ———
−273°C (Absolute zero)	0°K	(*d*) ———

6-A3. Normal atmospheric pressure is $1.013 \times 10^5 \, \text{N/m}^2$. (*a*) What is the pressure of the atmosphere in pounds per square foot? (*b*) What force does the atmosphere exert' against a windowpane one foot square? (*c*) Why does the window not break?

6-A4. Given that "a thermometer measures its own temperature," by what logic can a thermometer be used to measure the temperature of another object?

6-A5. Rub your hands vigorously together, then hold them against your face. Explain the observed rise in temperature of your hands.

6-A6. List some thermometric properties of matter that can be used to indicate temperature. Give some compelling reasons for the choice of a gas thermometer as the basis for defining a temperature scale.

6-A7. Your roommate tells you that "thermal energy" is a fiction invented by some professor to save the Law of Conservation of Energy. Can you find any merit in this idea? What argument would you use to convince your roommate that thermal energy is just as valid a concept as mechanical energy?

6-B1. The gas in a constant volume gas thermometer exerts a pressure of $1.020 \times 10^5 \, \text{N/m}^2$ against the walls of the confining vessel when the gas is maintained at the ice point. What pressure will the gas exert if its temperature is raised to the steam point?

6-B2. In an experiment with a constant volume gas thermometer, a student observes that the confined gas exerts normal atmospheric pressure ($1.013 \times 10^5 \, \text{N/m}^2$) when the thermometer is in contact with melting ice. When this gas thermometer is placed in liquid nitrogen at its boiling temperature, the pressure falls to $0.286 \times 10^5 \, \text{N/m}^2$. (*a*) What is the boiling temperature of nitrogen in degrees Kelvin according to this measurement? (*b*) What is the temperature in degrees Fahrenheit?

6-B3. A cylindrical vessel with a piston encloses 1500 cm³ of Ne gas at normal atmospheric pressure and at a temperature of 27°C. (*a*) What is the numerical value of pV/T? (*b*) What pressure will this gas exert if the volume is reduced to 1000 cm³ and the temperature is raised to 127°C?

6-B4. The gas constant (pV/T) for a certain specimen of gas equals 8000 N m/°K. (*a*) What volume will this specimen occupy under normal conditions ($p = 10^5$ N/m² and $T = 27°$C)? (*b*) What volume will the gas occupy if $p = 10^8$ N/m² and $T = 727°$C?

6-B5. How much work would one need to perform in order to raise the temperature of 5 kg of water by 3°K?

6-B6. An ice cream freezer without ice (Fig. 6.10*a*) is filled with pure water (4 kg) at 25°C. The motor is started, stirring the water vigorously; after 10 min the temperature is observed to be 29°C. (*a*) How much work was done by the motor, assuming that all the energy was changed into thermal energy in the water? (*b*) What is the likely rating of the motor in watts, given that a watt represents energy expended at the rate of 1 joule/sec?

6-B7. On a hot summer day approximately 1400 joules of energy fall on each square meter of the Earth's surface per second. If a child's outdoor pool has a cross section of 1 m² and contains 250 kg of water, what should be the rise in temperature in one hour of exposure to the noonday Sun?

6-B8. It is estimated that by the year A.D. 2000 electrical energy will be used in the United States at the rate of 10^{12} joules/sec (watts). Given that the Sun provides 1400 joule/m² sec and that the land area of the United States equals 8×10^{12} m², what fraction of the incident energy would we have to utilize if we were to depend on solar energy? (Remember, in making your estimate, that the Sun doesn't shine at night!)

THE CLASSICAL APPROACH TO THE UNSEEABLE

Early in the nineteenth century it was realized that if a complete understanding of macroscopic phenomena was to be attained, it would require insight into a submicroscopic world that might remain forever beyond direct observation and investigation. In order to probe this submicroscopic world, sundry ingenious instruments were invented; some extended the boundary line separating the macroscopic from the submicroscopic world, whereas others simply gave more precise data regarding the external effect of the collective behavior of objects that cannot be seen. A period of "model building" ensued from which physical structures of the unseeable world were inferred from these data.

Basic to all such model building was the assumption that the submicroscopic world is a world of waves and particles obeying laws and principles similar to those obeyed by ordinary objects. Eventually physical phenomena were to be observed that forced the reevaluation and finally the abandonment of this very plausible assumption, but a chronicle of the scientific work of the years 1800 to 1890 fills some of the brightest pages in the history of man's accomplishments in physics.

The
Structure of Matter

Most of our concern to this point has been centered on the behavior of objects under conditions in which all aspects of their performance were clearly visible. We have discussed force not only in terms of its effect on objects that can be seen and handled, but forces have themselves been expressed in terms of accelerations based on distances and observable amounts of time. Our discussion of force led to definitions of work and energy that were couched in terms which were equally susceptible to direct scrutiny and measurement. Even when the energy was partially dissipated in the form of "thermal energy," we contented ourselves with describing the thermal energy in terms of observations of visible changes in an instrument reading—in this case, a thermometer. In our study of wave motion we have concerned ourselves with waves that we can see. In other words, we have not dealt with the microscopic, or the submicroscopic, or the unseeable; the portion of physics that has occupied our attention has been the portion of classical physics that deals with macroscopic rather than microscopic systems.

A portion of classical physics and most of modern physics, however, is concerned with the unseeable; it searches for a detailed description of the structure of matter, the nature of light, and the precise form in which thermal energy appears when mechanical energy is "lost."

We study the unseeable by observing the effects that it produces on something we can see. In some cases, the effect may be an observable, visual effect of a single submicroscopic event, such as the track that a nuclear particle makes in a cloud or bubble chamber (to be described in Sec. 10-6). In other circumstances, a large number of submicroscopic particles in the form of a block of metal or a vessel filled with a liquid or a gas will be studied as a group. We observe the behavior of the conglomerate sample under pressure, in an electric field, or in combination with other materials. We may observe the light that the material either emits or absorbs. After collecting data on the effects produced by individual events, and on the behavior of a bulk sample of the material, we can begin to formulate a *model*.

A model is a sketch or a word description of that which may be unseeable. The model shows how we believe the unseeable object would

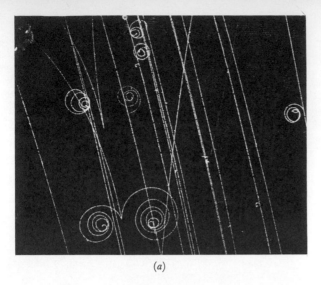

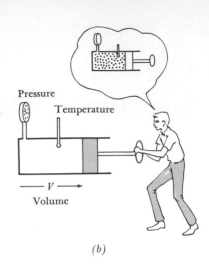

(a)

(b)

FIG. 7-1

Seeing the unseeable.

(*a*) Submicroscopic nuclear particles leave a trail of tiny bubbles in a bubble chamber. (*Lawrence Berkeley Laboratory, University of California, Berkeley, Calif*)

(*b*) The observer secures data on macroscopic characteristics of the gas but evolves theories depicting the gas to consist of millions of submicroscopic molecules.

appear if we could see it, but one does not expect that it will ever correspond with nature because it is certain to be based on less than complete information. The model summarizes what is known about this aspect of nature at the moment it is invented, but it will be changed or discarded as new information is gathered. The prestige of a given model is greatly enhanced if it successfully predicts events in advance of their observation in the laboratory. On the other hand, a single failure may destroy it. Since nature is never precisely like any of our models, we never say, "Nature is like this model," but rather "The experiments performed to date indicate that nature behaves as it would were it to consist of a structure like this model."

In discussing the structure of matter in this chapter, we will present a model that has been extremely useful not only in describing matter but in predicting its behavior in varied circumstances as well. The atomic model that we will use is not in its most modern form; in fact, because of its relative simplicity, we have chosen one that is now somewhat out of date.

7-1 SOLIDS, LIQUIDS, AND GASES

At a given temperature, water is almost certain to be found in one of three "phases." Below 273.15°K, it is a solid (ice); between 273.15°K and 373.15°K. it is a liquid; and above 373.15°K, at atmospheric pressure, it is a vapor (steam). Furthermore, water can be transformed from one of these phases to the other simply by changing its temperature.

Many familiar materials undergo these transformations. We are familar with iron as a solid; yet iron can be melted and, at a very high temperature, even vaporized. If air is cooled under proper conditions, it will condense into liquid nitrogen and liquid oxygen at temperatures of −196°C and −183°C respectively. If cooled further, the separate constituents will freeze into solids at −210°C and −218°C.

Not all materials, however, go through these stages. A block of wood under ordinary circumstances does not melt, and an egg, if heated, changes from a liquid into a solid instead of changing from a liquid into a vapor. Furthermore, in the case of the wood or the egg, the process is not reversible; one cannot recover the original physical characteristics by restoring the original temperature.

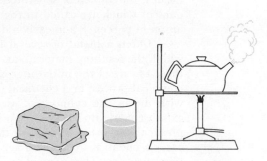

FIG. 7-2

Three phases of matter. Many common materials can be found in any of three phases: solid, liquid, or vapor.

FIG. 7-3

Some materials don't melt! When heated, some materials undergo a chemical reaction.

7-2 THE ATOMIC-MOLECULAR THEORY OF MATTER

Our present understanding of the structure of matter is based in large part on an atomic-molecular model that was developed during the nineteenth century. We will describe this model without attempting to present a chronology of the individual contributions of Lavoisier, Dalton, Gay-Lussac, Avogadro, Cannizzaro, Mendeleev, and others to its development; rather, in the interest of brevity, we will describe the most important features of the composite model that resulted from their efforts and, in the next section, state the essential steps in the logic through which the atomic-molecular view of matter was developed.

We all recognize that many familiar materials are mixtures of certain basic ingredients—that is, of components which, although randomly distributed through the bulk of the material, still retain their separate identities and physical properties. In many cases, one can devise a means of separating a mixture into its basic component materials. A handful of sand, for example, can be separated, if one uses sufficient care, into several well-defined piles, each of which will constitute a collection of particles with identical features of color, solubility, and crystal structure. As a second example, using more sophisticated techniques, the air can be separated into constituent gases, the most abundant of which are called nitrogen (78 percent), oxygen (21 percent), argon (1 percent), and carbon dioxide (0.03 percent).

Often a blend of several ingredients will undergo a process by which the resulting material takes on properties that differ greatly from those of any of the initial ingredients; such a process will be called a chemical process or a chemical reaction. Gasoline, for example, mixed with air in the cylinder of an automobile, burns and produces an exhaust made up of water vapor and other gases all having different physical properties from those that went into the process. In nearly all cases a chemical reaction is accompanied by the absorption or evolution of energy. The frying of an egg, for example, is a chemical reaction in which the physical properties of a material are changed by the addition of thermal energy; the explosion of a stick of dynamite, on the other hand, is a chemical reaction in which the physical properties of a mixture of ingredients are changed while thermal energy is generated.

Elements and compounds

A substance that has uniform properties throughout, that possesses a well-defined melting and boiling temperature, that can be returned to its initial conditions of volume, pressure, and temperature, and moreover, that cannot be separated into simpler ingredients through any chemical process whatsoever is called an *element*. Combinations of only 92 elements result in the seemingly endless variety of materials with which we are familiar; yet out of the 92 elements nearly a third play an insignificant or, at best, a very subtle role in nature. Some are very abundant or possess useful properties and therefore are very familiar— for example, carbon, hydrogen, iron, gold, and oxygen; others remain obscure—for example, praesodymium, osmium, gadolinium, and astatine.

Unless one has performed many experiments with familiar substances, it may be difficult to distinguish elements from other familar materials in our environment. Water, for example, possesses all the characteristics we have ascribed to an element except that it can be decomposed into known elements, namely, hydrogen and oxygen.

A substance that has uniform properties throughout, that possesses a well-defined melting and boiling temperature, that can be returned to its initial conditions of pressure, volume, and temperature, but that *can* be decomposed into simpler ingredients is called a *compound*. Since the number of distinguishable materials with different physical properties number in the hundreds of thousands, it is clear that most of nature consists of compounds and mixtures of compounds.

In the following sections we will trace the logic by which scientists have identified the 92 elements in the myriad of compounds in nature. Elements will be found to consist of tiny particles called *atoms*, whereas compounds consist of arrangements of atoms called *molecules*.

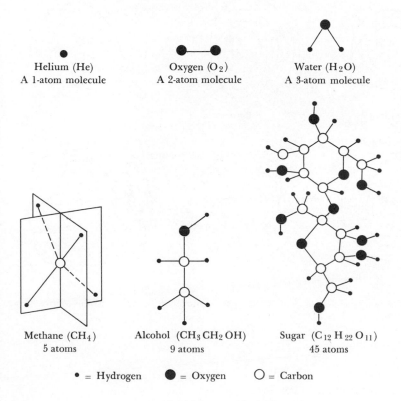

Helium (He)
A 1-atom molecule

Oxygen (O$_2$)
A 2-atom molecule

Water (H$_2$O)
A 3-atom molecule

Methane (CH$_4$)
5 atoms

Alcohol (CH$_3$ CH$_2$ OH)
9 atoms

Sugar (C$_{12}$ H$_{22}$ O$_{11}$)
45 atoms

• = Hydrogen   ● = Oxygen   ○ = Carbon

FIG. 7-4

Molecules. A given compound consists of nearly identical clusters of atoms called molecules.

In later chapters we will make use of further experimental information to develop a model of the structure of the atom. Although atoms of a given element all behave alike in chemical experiments, we will find that a given element may be made up of two or more *isotopes*, that is, of two or more kinds of atoms that differ in mass; hydrogen, for example, consists of two isotopes, one of which is nearly twice as massive as the other. We will also learn that each atom is a tiny planetary system in which electrons travel in orbits around a central, more massive nucleus.

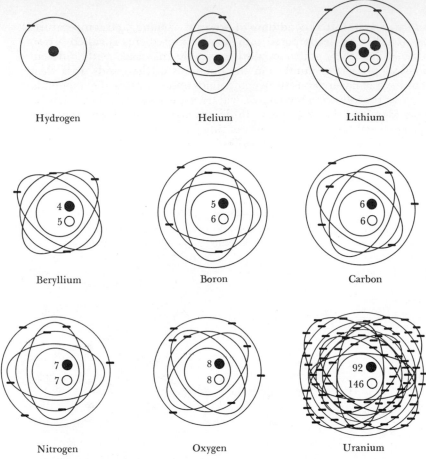

Hydrogen   Helium   Lithium

Beryllium   Boron   Carbon

Nitrogen   Oxygen   Uranium

An atomic model based on protons (●), neutrons (○), and electrons (—).

FIG. 7-5
Atomic structure.

7-3 A RATIONALE FOR
THE MOLECULAR THEORY OF MATTER

Our day-to-day experience with matter does not prepare us to accept the molecular model of matter that has been described in the preceding section. For example, when we examine a glass of water or a piece of iron, we see no evidence of atoms or molecules. Furthermore, if we subdivide either the water or the iron, each part appears to have the same physical characteristics that were possessed by the larger specimen. According to the molecular view of matter, were we to continue to subdivide either of the specimens, we would eventually secure one molecule of water or one atom of iron that, if further subdivided, would not possess the physical characteristics of the associated bulk specimen. However, it is not practical to carry out such extended subdivisions because the parts become smaller than the tools that we must use. In any case, the subdivided specimens would become submicroscopic long before we would reach a specimen of molecular size. Clearly our belief in atoms and molecules rests on less direct but, hopefully, equally convincing evidence.

A chemistry based on seven chemicals

Let us place ourselves in the frame of mind of a scientist of approximately the year 1800, and assume that the atomic-molecular view of matter is still only a promising hypothesis. We will assume that we have on hand seven materials bearing the names "hydrogen," "nitrogen," "oxygen," "hydrogen peroxide," "water," "nitric oxide," and "ammonia." Experiments show that they possess identifiably different physical properties; for example, they possess different melting points, different boiling points, and, at a particular pressure and temperature in the vapor phase, different densities. However, we know nothing of their structure, having been told that the names "hydrogen peroxide" and "nitric oxide" had been assigned to these materials simply because it had been learned that they could be prepared by processes in which hydrogen and oxygen, and nitrogen and oxygen, respectively, were used as raw materials. Furthermore, we are told that water and ammonia can be prepared from hydrogen and oxygen, and from hydrogen and nitrogen respectively. This advance knowledge is stated as entry (a) in Table 7-1; please ignore, for the moment, the bracketed chemical equations, [entry (d) in each case], because they represent conclusions we will eventually reach.

The laws of definite
proportions and combining gas volumes

Careful experiments are now carried out with the results shown in the table. Notice the cryptic entries (b) and (c) in the space designated

FIG. 7-6

A chemistry system based on seven chemicals. Our rationale for the atomic-molecular model is based on experimental data that can be secured using the seven materials shown below. We start by assuming that we know only that the materials possess identifiably different physical properties and that four of them can be produced from the other three.

The basic 3   The derived 4

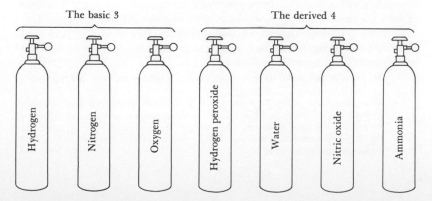

Hydrogen   Nitrogen   Oxygen   Hydrogen peroxide   Water   Nitric oxide   Ammonia

TABLE 7-1

Four reactions involving hydrogen, oxygen, and nitrogen

REACTION NUMBER	DESCRIPTION
I	(a) Hydrogen + oxygen = hydrogen peroxide (b) 1 unit mass + 16 unit masses = 17 unit masses (c) 1 vol + 1 vol = 1 vol (d) $[H_2 + O_2 = H_2O_2]$
II	(a) Hydrogen + oxygen = water (b) 2 unit masses + 16 unit masses = 18 unit masses (c) 2 vols + 1 vol = 2 vols (d) $[2H_2 + O_2 = 2H_2O]$
III	(a) Nitrogen + oxygen = nitric oxide (b) 14 unit masses + 16 unit masses = 30 unit masses (c) 1 vol + 1 vol = 2 vol (d) $[N_2 + O_2 = 2NO]$
IV	(a) Hydrogen + nitrogen = ammonia (b) 3 unit masses + 14 unit masses = 17 unit masses (c) 3 vols + 1 vol = 2 vols (d) $[3H_2 + N_2 = 2NH_3]$

as "Reaction I." The first entry (b) states that a unit mass of hydrogen is observed to combine with 16 unit masses of oxygen to produce 17* unit masses of hydrogen peroxide, where a "unit mass" can be any amount that seems convenient—a gram, a milligram, a kilogram, and so on. The second entry (c) states that if all volumes are measured in the vapor phase and at the same temperature and pressure, the volume of the unit mass of hydrogen is the same as the volume of the 16 unit masses of oxygen; furthermore, the 17 unit masses of hydrogen peroxide produced by their combination also occupy a unit volume. The combined volume is not equal to the sum of the two initial volumes!

The observations to which reference is made here are called the "Law of Definite Proportions," which states that elements always combine in definite mass ratios characteristic of the combining elements and the compound being formed (items b), and the "Law of Combining Gas Volumes," which states that when two gases combine to form a third gas, the volumes of the constituents and the product are related in the ratios of small whole numbers (items c).

A similar peculiarity is observed in two of the other reactions. Two volumes of hydrogen combine with one volume of oxygen to form not three but two volumes of water vapor (Reaction II,) and three volumes of hydrogen combine with one volume of nitrogen to form two volumes of ammonia (Reaction IV).

It would be well to note that the volume of gas identified as

---

* This statement would be more exact if we said that 1.008 unit masses of hydrogen combine with 16.000 unit masses of oxygen to produce 17.008 unit masses of hydrogen peroxide. The case for the molecular view of matter can be stated just as well when we use the nearest whole numbers and is much easier to read (and write).

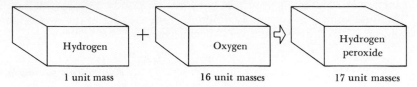

FIG. 7-7

Reaction I. A unit volume of hydrogen combines with a unit volume of oxygen to produce a unit volume of hydrogen peroxide. Mass, however, is conserved, at least within the accuracy of the best available chemical balances.

"1 vol" in Table 7-1 is not arbitrary if the mass relation and the volume relation in each case refer to the same chemical experiment. Experiments will show indeed that if we interpret the "16" in Reaction I as 16 kg of oxygen and "1" as 1 kg of hydrogen, the volume occupied by this amount of oxygen or hydrogen at atmospheric pressure and 0°C equals 11.2 m³, approximately the volume of the interior of a bus. However, there is little advantage in expressing masses and volumes in MKS units at this point, so for the present we will continue using the arbitrary units that we have called "unit masses" and "vols."

We can now go back over the data shown in Table 7-1 and compute the relative densities of the various gases. Starting with the definition of density,

$$\text{Density} = \frac{\text{mass}}{\text{volume}}$$

we can see that the density of hydrogen in these arbitrary units is equal to 1 whether one inspects Reactions I, II, or IV; oxygen has a density of 16 in the same units; nitrogen has a density of 14; and so forth. In summary, the relative densities are listed in Table 7-2.

We must now reflect on the magic that must be involved in the combination of these simple materials to form other materials while undergoing the remarkable volume changes we have noticed. Three volumes of hydrogen combine with one volume of nitrogen to produce

			TABLE 7-2			
			Density ratios*			
HYDROGEN	NITROGEN	OXYGEN	HYDROGEN PEROXIDE	WATER	NITRIC OXIDE	AMMONIA
1	14	16	17	9	15	8.5
2	28	32	34	18	30	17

* The data in the first row are obtained from Table 7-1 by applying the definition of density. The second row is obtained by multiplying each entry in the first row by 2; this was done in order to secure whole numbers. The relative densities remain unchanged.

exactly two volumes of ammonia, not some randomly different volume. The final volume is not simply different from the sum of the two input volumes; it is different in a very special way.

Avogadro's principle

The answer to the mystery of combining volumes was given by Avogadro in 1811. Starting with the assumption that gases consist of particles (or "molecules"), a hypotheses that had been used by a number of Avogadro's predecessors but without complete success, Avogadro's principle states

Equal volumes of all gases at a specified temperature and pressure contain equal numbers of molecules.

Although we have no way of knowing at the outset whether Avogadro's hypothesis is true or false, we can at least examine the consequences of accepting it as true. By definition, each of the densities tells the mass of gas in a unit volume of the gas in question. Since Avogadro's principle requires that a unit volume of all gases contain the same number of molecules, we see that *the densities stand in the same ratio as the masses of the corresponding molecules.* Thus the ratio of the mass of a molecule of nitric oxide to the mass of a water molecule equals $\frac{30}{18}$, the ratio of the mass of a water molecule to that of a hydrogen molecule equals $\frac{18}{2}$, and so forth.

We thus have a means of measuring the relative masses of all molecules. We need only measure the density of the gas in question and compare the measured density to the known density of hydrogen, nitrogen, or one of the other gases included in our discussion above.

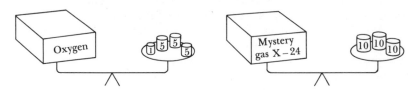

FIG. 7-8

Determining the relative mass of two molecules. One vessel contains oxygen gas at a given temperature and pressure; the other vessel has the same volume and contains an unknown gas at the same temperature and pressure. Since the two vessels contain the same number of molecules (Avogadro's principle), the ratio of the masses of the gases equals the ratio of the masses of the individual molecules.

Determining molecular structures

The data recorded in Table 7-1, items (b) and (c), may now be used to discover the most likely structures of the seven molecules that comprise the seven chemicals. Certain assumptions will be employed as follows:

1. Because experience shows that hydrogen, nitrogen, and oxygen cannot be disassociated into simpler substances, we will assume that they are elements.

2. We will assume that an element in the gaseous phase consists of a simple combination of like atoms. In developing a model of the molecules that make up an element we will favor a system in which the molecules are monatomic (one atom per molecule) over one in which they are diatomic (two atoms); we will favor diatomic molecules over triatomic molecules, and so forth.

3. We will assume that under ordinary circumstances all molecules formed by atoms of a given kind are identical; that is, we will assume that all hydrogen molecules are either atomic (H), or diatomic (H_2), or triatomic (H_3), etc., and not some mixture of H, H_2, and H_3.

Our procedure is to write every guess that comes to mind provided it is consistent with the data regarding combining masses and volumes. In conformity with assumption (2) above, we will start by limiting our guess to ones that involve monatomic and diatomic molecules for the elements. These guesses are

I-1	H	+	O	= HO
I-2	H	+	O_2	= HO_2
I-3	H_2	+	O	= H_2O
I-4	H_2	+	O_2	= H_2O_2
II-1	2H	+	O_2	= 2HO
II-2	$2H_2$	+	O_2	= $2H_2O$
III-1	N_2	+	O_2	= 2NO
IV-1	$3H_2$	+	N_2	= $2H_3N$

Certain "guesses" have been eliminated in advance and do not appear in the preceding list. For example, in Reaction II, were we to assume that "water" is formed from diatomic hydrogen and monatomic oxygen we would have written

$$2H_2 + O = 2H_2O_{1/2}$$

which would have each water molecule possess half an oxygen atom; this is clearly nonsense.

Looking at our guesses, we see that Reactions III and IV require hydrogen, nitrogen, and oxygen to be diatomic. Accepting this and applying assumption (3), we eliminate all guesses in Reactions I and II that would have either hydrogen or oxygen to be monatomic. This leaves us with a unique set of reactions: I-4, II-2, III-1, and IV-1, which have appeared as items (d) in Table 7-1.

Having identified the nature of all four reactions, we see that we have identified the atomic constituency of all seven molecules as fol-

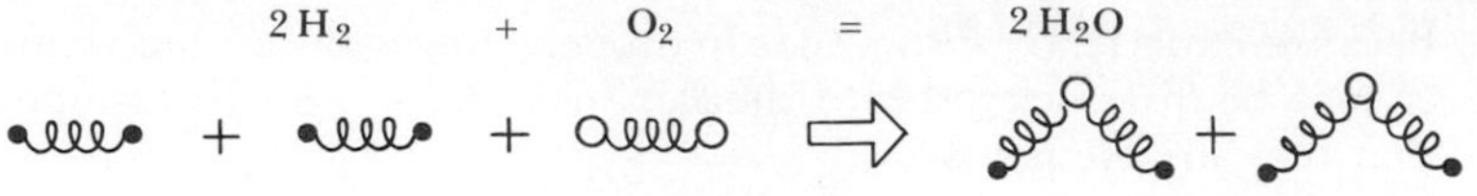

$$2\,H_2 \quad + \quad O_2 \quad = \quad 2\,H_2O$$

FIG. 7-9

Reaction II. Our analysis of mass and volume data leads to the above schematic diagram for the formation of water from hydrogen and oxygen. The triangular shape of the water molecule is known from other evidence.

lows: hydrogen (H_2), nitrogen (N_2), oxygen (O_2), hydrogen peroxide (H_2O_2), water (H_2O), and ammonia (H_3N).

The foregoing material, in miniature, is the chemical rationale for the atomic-molecular theory of matter. We have demonstrated the procedure by applying it to seven chemicals: our predecessors have already shown this procedure to be valid for all 92 elements and the thousands of compounds that may be formed from them.

Molecular and atomic masses

Since the relative masses of the molecules H_2, N_2, and O_2 are in the ratio 2:28:32 (Table 7-2), we can see that the masses of the atoms H, N, and O are in the ratio 1:14:16. Using the results of these and similar (and more accurate) experiments, one can set up a complete table of relative atomic masses. The relative atomic masses of a few of the well-known elements are listed in Table 7-3; other relative atomic masses will be found in the periodic table (Appendix 4).

TABLE 7-3			
Atomic masses of the elements			
(Based on C-12 = 12.00000)			
Hydrogen	1.00797	Neon	20.183
Helium	4.0026	Chlorine	35.453
Carbon	12.01115	Iron	55.847
Nitrogen	14.0067	Gold	196.967
Oxygen	15.9994	Lead	207.19
Fluorine	18.9984	Uranium	238.03

Atomic masses in this text will be quoted relative to the atomic mass of the most abundant isotope of carbon, which is arbitrarily assigned a mass of 12.00000 atomic mass units. As found in nature, however, carbon includes a small amount (1 percent) of carbon-13 ($M = 13.00335$ amu), thereby yielding the "average" value of 12.01115 shown in the table.

In most of our encounters with nature we deal with matter in the molecular rather than in the atomic state, because a molecule represents a more stable state of matter than an atom. Even if we were to start with a room full of an atomic gas, the atoms would surely combine quickly to form molecules, accompanied by a tremendous explosion!

We can utilize our list of relative atomic masses (Table 7-3) to generate a table of the relative masses of the molecules associated with a few common compounds (Table 7-4). Notice that the molecular mass in each case is just equal to the sum of the atomic masses of the constituent atoms in the molecule.

TABLE 7-4

Molecular masses of common compounds

Hydrogen (H_2)	2.01594	Oxygen (O_2)	31.9988
Methane (CH_4)	16.04303	Hydrogen peroxide (H_2O_2)	34.0147
Ammonia (NH_3)	17.0306	Hydrogen chloride (HCl)	36.461
Water (H_2O)	18.0153	Carbon dioxide (CO_2)	44.010
Nitrogen (N_2)	28.0134	Ethyl alcohol (CH_3CH_2OH)	46.0695
Nitric oxide (NO)	30.0061	Uranium hexafluoride (UF_6)	352.02

We now wish to reverse the logic utilized in our previous discussion of Avogadro's principle. The earlier approach is symbolized by the sketches in Fig. 7-8, where the relative mass of the molecules that constitute two different gases was stated to be the relative mass of equal volumes (i.e., the relative density) of the two gases at the same temperature and pressure. The reverse logic is to consider a quantity of each of several gases whose mass in kilograms in each case is equal to its molecular mass—for example, 18.0153 kg of water, 28.0134 kg of nitrogen, or 31.9988 kg of oxygen. In each case, this mass will occupy the same well-defined volume at a given temperature and pressure, provided, of course, that the material is in the gaseous phase.

Definition of the kmol

We will associate special significance with the quantity of gas whose mass is numerically equal to its molecular mass in kilograms, a "unit quantity" of gas that will be called a kilogram molecular mass or kmol (pronounced "kay-mole"). A *kmol* of oxygen therefore equals 31.9988 kg of O_2 gas, a kmol of hydrogen equals 2.01594 kg of H_2 gas, and so on.

A kilogram molecular mass (kmol) of any compound is a mass in kilograms of that substance numerically equal to its molecular mass.

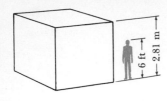

FIG. 7-10

A kmol of gas. A cubical vessel of the size shown in the above diagram will accommodate at standard temperature and pressure 2 kg of hydrogen, or 28 kg of nitrogen, or, in general, a mass of any gas numerically equal to its molecular mass. The vessel will contain 6.02×10^{26} molecules.

Experiments show that if one places 31.9988 kg of oxygen gas in a vessel that is large enough to accommodate it at atomospheric pressure (101,325 N/m²) and at a temperature of 273.15° Kelvin (0°C)—that is, at "normal pressure and temperature" —the volume of the gas will equal 22.4136 m³. The same vessel will accommodate 2.01594 kg of hydrogen, or 28.0134 kg of nitrogen, or 44.010 kg of carbon dioxide, etc., at normal pressure and temperature.

Avogadro's principle may now be reinterpreted to state that a kmol of any gas consists of a definite number (N) of molecules or, if only one kind of atom is present, of atoms. This number N, known as Avogadro's number, is numerically equal to 6.02252×10^{26} molecules/kmol.

Knowledge of Avogadro's number enables one to compute the mass of any atom or molecule from its atomic or molecular mass. For example, since the atomic mass of hydrogen equals 1.00783, we know that 1.00783 kg of hydrogen contains 6.02252×10^{26} atoms, hence that the mass of 1 atom of hydrogen equals 0.1675×10^{-26} kg. Similarly, since the molecular mass of ethyl alcohol equals 46.0695, one molecule of ethyl alcohol has a mass of 7.67×10^{-26} kg. In general, then,

$$ m = \frac{M}{N} \qquad (7\text{-}1) $$

where m equals the mass of one atom (or molecule) in kilograms, M is its atomic (or molecular) mass in atomic mass units, and N is Avogadro's number ($= 6.02252 \times 10^{26}$ atoms or molecules/kmol).

A procedure by which this important physical constant has been determined will be described in Chapter 8.

The general gas constant

In the previous chapter we noted that the product pV serves as an indicator of the temperature of the gas in a gas thermometer and, indeed, that Kelvin temperatures are defined to be proportional to this product [Eq. (6-7)]. The numerical value of the product pV depends, however, on the amount of gas present. Avogadro's principle may be employed at this point to prescribe the amount of gas present in such a way that the quantity pV/T is the same for all gases.

We have noted already that a kmol of any gas occupies a volume of 22.4136 m³ at atmospheric pressure (101,325 N/m²) and a temperature of 273.15° K. Therefore, for 1 kmol of any gas,

$$ \frac{pV}{T} = \frac{101{,}325 \times 22.4136}{273.15} = 8314.3 \ \frac{\text{joules}}{\text{kmol K}°} $$

Because this quantity is the same for all gases, it is to be regarded as one of the more important physical constants. The general gas law now takes the form

$$ \frac{pV}{T} = R \qquad (7\text{-}2a) $$

where $R = 8314.3$ joules/kmol K° and is called the *general gas constant*.

Equation (7-2a) relates to a situation in which one is conducting experiments on precisely 1 kmol of gas. If one possesses a smaller or larger quantity of gas, the equation must be modified to fit the new situation. The most straightforward way of doing so is to adjust the value of the gas constant in proportion to the number of kmoles under study. For example, if one were to have a specimen consisting of 0.45 kmol of gas at normal temperature and pressure, the volume of the specimen would be proportionally smaller and therefore the value of pV/T would equal $0.45R$ instead of R. In general, then, if we let n represent the number of kmols of gas under investigation, the general gas law takes the form

$$\frac{pV}{T} = nR \qquad \text{(for } n \text{ kmols)} \qquad (7\text{-}2b)$$

7-5 THE PERIODIC TABLE

If the elements are arranged in the order of their atomic masses, many chemical properties are found to be repeated at regular intervals in the list so formed. Helium, entry number 2 in the list, for example, appears to be chemically inert and never forms a compound with any other element. However, this property of chemical inertness is also found in neon (#10), argon (#18), krypton (#16) and xenon (#54). Hydrogen (#1), furthermore, forms molecules with most of the same elements as lithium (#3), sodium (#11), potassium (#19), rubidium (#37), and cesium (#55). Similar familylike relationships are found in the sequences of elements #9, 17, 35, and 53, and in other sequences of elements.

These familylike characteristics were first noted by Mendeleev in Russia and by Meyer in Germany a hundred years ago. Arranging the then-known elements in their order of increasing atomic mass in the rows of an eight-column table, it was found that elements of similar chemical properties arranged themselves rather naturally in the columns—that is, with lithium directly above sodium, which was directly above potassium, and so on. Such a "periodic table" will be found in Appendix 4; this version of the periodic table, however, reflects some of the progress that has taken place since Mendeleev's time. In order to form columns of materials with similar chemical properties, it was necessary for Mendeleev to leave some blank spaces; these blank spaces turned out to be the most exciting aspect of the table. For example, Mendeleev knew of no elements to fit into the spaces numbered 44, 68, 72, as well as many others. One could surmise that these empty spaces designated elements yet to be discovered. It is most remarkable to note that every empty space has now been filled with an element that has since been discovered. Furthermore, no element has been found that fails to fit into this amazing table.

The position of each element in the periodic table will be called the atomic number of the element in question. Ultimately (Chapter 13) we will see that this number corresponds to the electric charge on the nucleus of the atom in question and to the number of electrons in the charge cloud that surrounds the nucleus. Many years, however, lapsed between the first publication of the periodic table and a full comprehension of the structural factors that make this table possible.

7-6 THE KINETIC INTERPRETATION OF TEMPERATURE

If one places even a small amount of gas in an otherwise evacuated vessel, the gas immediately "fills" the vessel. We visualize the molecules that constitute the gas to be in continual motion, each molecule colliding with other molecules and with the walls of the vessel in a random and chaotic manner. At any instant, individual molecules are traveling in every possible direction with a wide range of speeds.

The pressure exerted by the gas against the walls of the confining vessel is regarded as the result of millions of collisions of the molecules against the walls (Fig. 7-11). If one heats the gas, the added energy constitutes an increase in the average kinetic energy of the molecules and manifests itself as a rise in the temperature. This view of the behavior of a gas is known as the *kinetic theory of gases*.

By making use of certain simplifying assumptions regarding the nature of a gas, we can derive a relationship between the average kinetic energy of the molecules in a vessel and the temperature. These assumptions are as follows:

FIG. 7-11

Force due to a stream of particles. The basic assumption of the kinetic theory of gases is that the pressure exerted by a gas on the walls of the confining vessel is due to the repeated impact of molecules against the interior surface. Unlike the collisions of the bullets in the sketch below, molecules make elastic collisions; that is, there is no loss of kinetic energy as a result of a collision.

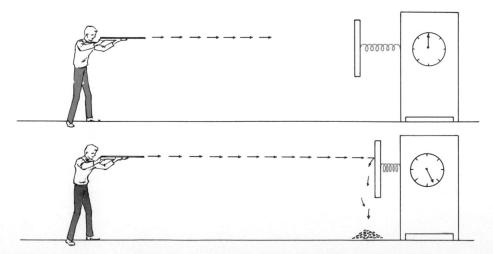

1. Although the speeds of individual molecules at a given temperature vary widely, we will deal with the gas as if all molecules of a given kind travel with the same effective speed v.

2. Although molecules travel in every conceivable direction we will treat the gas as if exactly one-third of the molecules are traveling in each of the three coordinate directions at any given instant.

3. Collisions are assumed to be perfectly elastic; that is, we assume that the total kinetic energy is the same before and after every encounter between two molecules or between a molecule and the walls of the confining vessel. In a collision with a wall, the molecule experiences a change in direction but no change in speed.

These assumptions, together with the general gas law Eq. (7-2a), lead to an equation that shows that *the average kinetic energy of the molecules is proportional to the temperature of the gas.* (For details of the derivation of this expression, see Appendix 6.) The equation is

$$\overline{KE} = \frac{3}{2}\frac{R}{N}T \tag{7-3a}$$

where $\overline{KE}$ is the average kinetic energy per molecule and T is the absolute temperature in degrees Kelvin. Since R and N are constants, we can write

$$\overline{KE} = \frac{3}{2}kT \tag{7-3b}$$

where $$k = \frac{R}{N} = 1.3805 \times 10^{-23} \text{ joules/molecule } °K$$

As an example of the application of this expression to a specific situation, we will compute the average kinetic energy and the velocity of molecules at room temperature (300°K or 27°C). The equations show that the average molecular kinetic energy equals 0.62×10^{-20} joules at this temperature. Although air is a mixture of nitrogen (78 percent) and oxygen (21 percent), plus small amounts of other gases, the equation draws no distinction between them, which means that the average kinetic energy of light and heavy molecules at a given temperature is the same.

On the other hand, the speeds of molecules depend on their mass. Making use of the expression

$$\overline{KE} = \tfrac{1}{2}mv^2$$

and using the above-computed value of the average kinetic energy, we find that

$$v(\text{oxygen}) = 483 \text{ m/sec}$$

and $$v(\text{nitrogen}) = 518 \text{ m/sec}$$

at room temperature. By way of comparison, sound has a velocity in air of 331 m/sec; the velocities of rifle bullets range from 300 m/sec to 1500 m/sec.

Equations (7-3) provide an entirely new and important meaning to the concept of temperature, namely,

The temperature of a gas is a measure of the average kinetic energy of the molecules that make up the gas.

The thoughtful reader may wonder if the assumptions that were made in deriving the equations on which this conclusion is based are realistic. Is it valid to assume that one-third of the molecules are traveling in each of the coordinate directions when we know that they are traveling in all directions? How can we associate one effective speed with molecules of a given kind when we know that they have a wide range of speeds? Are the conclusions we have reached for a gas in a cubical vessel also valid for gases enclosed in a box of irregular shape? Can we apply the same equations to gases that appear not to be enclosed at all—for example, the air in this room? Unfortunately, little can be done to satisfy the skeptic except to say that, by using more sophisticated mathematical techniques, one can relax some of these limiting assumptions yet arrive at the conclusion that temperature is a measurement of the average kinetic energy of molecules of the object whose temperature is being measured.

SUMMARY Elements and compounds differ in that compounds can be decomposed into elements, whereas elements cannot be separated into simpler substances.

When elements combine to form compounds, they conform to two laws:

1. The Law of Definite Proportions: elements always combine in definite mass ratios characteristic of the combining elements and the compound being formed.

2. The Law of Combining Gas Volumes: when two gases combine to form a third gas, the volumes of the constituents and the product are related in the ratios of small whole numbers.

Avogadro's principle starts with the assumption that elements and compounds consist of definite simple arrangements of atoms into molecules and states that

Equal volumes of all gases at a specified temperature and pressure contain equal numbers of molecules.

With knowledge of the mass ratios and the small-whole-number volume ratios for a selection of chemical reactions involving some of the same elements, one can set up a series of guesses regarding the molecular structure of the elements and compounds and eliminate all but a single internally consistent set of these guesses. In this way, chemists have determined the molecular structure of all elements and compounds.

A *kilogram molecular mass* (a kmol) of any compound is a mass in kilograms of that substance numerically equal to its molecular mass (M). Since 1 kmol contains 6.023×10^{26} molecules, it follows that the mass of one molecule in kilograms is

$$m = \frac{M}{6.023 \times 10^{26}} \qquad (7\text{-}1)$$

For a kmol of gas, the general gas law is

$$pV = RT \qquad (7\text{--}2)$$

where R is the general gas constant and equals 8314.3 joules/kmol °K.

According to the kinetic theory of gases, the pressure exerted by a gas on the walls of the confining vessel is due to the impact of the gas molecules. A mathematical analysis leads to an expression for the average kinetic energy per molecule

$$\overline{\text{KE}} = \frac{3}{2} \frac{R}{N} T \qquad (7\text{-}3)$$

This equation shows that temperature is a measure of the average kinetic energy of the molecules.

Tables of useful physical constants will be found in Appendices 4 and 5.

**QUESTIONS
AND
PROBLEMS**

7-A1. Prepare a list of the materials with which you have ever had experience in all three of their phases. Next, prepare a list of all the materials that you have encountered in two of the three phases.

7-A2. We have seen that some materials (eggs, for example) do not go through the three phases—solid, liquid, and gas—when they are heated or cooled. Explain what does happen in such cases.

7-A3. In a reaction involving hydrogen and nitrogen gas, 5000 cm³ of ammonia gas (NH_3) is produced (Reaction IV). What volumes of hydrogen and nitrogen were involved in this reaction?

7-A4. What is the mass in atomic mass units of each of the following molecules:

$$F_2; \ N_2O; \ CCl_4; \ CN; \ \text{and} \ CCl_3H?$$

7-A5. (*a*) How many kilograms of each of the following compounds constitute a kmol:

$$He; \ CH_4; \ CN; \ CO_2?$$

(*b*) What is the mass of one molecule of each of these compounds?

7-A6. Notice that, according to the periodic table (Appendix 4), lithium (Li), phosphorus (P), and sulfur (S) should have chemical properties similar to those of hydrogen (H), nitrogen (N), and oxygen (O) respectively. Starting with the information given in Table 7-1, prepare a list of a dozen possible molecules, each involving one or two of these six elements.

7-A7. Considering the information presented in this chapter, restate the argument for basing our temperature scale on a gas thermometer.

7-B1. In our analysis of the structures of the molecules involved in the reactions listed in Table 7-1, we did not include this guess:

$$N_2 + O_4 = 2NO_2$$

Which of our assumptions would have been violated by this guess?

7-B2. Oxygen gas (O_2) has a density of 1.43 kg/m³. Under certain circumstances, however (for example, in the presence of a high-voltage electrical discharge), oxygen may form another molecular form called ozone (O_3). (*a*) What should be the density of ozone? (*b*) How can we rationalize the existence of this molecule, given our assumption (page 133) that "all molecules formed by atoms of a given kind are identical"?

7-B3. Let us extend our "chemistry based on seven chemicals" to an eighth substance called carbon dioxide, which is found to be separable into oxygen (72.7 percent by mass) and a "new" element called carbon (27.3 percent by mass). On the basis of this much information, write two or more reaction equations and the associated best guess as to the atomic mass of carbon. What additional information would you need to eliminate some of these guesses?

7-B4. Let us assume that a system of atomic masses should be established based on the hydrogen isotope $_1^1H$ (see Appendix 8) instead of on carbon-12 ($_6^{12}C$). In this new system we would assign a value of 1.000000 to $_1^1H$. (*a*) What would be the mass of carbon-12 in this system? (*b*) What would be the numerical value of Avogadro's number?

7-B5. The best vacuum possible at present is approximately 10^{-15} atm. (*a*) What volume would a kmol of gas at this pressure occupy at the standard temperature of 0°C? (*b*) What would be the average distance between molecules at this pressure?

7-B6. Assume that a single molecule travels with a velocity of 500 m/sec, bouncing back and forth between two walls 0.10 m apart. How many collisions will this molecule have with a given face each second?

7-B7. Assume that a certain vessel contains a mixture of nitrogen, carbon dioxide, and ammonia molecules in the vapor state. Given that the average velocity of the nitrogen molecules equals 500 m/sec, (*a*) what is the average energy of each of the constituent molecules? (*b*) What is the average velocity of the carbon dioxide (CO_2) molecules and of the ammonia (NH_3) molecules?

7-B8. The gas at the center of the explosion of an atomic bomb may reach a temperature of 10,000,000°K. (*a*) What is the average velocity of nitrogen molecules under these conditions, assuming that nitrogen remains in the molecular configuration? (*b*) Assuming that the nitrogen is all dissociated into nitrogen atoms, what would be their average velocity?

Electricity

The forces that provide unity to the galaxies, that hold the planets in their orbits, and that hold man so nearly fixed on the surface of the Earth are gravitational forces. Every man and woman since antiquity has been aware of the existence of this force literally from the day of his or her personal cradle.

Built into the fabric of matter, however, are several other kinds of forces and force fields. One of these, the electrical force, in what would seem to be comparable circumstances, is billions of times stronger than the gravitational force. Electric forces permeate every facet of nature and play a dominant role in all life processes.

A major part of our knowledge of electricity has been acquired only recently. It is quite remarkable, in fact, that an aspect of nature so vital could have remained so nearly invisible through most of man's history. Indeed, as recently as a century ago most of man's knowledge of electrical phemomena bore a marked resemblance to magic and witchcraft.

Man uses what he understands. When he came to understand electricity less than a century ago, he put this knowledge to use in fashioning a technical revolution and, with it, a cultural revolution unparalleled by anything that had happened before. The most important effect of this understanding was to make enormous amounts of energy available to man, thereby providing him with food as well as with many necessary and not-so-necessary creature comforts and, when provoked, enhancing his ability to wage war.

8-1 ELECTRIFICATION BY CONTACT

Each of us has experienced an "electric" shock on touching some metal object after walking across a wool rug or has noticed that one's hair is attracted to the comb on a crisp winter morning. In each of these cases, two dissimilar materials, the leather of one's shoes and the

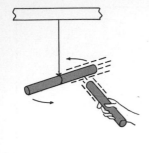

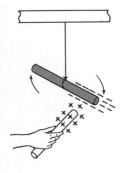

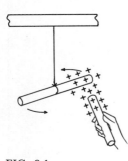

wool of the rug in the first case, and the comb and one's hair in the second, have been brought into contact; when they are separated, we see effects that we call "electrification." The objects are said to be "charged" or "electrically charged."

More diagnostic observations of electrification can be carried out in the laboratory. One may rub a glass rod with a piece of silk and find that the silk will be attracted to the glass after they have been separated. A similar effect will be observed with a hard-rubber rod and a piece of wool or fur. The phenomenon is observable whenever contact is made between two dissimilar materials but seems to be enhanced by rubbing the materials together; presumably rubbing produces more contact between the materials.

Two kinds of electrification

The electrification that appears on glass when rubbed with silk is different from that which appears on hard rubber when rubbed with wool. Convincing evidence of this fact can be seen if one suspends a charged hard-rubber rod and a charged glass rod as shown in Fig. 8-1. If a charged hard-rubber or a charged glass rod is brought near either of the suspended rods, a force of repulsion or a force of attraction will be evident, depending on whether the interacting charged rods are "like" or "unlike." The electrification on a hard-rubber rod is clearly "opposite" the electrification on a charged glass rod.

As we evolve theories regarding electrification, we will need names for these types of electric charges. We shall call the kind of electric charge that appears on glass "positive" and that which appears on hard rubber "negative." The designations, however, are quite arbitrary.

Further experiments will show that while glass receives a positive charge, the silk used in its production receives a negative charge; similarly, whereas the hard-rubber rod receives a negative charge, the wool or fur receives a positive charge.

The property of electrification can be transferred from one object to another. Let two light celluloid or pith spheres be coated with aluminum paint and be suspended, with silk or nylon thread, a few

FIG. 8-1

Two kinds of electrification. The electrified hard-rubber rod (gray) repels another electrified hard-rubber rod, but it is attracted to an electrified glass rod (clear). The experiment indicates that there are two kinds of electrification and that like charges repel each other while unlike charges attract each other.

FIG. 8-2

Pith-ball experiments. Charges can be transferred directly from a charged to a previously uncharged object.

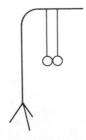
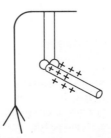
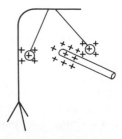

centimeters apart, as shown in Fig. 8-2. Initially these objects show no effect on each other. If the glass that has been rubbed with silk is brought into contact with both spheres simultaneously, and is then removed, the metalized spheres will not only repel each other but both will be repelled by the glass rod. On the other hand, they will be attracted by a charged hard-rubber rod.

FIG. 8-3

Charging by induction. (*a*) Two metal hemispheres are suspended by insulating fibers. (*b*) When a positively charged object is brought near one hemisphere, opposite charges appear on the two hemispheres. (*c*) If the hemispheres are separated while the external charge is still in the vicinity, the two hemispheres retain opposite charges after the external charge has been removed.

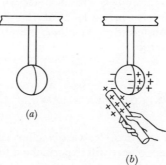

(*a*)

(*b*)

(*c*)

The arrangement shown in Fig. 8-3 provides additional insight into the nature of electrification. Two metal hemispheres are shown suspended by silk threads and in contact with each other. Initially the sphere so formed is uncharged. A positively charged glass rod is held near one of the hemispheres without making contact with it, and the hemispheres are separated before the glass rod is moved aside. Tests will show that both hemispheres are now charged, one positively and the other negatively. The experiment suggests that the uncharged metal sphere initially possessed both negative and positive charges, one or both of which are mobile and can be displaced by a neighboring electric charge.

The electroscope

A more sensitive device for the detection and study of electrification is the electroscope or electrostatic voltmeter shown in Fig. 8-4. The essential part of this device consists of a metal rod approximately 3 in. long, on the lower end of which a strip of gold foil has been attached. This metal rod, terminated at the upper end by a metal knob, is centered in a plastic or amber plug, which in turn is installed in a cylindrical metal container with glass windows. The plastic plug, like the silk threads of earlier experiments, serves to keep any electric charge that may be put on the metal rod-foil system from leaking off.

The gold-leaf electroscope is a very sensitive detector of electric charges. For example, if one brings a charged glass rod near the knob at the top of the electroscope, the gold leaf will deflect, presumably because positive charges are repelled to the bottom by charges on the glass or because negative charges are attracted to the top. If one touches the glass rod to the knob, the gold leaf will deflect and remain deflected, having received charge directly from the glass.

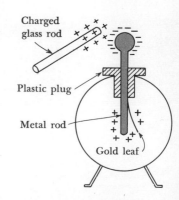

Charged glass rod

Plastic plug

Metal rod

Gold leaf

FIG. 8-4

A gold-leaf electroscope. The plastic plug is an insulator to prevent charges from leaking off the central metal parts. When a positively charged object is brought near the top of the electroscope, charges will flow along the rod, causing opposite charges to appear on the two ends of the rod.

Most materials possess the property of being able to transmit or "conduct" electricity to a certain extent; even the plastic plug used in isolating the metal rod-foil system in the electroscope conducts electricity well enough that the charge will eventually escape. Metals, however, are excellent conductors. If one joins two oppositely charged metal objects with a metal wire, both objects lose their charge immediately, precisely as if the excess charge on one sphere had been transferred to the other, thereby canceling that sphere's excess opposite charge.

8-2 COULOMB'S LAW

When two electrically charged objects are brought near each other, each object will experience a force. According to Newton's Third Law, these forces must be of equal magnitude but in opposite directions. If the charges are of the same sign—that is, if they are both positive or both negative—the objects will repel each other, whereas if the charges have opposite signs, the objects will attract each other.

Securing a set of charges having known relative magnitudes

Coulomb's law describes the force that one electric charge experiences when in the presence of another. However, because we have not yet defined a unit of electric charge, we must initiate our discussion with a description of a procedure by which we can change the amount of charge on an object in a predictable way.

Suppose that we have charged two identical conducting spheres while they were in contact with each other. Upon separating them, they should have identical charges. Let us represent the charge on each of these spheres by the symbol Q.

If we bring an uncharged sphere of the same size into contact with one of these charged spheres, it is reasonable to assume that the charge on the charged sphere will be shared equally with the sphere it touches; thus we will have three spheres with charges Q, $Q/2$, and $Q/2$ respectively. Continuing this procedure, we can secure additional charged spheres with charges $Q/4$, $Q/8$, and so on. Although we do not have a unit in which to express electric charge, we have a set of spheres possessing charges of known relative magnitudes.

The force law for electric charges

Measurements of the force experienced by any of these spheres in the presence of any one of the others reveal a force law identical in appearance to the Law of Universal Gravitation; that is, the force is proportional to the product of the charges and inversely proportional to the square of the distance between them. This relationship is called Coulomb's law and may be written in the form

$$F = +\frac{kQq}{r^2} \tag{8-1a}$$

where the interacting charges are Q and q and where k is a constant of proportionality that plays almost precisely the same role in electrification that G plays in gravitation. At this point, however, we have not defined a unit of charge nor have we selected a value for k, but it is well to note that the choice of the unit of charge and the numerical value of k are interrelated.

The positive sign in Eq. (8-1a) signifies that like charges repel each other as shown by the fact that, if Q and q are of the same sign, their product is a positive number; a positive force is associated with a force that tends to increase the magnitude of the radius vector. If Q and q have opposite signs, Eq. (8-1a) indicates a negative force, hence a force of attraction.

8-3 A TENTATIVE DEFINITION OF THE COULOMB AND THE AMPERE

We may now establish a working definition of the unit of charge by making use of the fact that, according to Eq. (8-1a), the choice of a unit of charge and the numerical value of k are interrelated. The situation, however, differs from the one we encountered in gravitation where the unit of mass had already been arbitrarily defined in terms of inertial properties and where the constant G was determined experimentally. The value that we select is 9×10^9 (or, more precisely, 8.9474×10^9)N m²/coul², where the unit of electrical charge determined by this choice is called the *coulomb*. Coulomb's law now becomes

$$F = +9 \times 10^9 \frac{Qq}{r^2} \tag{8-1b}$$

We refer to this as a tentative or working definition because, in fact, the coulomb is not defined by Coulomb law; the measurement of electric charges through the measurement of the forces they exert on each other is simply too difficult to carry out with the precision we require. Strictly speaking, the coulomb is defined by the effect of an electric current (electric charges in motion) on a nearby current. We will have another look at this aspect of electric charges in Sec. 9-5.

Equation (8-1) leads to the following operational definition:

A coulomb is that point charge which, when one meter from an identical point charge, experiences a repelling force of 9×10^9 newtons.

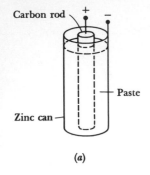

Carbon rod

Paste

Zinc can

(a)

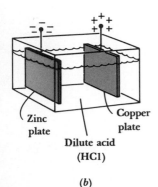

Zinc plate

Copper plate

Dilute acid (HCl)

(b)

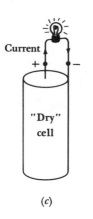

Current

+ −

"Dry" cell

(c)

FIG. 8-5

Electric cells. An electric cell consists of two dissimilar materials separated by a paste (a) or liquid (b) electrolyte. If the electrodes of a cell are connected by a conductor as in (c), a current of charges flows from one terminal to the other.

Up to now we have dealt only with electrification as produced by contact between two dissimilar materials in experiments that suggest that it is a difference in the "chemistry" of the two materials that produces the transfer or separation of charge. Most readers will be equally familiar with the phemomenon of electrification as produced by a device known as an *electric cell*, in which two conductors, typically a carbon rod and a zinc can, are put into effective contact through the agency of a liquid or paste "electrolyte." In the familiar dry cell (not strictly dry but sealed), one terminal, connected to a carbon rod, becomes positively charged while the other terminal, connected to the zinc can, takes on a negative charge. A combination of electric cells is called a battery.

An electric cell has a distinct advantage over electrification by contact or by induction (Fig. 8-3), because the electric cell will continue to provide electric charge steadily and almost indefinitely, in fact, until the zinc has been largely consumed by the chemical reactions that accompany the transfer of charge.

If the terminals of a battery are joined by a conductor as in Fig. 8-5(c), a steady transport of charge, which we shall call an electric current, will be established. The unit of electric current is the *ampere* and is defined as follows:

An electric current of 1 ampere flows in a conductor if 1 coulomb of electric charge is transported across any cross section of the conductor each second.

8-4 ELECTRIC FIELDS

Let us place a static, or stationary, electric charge on a metal sphere as in Fig. 8-6. Any small positive charge, which we shall call a test charge, brought into the vicinity will experience a force, thus serving to demonstrate that the space is affected by the charge on the sphere. We will ascribe this force to an "electric field" in the region.

The *direction* of an electric field will be defined as the direction of the force on a positive test charge. The test charge thus enables us to verify the presence of an electric field, to define its direction, and, as we shall see, to express its magnitude or its intensity.

The electric field intensity

The *intensity* of the electric field E at any point is defined in terms of the force experienced by a test charge. In equation form, this definition is

$$E = \frac{F}{q} \tag{8-2}$$

148

FIG. 8-6

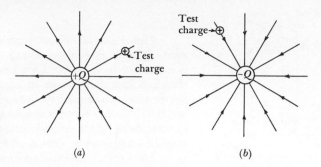

A test charge serves to detect an electric field. The direction of an electric field is the direction of the force on a positive test charge. A "line" represents the path along which an unrestrained positive charge will travel.

(a)   (b)

where F is the force experienced by a test charge q. The electric field intensity E is expressed in newtons per coulomb.

There is a striking parallel between the appearance of the equations that describe electric and gravitational fields. In fact, one can find an equation in gravitation analogous to nearly every equation used to describe electric charges (Appendix 7). They are, however, entirely different phenomena. An electric charge experiences no force in a gravitational field (unless the charge resides on an object that possesses mass); conversely, an uncharged mass experiences no force that is ascribable to an electric field.

The intensity of a Coulomb force field

We can secure an expression for the electric field intensity due to a point charge Q by applying our definition to Coulomb's law. If a test charge q is brought to a point in the vicinity of Q, it experiences a force F. The field is, then,

$$E = \frac{F}{q} = \frac{kQ}{r^2} \tag{8-3}$$

where r is the distance of the point in question from the center of the charge Q. A graph of the electric field intensity near a point charge is shown in Fig. 8-7.

FIG. 8-7

The electric field intensity due to a negatively charged sphere. The shaded area will be discussed in Sec. 8-5 (Fig. 8-13).

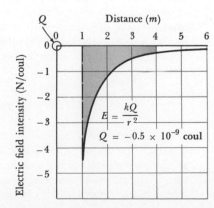

$$E = \frac{kQ}{r^2}$$

$$Q = -0.5 \times 10^{-9} \text{ coul}$$

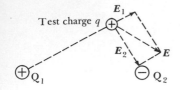

FIG. 8-8

The electric field due to
two point charges. The
field at a given point in
the vicinity of two point
charges equals the vector
sum of the two forces that
a unit positive charge
would experience at the
point in question.

Our definition of the electric field intensity can be extended to the
case of two or more point charges. Since the electric field at any point
is measured by the force on a positive test charge, it is clear that for a
situation like that shown in Fig. 8-8 the test charge will experience
vector forces from each of the individual charges when placed in their
vicinity. The net force is the vector sum of the individual forces; thus
electric fields due to many individual charges are vector quantities and
may be combined by vector addition.

In representing the resultant field due to an array of electric
charges, one draws lines of electric force in the region where the field
has an appreciable magnitude. The field due to each of several arrangements of charge is shown in Fig. 8-9. Although such sketches are very
useful in thinking about the field, it is well to keep in mind that the

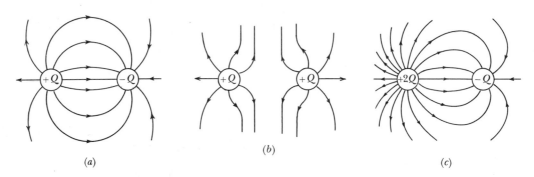

FIG. 8-9

Use of lines to represent the electric field due to two charges. Any line shown in these
sketches represents a path that a positive charge would tend to follow if placed in the
field.

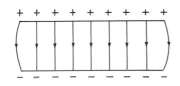

FIG. 8-10

A charged capacitor.
Two metal plates
separated by an insulator
constitute a capacitor.
The intensity of the
electric field in the space
between the plates is
uniform and is determined
by the amount of charge
on the plates.

lines shown in these figures only symbolize the paths along which
charges would tend to move if placed in the field. In each of the figures,
the lines appear to "leave" or "radiate from" the positive charge and
"terminate on" the negative charge. However, words like "leave,"
"radiate," or "terminate" should not be taken literally, because the
lines we have drawn are only a means of representing the electric field
in the region; according to the model we are describing, nothing
whatsoever "leaves," terminates on," or "radiates from" an electric
charge.

The uniform electric field of a capacitor

A capacitor is an electrical device consisting of two conductors
separated by an insulator; in its simplest form, it consists of two parallel
metal plates in a vacuum, separated by small insulating dividers. If
one of the plates is charged positively and the other is charged negatively, a test charge in the space between the plates will experience a

force. This force will be almost precisely the same at every point between the plates—that is, the field is uniform. The electric field, however, is zero both above and below the plates; it is as if the positive charge on one plate produces a field that is canceled by the field due to the negative charge on the other plate.

In nearly every electrical situation involving a capacitor, the electric charges on the two plates are equal but of opposite sign. This occurs because the battery used to charge the capacitor acts like an electric pump, removing charge from one plate and depositing it on the other. Since the battery does not create charge but only displaces it, the two plates must receive equal and opposite charges.

8-5 ELECTRIC POTENTIAL AND ELECTRIC POTENTIAL DIFFERENCE

The grandeur of the Niagara Falls arises both from the height of the Falls and the enormous quantity of water that spills over the Falls each second. These two properties contribute equally to the Falls as a provider of energy, since the total energy available each second is determined by the potential energy possessed by each kilogram (the gravitational potential in joules per kilogram) and by the number of kilograms arriving at the Falls each second (the "current" in kilograms per second).

In the equivalent electrical situation, the amount of charge transmitted past any point in a conductor per second is the counterpart of the amount of water flowing over the Falls per second; in either case it would be called the current. The gravitational potential of the Falls has its counterpart in electricity also; it is called the electrical potential and measures the energy possessed per coulomb of charge (joules per coulomb).

The difference in potential

The difference in potential ΔV_{BA} between two points B and A is equal to the work done by electrical forces on 1 coulomb of positive charge in moving from B to A.

This definition can be interpreted to say that the difference in potential between two points B and A is the change in potential energy of 1 coul of positive charge in going from B to A. For an arbitrary test charge then,

$$\Delta V_{BA} = \frac{\Delta W_{BA}}{q} \tag{8-4a}$$

or

$$\Delta W_{BA} = \Delta V_{BA} q \tag{8-4b}$$

where ΔW_{BA} is the difference between the energy of the charge q at point B and its energy at point A and, therefore, measures the work the field can do on the charge going from B to A. The difference in potential ΔV_{BA} is expressed in joules per coulomb or "volts."

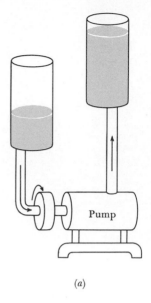

(a)

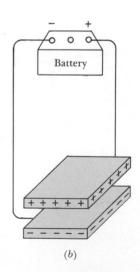

(b)

FIG. 8-11

Charging a capacitor. The act of charging a capacitor by means of a battery is analogous to the action of a pump. The energy provided by the "pump" appears as gravitational energy (a) or as electrical energy (b).

The potential energy of an object is always expressed relative to some position or condition at which its numerical value is defined to be zero. In the case of gravitation, this reference position was chosen to be a location at sea level, the floor of the laboratory, or a point at infinity, depending on which position served best to simplify the analysis of the problem at hand. A similar procedure is followed in dealing with electric charges and their associated electric fields.

> The potential at a point is equal to the work done by electric forces on 1 coulomb of positive charge in moving from that point to the position at which the potential is defined to be equal to zero.

Thus the potential V at a point is equal to the potential energy of 1 coul of positive charge if placed there. In equation form this definition is

$$V = \frac{W}{q} \tag{8-5a}$$

or

$$W = Vq \tag{8-5b}$$

where W represents the energy of the test charge q at the location in question. The potential V is expressed in joules per coulomb, or volts.

Potential difference and potential in a uniform field

When a *unit* positive charge travels from some point B to some other point A in a uniform electric field, since work equals force times distance, the work done by the field on the charge is

$$\Delta V_{BA}(\text{for uniform field}) = E \, \Delta d \tag{8-6a}$$

where Δd is the component of the displacement BA of the unit charge along the field direction. If we define A to be the reference position at which the potential is assigned a value of zero, we have

$$V = Ed \tag{8-6b}$$

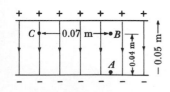

FIG. 8-12

Potential difference in a uniform field. If the electric field intensity is uniformly equal to 20,000 N/coul between the plates, then $\Delta V_{BA} = +800$ volts and $\Delta V_{BC} = 0$.

where d is the component of the displacement along the field direction.

Let us apply the concepts that have just been introduced to the case of a capacitor that has been charged by contact with the terminals of a 1000-volt battery (Fig. 8-12). For the specific case shown in the figure, the electric field in the space between the plates equals 20,000 N/coul. The differences in potential between the points shown in the figure, according to Eq. (8-6a), will be

$$V_{BA} = V_B - V_A = \quad 800 \text{ volts}$$

$$V_{AB} = V_A - V_B = -800 \text{ volts}$$

and

$$V_{BC} = V_B - V_C = \quad 0 \text{ volt}$$

Numerical values of the potential at the points indicated depend on the location of the point at which the potential is assigned a value of zero. Selecting the negative plate as the reference position, we may write

$$V_A = \quad 0 \text{ volt}$$
$$V_B = 800 \text{ volts}$$

and

$$V_C = 800 \text{ volts}$$

Potential and potential difference in an inverse square field

The energy of a charge in the electric field of another charge is analogous to the gravitational energy of an object in a gravitational field. In both cases, the force field is an inverse square field; hence the force on the object whose energy is under scrutiny is not constant but varies with distance from the other object. In all such cases, as was explained in Sec. 3-4, the work one must do to move the object from one point to another may be determined from a graph of force versus distance by measuring the area lying between the force curve and the displacement coordinate. Since the form of the Universal Law of Gravitation is identical to Coulomb's law, the work required to move a test charge q from a position at a distance r from another charge Q to infinity, by comparison to Eq. (3-11b), may be written

$$\text{Electrical potential energy} = \frac{kQq}{r} \qquad (8\text{-}7)$$

The potential at that position is the potential energy at a unit test charge if placed there, that is,

$$V = \frac{kQ}{r} \qquad (8\text{-}8a)$$

A graph of this expression for the specific case introduced in Fig. 8-7 is shown in Fig. 8-13. The arbitrary choice that the potential be set equal to zero when the charges no longer exert forces on each other—

FIG. 8-13

Potential due to a negatively charged sphere. The potential in the field described by Fig. 8-7 is negative everywhere, since a unit positive charge would have less energy at any finite location than it would have at infinity. The shaded area in Fig. 8-7 equals the difference between the potentials at $r = 1$ m and $r = 4$ m.

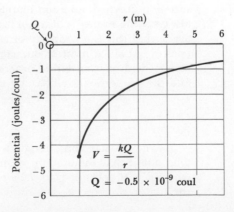

that is, when $r = \infty$—is reasonable. The work done by electrical forces on 1 coul traveling from $r = 4$ m to $r = 1$ m equals the shaded area in Fig. 8-7. The same result may be secured much more easily from Fig. 8-13, where one observes that $V(r = 1$ m$) = -4.5$ joules/coul and $V(r = 4$ m$) = -1.13$ joules/coul; hence $\Delta V = 3.37$ joules/coul.

The difference in potential between any two points B and A in the electric field due to a point charge Q is

$$\Delta V_{BA} = V_B - V_A = kQ\left(\frac{1}{r_B} - \frac{1}{r_A}\right) \tag{8-8b}$$

where r_B and r_A are the distances from the charge Q to the points.

A statement of the potential or of the potential difference describes a property of points in a region resulting from the presence of an electric field. The potential tells us the energy that would be possessed by a unit charge if placed at a given point in the field; the difference in potential tells us the amount of energy involved in causing a unit charge to travel from one point to another. Thus the existence of a potential at a point or of a difference in potential between two points does not mean that energy is present; energy is involved only if a charge is placed at the point or is caused to travel from one of the points to the other.

8-6 THE PARTICLE NATURE OF ELECTRIFICATION

The work of chemists in the nineteenth century showed that matter cannot be subdivided indefinitely but instead consists of basic structural units that we call atoms. The essential features of the argument leading to this hypothesis and its confirmation have been presented in Sec. 7-3 and were based on observations of ratios of masses and volumes of materials involved in chemical reactions.

Some of the experiments of these early chemists contained the germ of a parallel particle view of electricity. These experiments, and others performed specifically to test the hypothesis, show that in any situation in which an atom manifests an electric charge, that charge is always some multiple of a fundamental elementary charge. One may argue that this is a property of matter rather than a property of electricity, but since most of our contact with nature takes place in our contact with matter, for all practical purposes, the properties of matter and the properties of electricity are synonymous.

An experiment in electrolysis

The crucial experiments were conducted by Faraday in 1833, who noted that the amount of matter deposited or released by "electrolysis" is proportional to the amount of charge that is passed through the material. Experiments similar but not identical to Faraday's experiments are sketched in Fig. 8-14. (Faraday did not use the same chemicals as those indicated here.)

In the figure, we show a battery that serves as a voltage source or "pump" to send electric charge through a series of vessels. Each vessel contains two electrodes separated by a liquid. Vessel 1 with graphite electrodes contains a dilute water solution of sulfuric acid (H_2SO_4); vessel 2 with silver electrodes contains a solution of silver nitrate ($AgNO_3$) in water; vessel 3 with copper electrodes contains a solution of copper sulphate ($CuSO_4$) in water. When an electric current is passed through these vessels, hydrogen bubbles off electrode A, oxygen bubbles off electrode B, silver leaves electrode D and is deposited on electrode C, and copper is removed from electrode F to be deposited on electrode E. When the amounts of the materials that are either released or deposited are measured, the masses are found to stand in the ratio $1:8:108:32$ for hydrogen, oxygen, silver, and copper, respectively, as compared to the relative atomic masses of $1:16:108:64$ for the same elements (see Table 7-3). Thus the relative masses released or deposited electrically are the same as the relative atomic masses except for two obvious and interesting exceptions: the amount of oxygen and of copper are exactly half the amounts one would have expected on the basis of mass data alone.

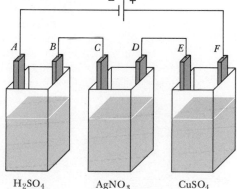

FIG. 8-14
The essence of Faraday's experiments with electrolysis. Hydrogen and oxygen bubble from electrodes A and B respectively, silver is deposited on electrode C, and copper is deposited on electrode E. The mass ratios of materials released or deposited suggest that the ions each possess some small whole number of elementary charges.

The conclusion one can reach from these observations is quite straightforward. The battery establishes an electric field between the respective electrodes in each of the liquids. The materials that appear ultimately at the electrodes are present in the liquid in the form of charged particles called "ions." The positive ions are drawn to the nearest negatively charged electrode where they are either deposited as a solid or released as a gas. A hydrogen ion and a silver ion each possess a certain charge; an individual oxygen or copper ion, however, possesses exactly twice that charge. Thus a given total charge releases or deposits only half as much material as it would have if the ions were singly charged.

Measurements show that the release of 1 kmol of hydrogen (2.016 kg) requires 192,974,000 coul, which may be interpreted to mean that the release of N atoms of hydrogen possesses 96,487,000 coul of electric charge, where N is Avogadro's number. Thus, letting the charge carried by a hydrogen ion be represented by e, we have

$$Ne = 96{,}487{,}000 \text{ coul} \qquad (8\text{-}9)$$

where N (Avogadro's number) is the number of atoms in a kmol. Twice as much charge would be required to deposit or release N atoms of a substance whose ions are doubly charged because twice as much charge is involved in depositing each ion.

8-7 CATHODE RAYS

The discharge of electricity through gases

Everyone is familiar with the beautiful and impressive possibilities of the discharge of electricity through a gas, a neon sign being the most frequently observed example. In order to construct a discharge tube, one installs a pair of metal electrodes in the two ends of a long glass tube. The tube may be partially evacuated if one wishes to observe the gaseous discharge of electricity in air, or it may be evacuated of essentially all residual gas and then "filled" with neon, hydrogen, or another gas at low pressure, perhaps at $\frac{1}{100}$ of normal atmospheric pressure. When an intermittent* high-voltage supply, with peak voltages up to 12,000 volts, is connected to the two electrodes, the tube will emit light characteristic of the gas remaining in the tube.

Let us suppose that the voltage source is applied to a discharge tube with air at atmospheric pressure; no evidence of conduction or emission of light will be seen. If the pressure is decreased, purple-pink

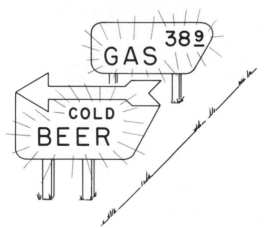

FIG. 8-15

Discharge tubes. "Everyone is familiar with the beautiful and impressive possibilities. . . ." Well, possibilities maybe!

---

* In the discussion that follows one should think in terms of an "intermittent" rather than an alternating voltage source. Whereas an alternating current (ac) simply reverses direction periodically, a truly intermittent current flows in one direction only. An induction coil (like the arrangement that energizes the spark plugs in a typical automobile) provides a much higher surge in one direction than in the other; thus it behaves like a high-voltage battery that is being turned on and off.

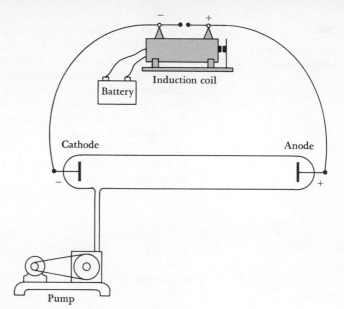

FIG. 8-16

A discharge tube. The induction coil provides an intermittent dc voltage as high as 12,000 volts. The electrodes are metal probes that are sealed in the ends of the glass tube.

streamers will begin to appear, to be replaced at even lower pressure by a steady, uniform glow. The stages in the appearance of a discharge tube containing successively smaller amounts of air are shown in Fig. 8-17.

The region identified as the Crookes' dark space does not begin to appear until the pressure falls below about $\frac{1}{1000}$ of normal atmospheric pressure. If one continues to decrease the pressure, the striations

FIG. 8-17

Successive stages of evacuation of an air-filled discharge tube. The sketches show successive stages of the appearance of a discharge tube as the pressure falls from approximately $\frac{1}{100}$ of atmospheric pressure to approximately $\frac{1}{1000}$ of atmospheric pressure. If the pressure is reduced further, the Crooke's dark space becomes larger and ultimately fills the entire tube.

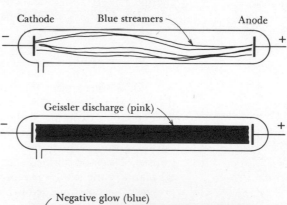

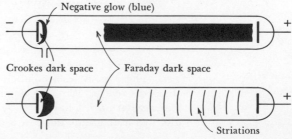

and the glow become fainter, and at a pressure of approximately 10^{-5} atm the Crookes' dark space will fill the entire tube. At this stage little or no light that can be associated with the gas in the tube will be observed; instead the glass will be observed to emit a yellowish-green light, particularly in the vicinity of the positive electrode. The activation of the glass appears to be due to some invisible radiation that comes from the cathode (the negative electrode); therefore the radiation has been called "cathode rays."

These effects were first observed a very long time ago. The first evidence of a discharge of electricity through a gas was reported as early as 1752 by Watson; the Faraday dark space was observed in 1838.

Cathode rays

Although studies of the glow discharge constitute an important and fascinating aspect of the history of physics, our concern with the subject at this point lies with phenomena taking place at high vacuum when the effect of the gas is no longer obvious. The names of Hittorf, Crookes, and Lenard are particularly important in this phase of the story.

Four particularly significant high vacuum tubes constructed in this early period are shown in Fig. 8-18. These tubes indicate that cathode rays emerge from the metal surface of the negative electrode in a direction normal to the surface [Fig. 8-18(a)] and proceed to travel in straight lines forming [in Fig. 8-18(b)] a sharp shadow of the cross-shaped obstruction. Because a pinwheel in their path [Fig. 8-18(c)] is caused to rotate, cathode rays appear to possess momentum, and since they can heat a spot on a piece of foil [Fig. 8-18(a)], they must transport energy. Finally, when deflected into a small metal cage built into the tube [Fig. 8-18(d)], cathode rays manifest a negative charge.

The discovery of the electron

Of all the experiments conducted with cathode rays, undoubtedly the most significant one was conducted by Sir J. J. Thomson in 1897. In Thomson's experiment the behavior of a beam of cathode rays was observed when it was sent through electric and magnetic fields. By assuming that cathode rays consist of a stream of negatively charged *particles*, Thomson verified that all these particles are alike in that they all manifest the same ratio of their charge (e) to their mass (m). The presently accepted value of this ratio is

$$\frac{e}{m} = 1.758796 \times 10^{11} \text{ coul/kg} \tag{8-10}$$

These particles are called *electrons* and Thomson is credited with their discovery. We will defer a detailed discussion of Thomson's experiment until Chapter 9, by which time we will have discussed the effect of a magnetic field on a moving charge.

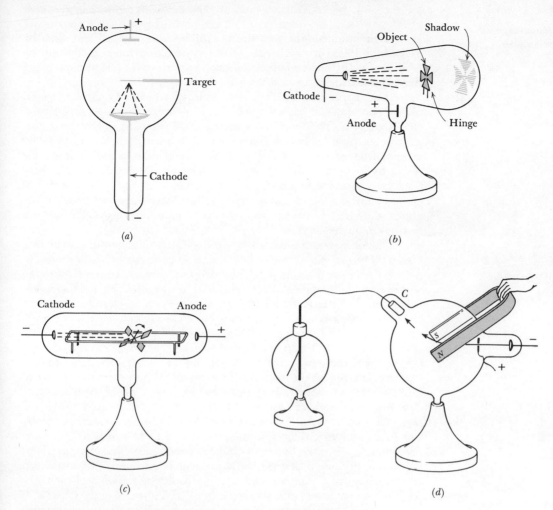

FIG. 8-18

Four landmark discharge tubes. (*a*) Rays emerge normal to a surface and hence can be concentrated at a point to produce heating (Crookes, 1879). (*b*) Rays seem to emanate from the cathode since they produce a shadow (Hittorf, 1869). (*c*) Cathode rays appear to have momentum; they cause a small wheel to spin (Crookes, 1870). (*d*) When the cathode rays are deflected into the cylindrical cage *C* by the magnet, a negative charge appears on the electroscope (Perrin, 1895).

8-8 MILLIKAN'S EXPERIMENT

We have seen that Thomson's experiment led to the identification of cathode rays as a stream of negatively charged particles and to a determination of the ratio of the charge on one of these particles to its mass. Clearly, if one were to devise an experiment that would lead to an independent measurement of e, he could combine this measurement of e with Thomson's measurement of e/m to secure a numerical value of m.

The independent measurement of e was carried out first by Wilson in 1903, using charged water drops, and more accurately by Millikan in 1910 in his famous oil-drop experiment. The apparatus for this experiment is shown schematically in Fig. 8-19.

The apparatus consists of a pair of insulated, parallel metal plates, one of which has a tiny hole at its center; the plates have a surface area of a few square centimeters and typically might be a few millimeters apart. The plates are located in a sturdy box to protect the interior from air currents. An electric field is established between the plates by a battery whose potential difference can be varied smoothly from zero to a few hundred volts. Two windows open into the box— one, to admit a sharp beam of light to illuminate the space between the plates, the other to permit an observer to examine this region using a low-power microscope.

The observer can introduce a mist of tiny oil drops into the space above the upper plate; one or more of these drops may fall through the hole in the upper plate. A typical oil drop will possess an electric charge whose sign and magnitude are relatively unpredictable. Furthermore, the amount of charge carried by an oil drop can be changed, but in a rather haphazard way, by turning on an x-ray tube whose beam is directed into the space between the plates.

If a given oil drop is charged negatively, it will experience two forces—one, the downward force Mg due to the gravitational field of the Earth acting on the mass M of the oil drop, the other, the electrical force qE due to the action of the electric field E of the parallel-plate capacitor on the charge q of the oil drop.

Although the oil drops are very small, their mass can be determined by studying their rate of fall when the electric field is turned off—that is, when the battery voltage is set at zero. This measurement is possible because a precise mathematical relationship exists between the "fluidity" of the air, the radius of a droplet, the density of the oil of which it is composed, and the velocity at which it descends. Clearly, if this is the case, measurements of the velocity of fall of a given drop and

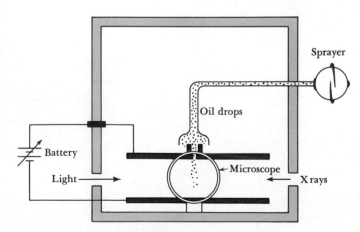

FIG. 8-19

Millikan's oil-drop experiment. By measuring the speed of fall of an oil drop under gravity, Millikan determined the mass of the oil drop; by measuring the electric force experienced by the same oil drop, he determined its charge. The charge was always a multiple of some elementary charge.

FIG. 8-20

A parachute analogy. The mass of a man could be determined by measuring his speed of descent in a parachute.

of the density of the oil of which it is composed will suffice to determine the radius of the drop and thus will yield a measurement of its mass.

It is important to note that this is not a case of "free fall," which was discussed in Chapter 1. The oil drop experiences a "drag" due to the resistance of the air and therefore falls at a steady speed, which is determined by the size of the droplet and the net force acting on it. The experiment resembles one in which an observer might seek to infer the mass of a man by observing the speed at which he descends in a parachute. To relate the matter to another familiar situation, recall that fog descends more slowly than rain; presumably one could determine the mass of a drop of rain or of a fog droplet by observing its speed of fall.

Suppose that one were to turn on the electric field and adjust its intensity to such a value that the oil drop stands motionless* in the space between the plates. This situation will be achieved only if the downward force due to gravitation is exactly equal to the upward force of the electrical field, that is, when

$$qE = Mg \qquad (8\text{-}11)$$

Since M and g are known and E can be determined from

$$V = Ed \qquad (8\text{-}6b)$$

* The procedure described here is a slightly simplified version; Millikan compared the drift velocities of an oil drop bearing various amounts of electric charge.

where V is the difference in potential between the plates and d is their distance apart, one can determine a value for the amount of charge on the oil drop; that is,

$$q = \frac{Mg}{E} = \frac{Mgd}{V} \qquad (8\text{-}12)$$

Identifying the elementary charge

The experiment can be repeated using the same oil drop but with a different charge, for one can change the charge on the oil drop simply by turning on the x-ray tube for a few seconds. One can also measure the charge on different oil drops, remembering, of course, that he must redetermine M, the mass of the droplet. Typical data for the charge observed to be carried in various separate trials are shown in Table 8-1.

These data and similar additional data secured by many observers over the years since Millikan's early experiments point to a rather straightforward conclusion: *the electric charge on an oil drop is some whole number multiple of a basic elementary charge.* Electric charge is "quantatized"; that is, it appears in nature in the form of particles of charge, each having an identical magnitude of approximately 1.6×10^{-19} coul. More recent measurements indicate that the elementary charge equals

$$e = 1.60210 \times 10^{-19} \text{ coul} \qquad (8\text{-}13)$$

TABLE 8-1		
TRIAL NUMBER	MEASURED CHARGE ON OIL DROP (coul)	NUMBER OF ELEMENTARY CHARGES
1	6.35×10^{-19}	4
2	14.44×10^{-19}	9
3	11.22×10^{-19}	7
4	8.07×10^{-19}	5
5	3.16×10^{-19}	2
6	4.80×10^{-19}	3
7	11.21×10^{-19}	7
8	1.61×10^{-19}	1
9	14.47×10^{-19}	9
etc.	etc.	etc.

8-9 THE FARADAY CHARGE, THE THOMSON CHARGE, AND THE MILLIKAN CHARGE

It would be well for us to review the situation to this point in order to ascertain how much we can honestly say we know and to separate this knowledge from any conjectures that we might wish to make.

1. *Faraday's Observations* (see Sec. 8-6). In experiments involving solutions of various salts in water, it was found that the amount

of electric charge required to deposit or release N atoms of matter is some whole number multiple of 96,487,000 coul—that is, 96,487,000 coul, or $2 \times$ 96,487,000 coul, or $3 \times$ 96,487,000 coul, etc., depending on whether the ion is singly, doubly, or triply charged. Thus if we let the charge on each ion be represented by e, $2e$, $3e$, etc., respectively, we have

$$Ne = 96{,}487{,}000 \text{ coul} \qquad (8\text{-}9)$$

2. *Thomson's Observations* (see Sec. 8-7). In Thomson's experiments, cathode rays were sent through a region in which an electric and/or a magnetic field could be established at will. Treating cathode rays as a stream of identical particles, Thomson demonstrated that the cathode particles each possess a charge e and a mass m such that their ratio is

$$\frac{e}{m} = 1.758796 \times 10^{11} \text{ coul/kg} \qquad (8\text{-}10)$$

3. *Millikan's Observations* (see Sec. 8-8). Millikan observed the electrical force on oil drops and found that each oil drop behaved as if it possessed a whole number multiple of elementary charges e, where

$$e = 1.60210 \times 10^{-19} \text{ coul} \qquad (8\text{-}13)$$

Do all three experiments deal with the same charge?

It must be noted that these three experiments deal with very different situations. Although we have used the same symbol e in describing the charge in each experiment, it is not at all obvious that the Faraday elementary charge, the Thomson elementary charge, and the Millikan elementary charge are identical. In none of the three experiments does the charge permit other and more direct observation of its properties.

A reading of the history of this period (1900 to 1917) indicates that scientists accepted the hypothesis that the Faraday, the Thomson, and the Millikan elementary charges are identical on what, in retrospect, would seem to be considerably less than conclusive evidence. It was a reasonable working hypothesis, but, like any hypothesis in physics, it has always been subject to change at any time experimentation could produce a result that contradicts it. A half-century of diligent experimental work has reinforced the assumption that Eqs. (8-9), (8-10), and (8-13) do, in fact, all relate to the same elementary charge. While no date can be set at which the identity of the charges represented by these equations became "certain," it has not been the subject of debate by scientists for many years.

Avogadro's number

Remarkable conclusions are possible when one assumes the symbols e in Eqs. (8-9), (8-10), and (8-13) to be identical. Introducing the Millikan-determined value of e into Eq. (8-9) yields a value of Avogadro's number; namely,

$$N = \frac{96,487,000}{1.6021 \times 10^{-19}} = 6.02252 \times 10^{26} \text{ atoms/kmol} \quad (8\text{-}14)$$

The mass of an electron

Using the value of e in Eq. (8-10), one secures a numerical value of m, the mass of an electron,

$$m = \frac{e}{e/m} = \frac{1.60210 \times 10^{-19}}{1.758796 \times 10^{11}} = 9.1091 \times 10^{-31} \text{ kg} \quad (8\text{-}15)$$

Without seeing either an atom or an electron, we have counted the atoms and determined the mass of an electron!

SUMMARY There are two kinds of electrification: (1) positive, which appears on glass that has been rubbed with silk and (2) negative, which appears on hard rubber after being rubbed with fur or wool. Like charges repel each other, whereas unlike charges attract each other with a force whose magnitude is given by Coulomb's law,

$$F = \frac{kQq}{r^2} \quad (8\text{-}1)$$

The Coulomb constant k equals 9×10^9, and the charges Q and q are expressed in coulombs.

The electric field intensity E at a point is defined as the force experienced by 1 coul of positive charge if placed there; thus

$$E = \frac{F}{q} \quad (8\text{-}2)$$

where q is a test charge used to measure the field.

The potential V at a point is equal to the work done by electric forces on 1 coul of positive charge in moving from that point to a point at which the potential is defined to be equal to zero; that is,

$$V = \frac{W}{q} \quad (8\text{-}5)$$

The electric field intensity and the potential at a position at a distance r from a small (or "point") charge are given by

$$E = \frac{kQ}{r^2} \quad (8\text{-}3)$$

and

$$V = \frac{kQ}{r} \quad (8\text{-}8a)$$

respectively.

The electric field intensity E in the space between two oppositely charged metal plates (a capacitor) is related to the difference in potential between the plates ΔV by

$$\Delta V = E \, \Delta d \qquad (8\text{-}6a)$$

where Δd is the separation of the plates.

The smallest known particle of electrification is the electron with a charge of -1.6×10^{-19} coul and a mass of 9.11×10^{-31} kg.

In the following problems reference will be made to charged particles called protons, electrons, and alpha particles. We will have much more to say about them in later chapters; at this point they may be treated simply as tiny particles having the following properties:

QUESTIONS AND PROBLEMS

	MASS (kg)	CHARGE (coul)
Electron	9.11×10^{-31}	-1.6×10^{-19}
Proton	0.167×10^{-26}	$+1.6 \times 10^{-19}$
Alpha particle	0.667×10^{-26}	$+3.2 \times 10^{-19}$

8-A1. Assume one proton to be held in a fixed position while another proton is brought into its vicinity. What are the direction and the magnitude of the force experienced by each of these protons when they are 10^{-10} m apart?

8-A2. Assume that a hydrogen atom consists of an electron traveling about a proton in a circular orbit whose radius equals 5.28×10^{-11} m. What must be the speed of such an electron?

8-A3. An electron at given position in an electric field experiences an acceleration a million billion (10^{15}) times the acceleration of gravity. What is the intensity of the electric field at that location?

8-A4. Rewrite the caption that accompanies Fig. 8-3 making use of hindsight, assuming that the metal hemispheres possess fixed positive charges and mobile negatively charged electrons.

8-A5. Rewrite the caption that accompanies Fig. 8-18, introducing the word "electron" where appropriate.

8-A6. A coil of wire is submerged in a vessel of water and an electric current of 5 amp is established in the wire for 3 min. From the rise in temperature (recall Sec. 6-6 and Fig. 6-10d) it is found that 10,000 joules of energy were added to the water. What was the difference of potential at the terminals of the battery or generator being employed?

8-A7. If we assume in an arrangement such as that shown in Fig. 8-18(a) that a billion electrons strike the target each second and that each electron has fallen through a difference of potential of 5000 volts, how many joules of energy are absorbed each second by the target?

8-A8. The table shown in Appendix 7 summarizes the analogous nature of gravitation and electrification. Examine this table carefully, then list at least two important ways in which electrification and gravitation differ.

8-B1. A proton in the space between two capacitor plates experiences a force of 1.44×10^{-14} N. (*a*) What is the electric field intensity in the space between the plates? (*b*) Assuming the plates to be 0.001 m apart, how much work would be required to carry the proton from one plate to the other? (*c*) How much work would be required to carry a coulomb of positive charge from the negative plate to the positive plate of this capacitor?

8-B2. The essential components of an electron gun are a filament that emits electrons and a metal plate with a hole at its center. When sealed in an evacuated tube, the electrons may be accelerated toward the plate by an external voltage supply as in Fig. 8-21. (*a*) If the accelerating voltage equals 1000 volts, how much kinetic energy will each electron acquire in falling through the stated difference of potential? (*b*) What will be the velocity of the electrons that pass through the hole? (*c*) Will these electrons be attracted back to the plate? Explain.

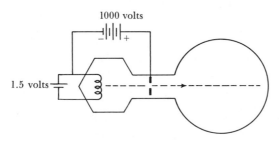

1000 volts

1.5 volts

FIG. 8-21

8-B3. Given that the electric field intensity in the space between the plates (Fig. 8-22) equals 100,000 N/coul, compute the work required to carry a proton (*a*) from *A* to *B* in the figure, (*b*) from *B* to *C*, (*c*) from *A* to *C*.

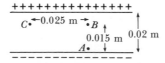

C $\leftarrow$ 0.025 m $\rightarrow$ B

0.015 m   0.02 m

A

FIG. 8-22

8-B4. (*a*) In a laboratory experiment a charge of $+10^{-7}$ coul is placed on an insulated metal sphere having a diameter of 0.10 m. Prepare a graph showing the electric field intensity versus distance in the range from $r = 0.1$ m to $r = 1.8$ m.

(*b*) By counting the appropriate squares in the preceding graph, determine the work required to carry a coulomb of positive charge from a position at $r = 1$ m to the surtace of the sphere. (Assume that the charge on the sphere remains centered on the center of the sphere.)

8-B5. After measuring the charge on an electron, a student performs an electrolysis experiment and observes that when a total charge of 90 coul is passed through a series of electrolytes as shown in Fig. 8-14, 10^{-5} m³ of hydrogen at normal temperature and pressure is released from electrode B. (a) How many atoms of hydrogen are released? (b) How many molecules of hydrogen are released? (c) What is the student's computed value of Avogadro's number?

8-B6. In one of Millikan's experiments an oil drop weighing 80×10^{-14} N was seen to fall through air at a constant velocity of 0.860 mm/sec when no electric field was employed. When an electric field of 318,000 N/coul was employed, the oil drop rose with a constant velocity of 0.125 mm/sec. (a) What was the net upward force on this oil drop? (b) What was the force on the oil drop due to the electric field? (c) What was the charge on the oil drop? (d) In the light of the conclusions reached by Millikan, how many electrons were on this particular oil drop?

8-B7. A student wishes to determine the mass of an electron by repeating certain experiments mentioned in this chapter. Performing Thomson's experiment, he found e/m to be equal to 1.81×10^{11} coul/kg; repeating Millikan's experiment, he found e to be 1.64×10^{-19} coul. What was his measurement of the mass of the electron?

Magnetism

In previous chapters we have discussed two distinct fields that appear in nature. The gravitational field as described in Sec. 2-12 was introduced to describe the "region of influence" that accompanies any object possessing mass. The electric field was introduced in Sec. 8-4 and served a similar purpose; it enabled us to describe a completely different "region of influence" that accompanies an object possessing an electric charge. In this chapter we will be concerned with a third kind of field, a *magnetic* field.

9-1 INTERACTION OF TWO MAGNETIZED OBJECTS

Magnetic phenomena have been known for thousands of years. Even the name we associate with magnetism speaks of the antiquity of the concept, for it derives from the name of a region in Asia Minor (Magnesia) where "magnetic" materials called lodestones (actually pieces of ferric oxide, Fe_3O_4) were discovered centuries ago. These materials possess the property of attracting (or repelling) other lodestones and, furthermore, of transmitting this property to pieces of iron. These "magnetized" pieces of iron are called magnets.

Consider a bar magnet suspended by a thread and free to rotate about a vertical axis as in Fig. 9-2. When so mounted, the magnet may oscillate for a considerable length of time about the vertical axis but ultimately will come to rest oriented lengthwise along a north-south line. If disturbed at any later time, it will return to the same orientation with the same end "pointing" north. One labels this end the north-seeking or, more simply, the north pole (or the N-pole) and the other end the south-seeking or the south pole (the S-pole). A magnetic compass is a miniaturized version of this arrangement in the form of a tiny magnet mounted on a smooth bearing in a watch-sized box (Fig. 9-3).

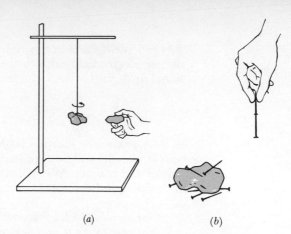

FIG. 9-1

A lodestone experiment. (*a*) Two lodestones. One region of the suspended lodestone manifests a force of attraction; another region manifests a force of repulsion, when a second lodestone is brought near. (*b*) A nail that has been left in contact with a lodestone may be attracted to, or repelled by, another such nail.

(*a*)

(*b*)

Let us label the north-seeking and the south-seeking poles of two identical bar magnets. We suspend one of them so that it can swing freely about a vertical axis and hold the other magnet in its vicinity. We will observe a force of repulsion whenever the N-pole of one is close to the N-pole of the other or whenever the S-pole of one is near the S-pole of the other. Conversely, we will observe a force of attraction if an N-pole is brought near an S-pole. This behavior can be summarized as follows:

Like magnetic poles repel each other; unlike magnetic poles attract each other.

9-2 THE MAGNETIC FIELD OF A MAGNETIZED OBJECT

We can say that a magnet affects the space around it because we observe that another magnet experiences a force when placed in its vicinity. We therefore have the same justification for saying that a

FIG. 9-2

A magnet lines up along a north–south line. The north-seeking pole of the magnet is called the north pole (or N-pole) of the magnet.

FIG. 9-3

A magnetic compass. A tiny magnet pivoted on a needle mount indicates direction.

FIG. 9-4

The law of magnets. When the N-pole of one magnet is brought near the N-pole of another, each experiences a force of repulsion.

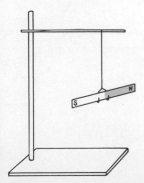

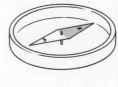

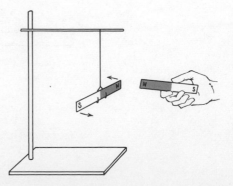

magnetic field exists in the vicinity of a magnet as we had for saying that an electric field exists in the vicinity of a charge or that a gravitational field exists in the vicinity of a mass.

The direction of a magnetic field

If a magnetic compass is placed near a magnet, the N-pole of the compass needle will be repelled by the N-pole and attracted by the S-pole of the magnet. We will use this simple experiment as the basis for defining the *direction* of a magnetic field as follows:

The direction of the magnetic field at a point in space is the direction indicated by the N-pole of a compass needle when placed there.

This definition of the direction of a magnetic field suggests a procedure by which one can map the magnetic field in any region where a magnetic field is believed to exist. As described in Fig. 9-6, a map of the field may be prepared by successively moving a small magnetic compass in such a way that at each new position the S-pole points at a dot which marks the previous location of the N-pole. A smooth curve is drawn through these dots; this line has the property that a compass needle placed anywhere on it will surely be tangent to it. The line therefore indicates the direction of the magnetic field.

The magnetic field of an object can also be mapped by using iron filings—that is, tiny elongated particles of iron that can be dispensed from a "salt" shaker. One places a large white card on top of the magnet to keep the filings from sticking to the magnet, then sprinkles iron filings uniformly over the top surface of the card. If the card is tapped gently, the iron filings, which have become magnetized because of their proximity to the bar magnet and thus are the equivalent of tiny compasses, will line up to reveal the shape of the magnetic field.

FIG. 9-5

The direction of a magnetic field. We define the direction of the magnetic field at any point as the direction indicated by a north-seeking pole of a magnetic compass when placed there.

FIG. 9-6

Mapping the field of a bar magnet. (*a*) A small magnetic compass is placed beside the magnet; its N-pole points at dot 1. When it is moved to a location where the S-pole points at 1, one marks the new location of the N-pole at 2. Many such sequences of dots yield a map of the magnetic field (*b*).

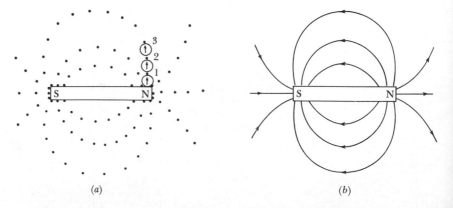

(*a*)   (*b*)

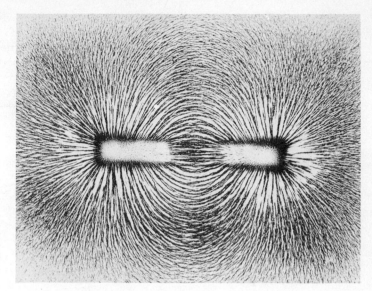

FIG. 9-7

Magnetic field revealed by iron filings. A piece of cardboard on top of a magnet sprinkled with iron filings produces a pattern suggesting the "lines" of the magnetic field. (*Phillips Hall*)

The experiment with iron filings might lead us to declare that we now have proof that "lines of magnetic force" emerge from the N-pole and converge into the S-pole. This would be an overinterpretation of the experiment because, regardless of the number of lines we see in the pattern of iron filings or have drawn by means of a compass needle, a magnetic field exists at all points in the space between the lines as well. The lines in either experiment only describe the direction of the magnetic field in the vicinity of the magnet; we will avoid the temptation to assign any more "reality" to them than that.

The magnetic field of the Earth

The Earth possesses a magnetic field whose direction in its vicinity permits one to regard the Earth as the magnetic equivalent of an enormous bar magnet. The S-pole of the Earth is located near the geographic North Pole, while the N-pole is near the geographic South Pole. This seeming contradiction arises, of course, from the way we have elected to name the magnetic poles of magnets and compass needles.

9-3 THE INTENSITY OF A MAGNETIC FIELD

Experience shows that an electrically charged particle at rest in a magnetic field experiences no force attributable to the magnetic field. The electrically charged pith-ball shown in Fig. 9-8, for example, makes no response when an intense magnetic field is established in the region where it is located.

An electric charge in motion in a magnetic field, however, does

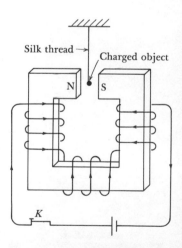

FIG. 9-8

An electric charge experiences no force when at rest in a magnetic field. A charged object in the space between the poles of an electromagnet is not affected when the magnetic field is established by closing the switch *K*.

experience a force; this force, furthermore, has a magnitude that depends both on the direction of motion of the charge and on its speed of travel. In addition, the magnetic interaction force has the peculiarity of acting in a direction perpendicular both to the direction of motion of the charge and to the direction of the magnetic field at the immediate location of the charge. This effect has already been shown in Fig. 8-18(*d*).

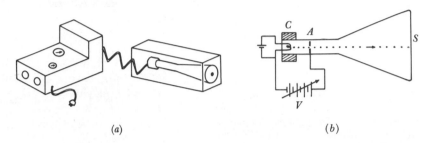

(*a*)   (*b*)

FIG. 9-9

A cathode-ray tube showing schematic of an electron gun. (*a*) The power supply makes electrical connection to the electron gun through a flexible electric cable. (*b*) The electron gun consists of a heated filament (*C*) and an anode (*A*) in the form of a disk with a hole in its center. Electrons "boiled off" the filament are accelerated by the voltage supply *V*; those that pass through the anode travel the length of the tube and strike the fluorescent screen (*S*).

FIG. 9-10

The deflection of an electron beam by a magnetic field. (*a*) If the direction of the magnetic field is parallel to the direction of motion of the electrons, the beam is not deflected. (*b*) If the magnetic field is oblique to the direction of motion, the beam is deflected in a direction perpendicular to both. (*c*) A maximum deflection is observed if the magnetic field is perpendicular to the direction of motion.
(*Phillips Hall*)

(*a*)   (*b*)   (*c*)

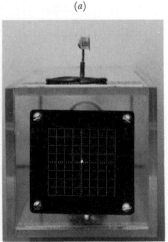

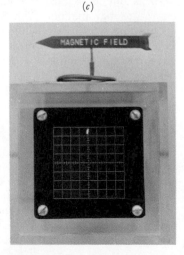

A modern version of the arrangement of Fig. 8-18(d) is shown in Fig. 9-9. This figure shows a cathode-ray tube attached to its power source by means of a flexible wire cable to facilitate the manipulation of the tube in the magnetic field. The electron beam is supplied by an "electron gun," which, in this context, means only that the cathode is heated by a hot tungsten filament to enhance the emission of electrons from the cathode, an effect known as the Edison Effect. The inside surface of the tube at S is coated with fluorescent paint as in a television tube to enable one to see the point of arrival of the electron beam.

A freely pivoted compass needle is used to indicate the direction of the magnetic field. When the tube is oriented so that the electron beam travels in a direction parallel to the field, the beam is not affected [Fig. 9-10(a)]. On the other hand, if the tube is pointed in any other direction, the beam will be deflected [Fig. 9-10(b)] by an amount determined by the orientation of the tube relative to the magnetic field, reaching a maximum when the axis of the tube is exactly perpendicular to the field direction [Fig. 9-10(c)]. Furthermore, the deflection is perpendicular both to the direction of motion of the charge and to the direction of the magnetic field.

The Q-v-B Rule

In stating a rule for the direction of the force acting on a moving charge, it should be remembered that all rules in electricity are formulated in terms of the behavior of positive, rather than negative, charges. Experiments like those described above for a moving electron show that a moving positive charge experiences a force in a direction described by a mnemonic, which, because of its relationship to Eq. (9-1a) below, will be known as the "Q-v-B" Rule. This rule is as follows:

One extends his right hand as for a handshake with his extended fingers pointing in the direction of the motion of the moving positive charge. When he has oriented this outstretched hand in such a way that it is possible for him to bend his fingers into the direction of the magnetic field, his thumb will be pointing in the direction of the force on the moving positive charge.

A moving negative charge experiences a force in a direction exactly opposite the force on a moving negative charge, other things being the same.

The force on a moving charge

Measurements show the force on a moving charge to be proportional both to the magnitude of the charge Q and to its velocity v. We use this observation to define and measure the intensity of the magnetic field through which the charge is traveling. Letting B represent the intensity of the magnetic field at a given point in *webers per square meter* and letting F represent the force in newtons acting on a charge Q when passing through that point with a velocity v in a direction perpendicular

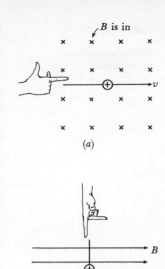

(a)

(b)

FIG. 9-11

The direction of the force on a moving positive charge. One holds his right hand with his fingers pointing in the direction of the moving positive charge (or in the direction opposite that of the moving negative charge). When he has oriented this outstretched hand so that it is possible to bend his fingers into the direction of the magnetic field, his thumb will be pointing in the direction of the force on the moving charge.

to the field, we write

$$F = QvB \qquad (9\text{-}1a)$$

Equation (9-1a) is the defining equation for the magnetic field intensity.

One should note the subtle difference between the logic by which B, in contrast to Q and v, enters into this relationship. *Experiments* show that the force F on a given charge is proportional to the magnitude of the charge Q and its velocity v. No further experiment is involved in the introduction of B into this equation; the equation defines the magnetic field intensity, and thus the relationship of B to the force F is not subject to experimental proof.

If a charge travels in a direction other than perpendicular to the magnetic field, the force on the charge is given by

$$F = Qv_\perp B \qquad (9\text{-}1b)$$

where $v_\perp$ represents the component of the velocity v in a direction perpendicular to the field direction. Clearly, if the charge travels along the field direction, whether parallel or antiparallel to the field, the component equals zero, and therefore Eq. (9-1b) correctly predicts that the charge should experience no force.

9-4 THE FORCE ON A CURRENT-CARRYING WIRE

Our discussion has been limited up to this point to consideration of the force on an individual charge in motion through a magnetic field. The more familar situation to most readers, however, is that in which charges are constrained to travel inside a conductor. When the charges are so constrained, the transverse force on them due to an external magnetic field is transmitted to the conductor; such forces make possible the operation of hundreds of practical devices, such as relays, meters, automatic switches, and electric motors.

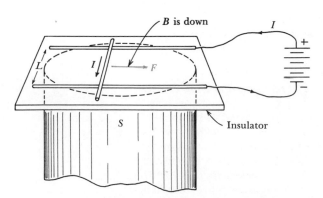

FIG. 9-12

Force on a conductor. A length L of wire carrying a current I through a uniform magnetic field B experiences a force F given by $F = ILB$.

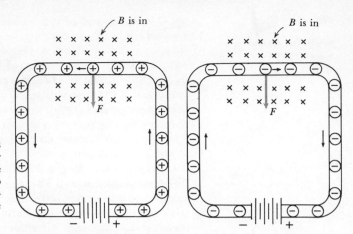

FIG. 9-13

The direction of the force on a conductor is independent of the sign of the elementary charges. One may not know whether the battery causes positive charges to travel to the left through the magnetic field or negative charges to travel to the right. The force on the wire is the same in either case.

One can write an alternate definition for the magnetic field intensity by defining it in terms of the force on a current-carrying wire. The defining equation is

$$F = ILB \tag{9-2}$$

where F is the force in newtons experienced by a wire of length L meters carrying a current of I amperes (coul/sec) in a direction transverse to a magnetic field of B webers per square meter. These alternate definitions of B are exactly equivalent in that it can be shown that Eq. (9-1) leads to Eq. (9-2), and vice versa. The proof, however, involves some analysis of the details of how conduction takes place which will not be undertaken here.

The charge carriers responsible for the transport of electric current in conductors may be positive or negative; in some conductors, electrolytes for example, both positive and negative carriers may be involved in the conduction process. It is of some importance that we note that the direction and magnitude of the force on a given conductor are independent of the sign of the charge carriers.

To verify this statement, let us connect a battery to the two ends of a conductor in a magnetic field as in Fig. 9-13. If the current consisted of positive charges in motion, we would conclude that the force on the wire should be a downward force. However, if the current actually consisted of negative charges moving in the opposite direction, we still expect to observe a downward force.

9-5 INDUCED VOLTAGE

We have seen that a current-carrying wire experiences a lateral force when located transverse to a magnetic field. Let us now examine another manifestation of this phenomenon that occurs when a wire is moved laterally across a magnetic field by an external agency.

Let a loop of wire be oriented so that one portion of the wire is located transverse to the magnetic field of a magnet (Fig. 9-14). If the magnet and all portions of the wire are at rest, no current flows in the wire. However, if the portion of the wire located in the field is moved quickly to the left, all charges residing on particles that constitute the substance of the wire will travel with the wire from right to left but at the same time will experience a force along the wire. If either the positive or the negative charges are mobile—that is, if the wire is a conductor—an electric current will flow around the circuit. This is called an *induced current*.

We can use the *Q-v-B* Rule quoted in Sec. 9-4 to determine the direction of this current. This rule will show that positive charges should experience a force down the page and, if mobile, will constitute a positive current. On the other hand, negative charges will experience a force up the page and, if mobile, will travel in that direction.

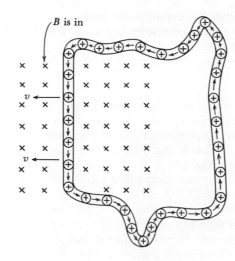

FIG. 9-14

An induced voltage. When the wire is moved across the field, individual positive charges experience a force along the wire, and, if mobile, they produce a positive current as shown.

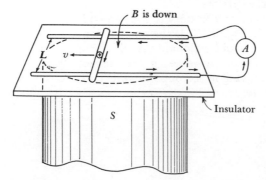

FIG. 9-15

Proof that induced voltage equals *BLv*. In traveling a distance *L*, a unit charge gains an amount of energy equal to $QvBL/Q$ or BLv.

The charges that are set into motion in this manner are capable of lighting lamps, causing motors to operate, and doing other practical things. Thus we have described a primitive form of electric generator.

In order to put the observations we have just made on a quantitative basis, let us replace the circuit shown in Fig. 9-14 by the one shown in Fig. 9-15. In the modified version, the movable portion of the circuit consists of a straight wire that slides on two parallel conducting bars.

Let the straight wire travel through the uniform field of the magnet at a constant velocity v. According to Eq. (9-1), each elementary positive charge experiences a force equal to QvB. In traveling the length of the wire, such a charge should gain energy equal to the force acting on it times the distance L that it travels under the action of that force; that is,

$$\text{Energy gained by a charge} = (QvB)L = (BLv)Q$$

In Sec. 8-5 a potential difference was defined as the amount of work an electric field would do on a coulomb in moving from one place to another in the field. We now have found another rather different situation in which a charge can attain energy in moving from one point to another. The energy gained by a unit charge under these conditions will be called an *induced voltage*. From the preceding expression we see that

$$\text{Induced voltage} = \frac{\text{energy gained by charge}}{\text{magnitude of charge}}$$

$$= BLv \qquad (9\text{-}3)$$

It is important to observe that we have not written a prescription for perpetual motion in the foregoing paragraphs. One does not get energy from any system unless he either extracts energy already possessed by the system or does work on the system. The current established in a wire when it is moved across a magnetic field constitutes a stream of moving charges traveling along the wire (Fig. 9-16). If we apply the Q-v-B Rule to this stream of charges, we find that the induced current is accompanied by a force F that tends to resist the motion of the wire. To keep the wire in motion, then, some external agent must continue to push against the wire exerting a force F' equal but opposite to F; hence the external agent must do work on the wire in order to generate electrical energy. It is the work done by this external agent that is translated into the electrical energy that lights the lamp or operates the motor.

FIG. 9-16

One must do work to generate electrical energy. If motion of the wire produces a current, the charges will experience a transverse force F. Some external agent, therefore, must exert an equal but opposite force F' to maintain the motion, hence doing work on the system.

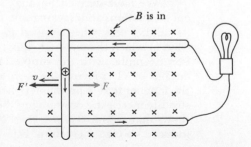

Equation (9-3), which expresses the voltage induced when a wire is carried across a magnetic field, will now be put in a form in which it has more general utility. The arrangement shown in Fig. 9-16 has been redrawn in Fig. 9-17 to show the wire traveling to the left with a velocity v across the magnetic field. The product Lv expresses the area swept out by the moving wire each second. The induced voltage, which we have seen is given by BLv, is therefore equal to the magnetic field intensity multiplied by the rate of change of the area of the magnetic field inside the loop.

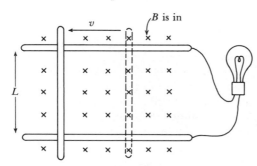

FIG. 9-17

Change of magnetic flux through a circuit. In one second the wire travels a distance v across the magnetic field, "sweeping out" an area Lv. The change in flux taking place in one second, therefore, equals BLv, which is numerically equal to the induced voltage.

An alternate interpretation of BLv is now possible. If we let A represent the sectional area of the magnetic field inside the loop, the product BA becomes a measure of the "quantity" of magnetic field enclosed by the loop. We will define a property that we will call the *magnetic flux* by this product —that is,

$$\phi = BA$$

where ϕ is the magnetic flux in webers, B is the magnetic field intensity in webers per square meter, and A is the area of the enclosed field in square meters. The induced voltage thus becomes equal to the rate of change of the magnetic flux through the loop; that is,

$$V = -\frac{\Delta\phi}{\Delta t} \qquad (9\text{-}4)$$

This is Faraday's Law of Electromagnetic Induction.

We have derived Faraday's law from our knowledge of a particular situation, and, lacking any further information, we can be sure of its validity only when applied to that case. There are, however, many ways in which the magnetic flux through a closed loop can be changed. For example, with the simple loop shown in Fig. 9-14, we have the options of (1) moving the loop to the right or to the left, (2) moving the magnet that is producing the field to the left or to the right, or (3) steadily increasing or reducing the magnetic field intensity. Experience shows that *regardless of the means by which the magnetic field is changed*, the induced voltage is given by Eq. (9-4).

The negative sign in Eq. (9-4) is a mathematical way of describing the direction of the induced voltage. The verbal expression of the direction of the induced voltage is known as Lenz's Rule:

The induced voltage in any closed path is in such a direction that any flux produced by it will be in a direction opposite the change in the inducing flux.

9-6 MAGNETIC FIELD PRODUCED BY AN ELECTRIC CURRENT

The account we have given, strictly speaking, is not a historical report presenting the facts as we now understand them in the order of their discovery. Although experimentation and theory (an explanation we call "understanding") tend to advance together, some remarkable exceptions can be found. A theory may be developed that remains without completely satisfactory experimental verification for many years, as for example the general theory of relativity. Similarly, an experimental observation may remain unexplained—that is, it may stand unrelated to general theories about nature—for many years after it is first announced. Such an experimental observation could only be said to be "ahead of its time" because such an observation makes little immediate contribution to the general theory until more information becomes available to enable some inspired theorist to fit the observation into a general view of nature. An observation of this sort, however, serves as a remarkable stimulus to research.

Oersted's discovery

Such an observation took place in July 1820 when the Danish physicist H. C. Oersted wrote a short note to the learned societies of Europe announcing the discovery of electromagnetism. He had suspected some connection between electricity and magnetism, but possibly because he labored under some incorrect preconceptions as to what to expect, Oersted had never made the appropriate tests. In fact, he first noticed that an electric current can produce a deflection of a compass needle while presenting a demonstration lecture.

Oersted's discovery is described by the sketch in Fig. 9-18. In this apparatus a positive current flows from right to left in a long straight wire. A compass needle held directly under this wire indicates that the current produces a magnetic field into the page; a compass needle above the wire shows that the same electric current produces a magnetic field out of the page.

A somewhat more revealing display of the field due to a current in a long straight wire utilizes the arrangement shown in Fig. 9-19. In this case, a wire is threaded through a hole in a large, horizontal, non-magnetic sheet (a piece of plywood or a sheet of plastic, for example).

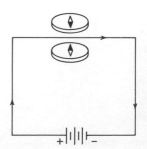

FIG. 9-18

Oersted's experiment. An electric current produces a magnetic field. The point end of the arrow represents the N-pole.

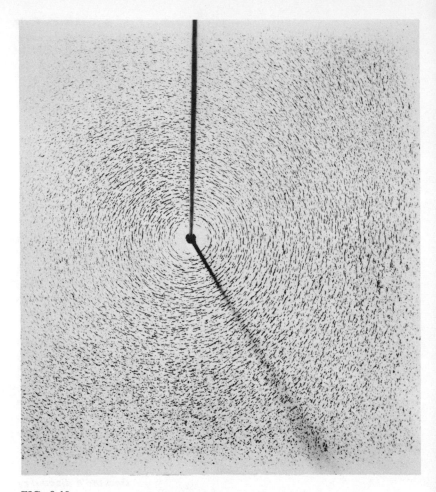

FIG. 9-19

Magnetic field due to steady current in a long straight wire. The magnetic field in a plane perpendicular to such a wire is in the form of closed circular paths around the wire. (*From PSSC* Physics, *D. C. Heath & Co., Lexington, Mass. 1965*)

The battery is located at a substantial distance from the sheet; the wire from one terminal of the battery runs along the floor, straight to the ceiling through the hole in the sheet, and then along the ceiling to the other terminal of the battery. A current of 20 amp or more is established in the wire. If one traces or maps the field in the vicinity of the wire, using the technique described in Sec. 9-2, the field direction from point to point suggests circles surrounding the wire. These circular paths can be seen in the pattern formed by iron filings sprinkled on the horizontal sheet. If one reverses the direction of flow of electric current, the direction of the associated magnetic field will also reverse.

The relationship between direction of current flow and the direction of the magnetic field it produces is described by the following rule.

If one grips a current-carrying wire with his right hand in such a way that his thumb points in the direction of flow of positive current, his fingers will point around the wire in the direction of the magnetic field produced by the current.

As before, we do not need to know whether the actual carriers of charge are positive or negative. The positive terminal of the battery utilized in the arrangement of Fig. 9-19, for example, was connected to the lower end of the wire while its negative terminal was connected to the upper end. We may never know whether this resulted in positive charge traveling out of the surface or negative charge traveling into the surface; either will produce the observed magnetic field in a counter-clockwise direction around the wire.

Experiments show that the intensity of the magnetic field B due to a long straight wire is proportional to the current in the wire I and is inversely proportional to the distance r from the wire to the point in question. The numerical relationship between B, I, and r that emerges from these measurements can be summarized by the equation

$$B = 2 \times 10^{-7} \frac{I}{r} \qquad (9\text{-}5)$$

where B must be expressed in webers per square meter, I in amperes, and r in meters.

Throughout this and much of the previous chapter, our discussion of electric charges has been based on the tentative definition of the coulomb introduced in Sec. 8-3. It should now be pointed out that the system of units which we are using utilizes Eqs. (9-2) and (9-5) to define the ampere; the coulomb is then defined as the amount of charge transported by one ampere in one second.

This nuance regarding the order in which one defines the coulomb and the ampere does not need to change our thinking about either charge or current. Both Eq. (8-1) and Eq. (9-5) correctly state the numerical relationship between the quantities in those equations; it is not important to us which of these equations is the more fundamental.

The direction of a magnetic field at a point in space is the direction indicated by the N-pole of a magnetic compass needle when placed there.

The magnitude of the magnetic field at a point is measured by the force experienced by a charge traveling through that point in a direction perpendicular to the magnetic field according to the equation,

$$F = QvB \tag{9-1a}$$

The direction of the force on a charge traveling transverse to a magnetic field is described by the Q-v-B Rule.

A current-carrying wire of length L transverse to a magnetic field B experiences a force equal to

$$F = ILB \tag{9-2}$$

The magnetic flux ϕ through an element of area A is defined as

$$\phi = BA$$

The induced voltage in any closed circuit is given by Faraday's Induction Law:

$$V = -\frac{\Delta\phi}{\Delta t} \tag{9-4}$$

According to Lenz's rule, the induced voltage in any closed path is in such a direction that any flux produced by it will be in a direction opposite the change in the inducing flux.

A current-carrying conductor possesses a magnetic field that surrounds the conductor. The intensity of the magnetic field at a distance r from a long straight wire is given by

$$B = 2 \times 10^{-7}\frac{I}{r} \tag{9-5}$$

In the following problems, the mass, charge, and ratio of charge to mass of the various particles may be taken to have the following values:

	MASS (kg)	CHARGE (coul)	RATIO (coul/kg)
Electron	9.11×10^{-31}	-1.6×10^{-19}	-0.176×10^{12}
Proton	0.167×10^{-26}	$+1.6 \times 10^{-19}$	$+0.96 \times 10^{8}$
Alpha particle	0.667×10^{-26}	$+3.2 \times 10^{-19}$	$+0.48 \times 10^{8}$

9-A1. Explain the apparent contradiction in the statement, "The magnetic pole near the north pole is a south magnetic pole."

9-A2. Explain why iron filings produce visible lines when sprinkled on a surface located in a region where there is a magnetic field. (See Figs. 9-7 and 9-19.)

9-A3. A proton traveling with a velocity of 2×10^6 m/sec transverse to a certain magnetic field experiences a force of 0.4×10^{-12} N due to its interaction with the field. What is the intensity of the magnetic field through which the proton is traveling?

9-A4. State in your own words the argument that the external effects of an electric current supplied by a given voltage source are independent of the sign of the charge carriers. Are there any exceptions to the argument?

9-A5. The magnetic field intensity in the space between the poles of an electromagnet equals 3.5 wb/m². If a wire loop having a diameter of 0.05 m is suddenly thrust into the field, what will be the change in magnetic flux through the loop?

9-A6. We see in Appendix 7 that electrification and gravitation have many analogous properties. What similarities have you noticed between magnetization and either electrification or gravitation? Why do we not attempt to pursue an analogy between magnetization and either electrification or gravitation?

9-B1. The magnetic field in the space between the pole pieces of a magnet equals 0.4 wb/m². (*a*) What force in newtons will be experienced by a proton traveling transverse to this field with a speed of 3×10^6 m/sec? (*b*) What will be the acceleration of this proton? (*c*) What will be the radius of the path of this proton?

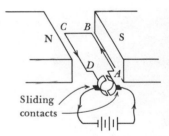

FIG. 9-20

9-B2. Figure 9-20 shows the basic components of an electric motor. A positive current equal to 2 amp flows as shown through a magnetic field of 0.5 wb/m²; $BA = CD = 0.10$ m. (*a*) What forces will BA and CD experience? (*b*) What will be the direction of these forces?

9-B3. A wire 0.2 m long is carried sidewise across a magnetic field of 0.4 wb/m² at a velocity of 0.3 m/sec. (*a*) What are the magnitude and direction of the force on a typical electron in the wire while the wire is in motion? (*b*) If this electron travels the length of the wire because of the action of this force, what energy would it gain? (*c*) How much energy would be given to a coulomb under the action of this force? (*d*) How would your answer be changed if we were to say that the charge carriers are positive instead of negative charges?

9-B4. Figure 9-21 depicts a wire loop 0.10 m long being rotated in a magnetic field whose intensity equals 0.5 wb/m². At the instant shown, BA is moving down and CD is moving up with a velocity of 5 m/sec. (*a*) In what direction will the induced current flow through the lamp? (*b*) What voltage will be impressed across the lamp?

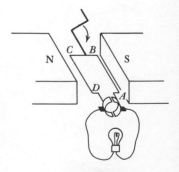

FIG. 9-21

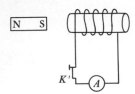

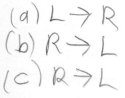

FIG. 9-22

(a) $L \rightarrow R$
(b) $R \rightarrow L$
(c) $R \rightarrow L$

9-B5. An airplane with a wingspread of 30 m is flying north with a velocity of 540 km/hr (= 150 m/sec) in a region in which the vertical component of the Earth's magnetic field equals 0.20×10^{-4} wb/m². (a) What is the induced voltage in the wings of the plane? (b) Does this difference in potential produce charge displacements, currents, heating, etc.? Explain.

9-B6. A voltage of 0.04 volts is induced in a single loop of wire whose radius is 0.06 m. (a) At what rate is magnetic flux changing in the loop? (b) What is the rate of change of the magnetic field intensity?

9-B7. A bar magnet is located in the vicinity of a wire solenoid as in Fig. 9-22. Switch K' is kept closed. Apply Lenz's rule to determine the direction of the induced current in ammeter A (a) when the bar magnet is carried toward the solenoid, (b) when the bar magnet is carried away from the solenoid, and (c) when the magnet is suddenly rotated through 180 degrees.

9-B8. A student places a magnetic compass at a distance of 0.10 m directly north of a long straight wire oriented as in Fig. 9-19. He then adjusts the current in the wire until the compass points in a direction exactly 45° west of north. In which direction is the current flowing through the horizontal sheet? If an ammeter in the circuit reads 10 amp, what is the magnetic field intensity of the Earth at that location?

9-B9. Apply Lenz's rule to the situation depicted in Fig. 9-22 to determine the electric and/or magnetic effects that occur in the secondary coil when (a) the magnet is spun about a vertical axis with the switch K' closed, (b) the magnet is spun about a vertical axis with the switch K' open, (c) the solenoid is carried toward the magnet with K' closed, and (d) the solenoid is carried toward the magnet with K' open.

Electric
Charges in Motion

An abbreviated statement of basic concepts of physics must limit its attention to the kind of fundamental considerations that have characterized the first nine chapters, avoiding the temptation to dwell on the uses that have been made of physical principles. Some practical developments, however, are the very tools that make research in physics possible; hence no statement of the facts of physics would be complete without mention of the means by which these facts are obtained. Accordingly, this chapter will be devoted to a brief look at a number of electrical devices that have played and continue to play a vital role in research. Incidental to our discussion of these devices will be a description of the behavior of charges in electric and in magnetic fields.

10-1 THE TRAJECTORY OF A CHARGED PARTICLE IN A UNIFORM ELECTRIC FIELD

Figure 10-1 shows two objects that have been projected into a region where they experience constant forces in a direction in each case perpendicular to the object's original direction of motion. The first figure shows the familiar situation of an object in the nearly uniform gravitational field of the Earth. The second figure shows a charged particle that has been fired from a "proton gun" into the space between two electrically charged capacitor plates. The two cases have much in common. Each object possesses mass (i.e., each possesses inertia), and thus each object will continue forward at constant speed. Each object will be accelerated in the direction of the force acting on it, one by the nearly uniform force of the gravitational field, the other by the uniform force of the electric field.

Let us consider the motion of a charged particle in a uniform electric field in some detail. Since the force acting on the particle equals qE [Eq. (8-2)], upon applying Newton's Second Law, we have

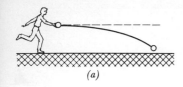

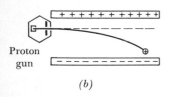

Proton
gun

(b)

FIG. 10-1

Trajectory of an object in
a uniform force field. An
object in a uniform force
field travels in a parabolic
path. (*a*) An object in a
uniform gravitational
field. (*b*) A charged
object in a uniform
electric field.

$$qE = ma$$

or

$$a = \frac{qE}{m}$$

The trajectory of an object projected into a uniform transverse force field can now be determined. We start by treating the motion of the object in the x direction (perpendicular to the force field) and its motion in the y direction (parallel to the force field) separately. Since no force acts in the x direction, the x velocity of the object is constant. Letting V_{ox} represent the constant velocity in the x direction, we have

$$x = v_{ox}t$$

The motion in the downward or y direction is not affected by the forward motion, and since the downward acceleration is constant, the average velocity during any time interval is determined by averaging the initial and final velocities. Letting v_y represent the downward velocity attained in time t, we have

$$v_{y(\text{av})} = \frac{0 + v_y}{2} = \frac{v_y}{2}$$

Since the velocity attained is just equal to the acceleration times the time, the displacement in the y direction becomes

$$y = v_{y(\text{av})}t = \frac{v_y}{2}t = \frac{1}{2}at^2$$

We now have two equations, one describing the motion in the x direction and another describing the motion in the y direction in the same amount of time t. Combining these equations, we have

$$y = \left(\frac{a}{2v_{ox}^2}\right)x^2 \qquad (10\text{-}1a)$$

which shows that the trajectory is a parabola. Introducing the expression for the acceleration of a charged particle in an electric field, we have

$$y = \left(\frac{qE}{2mv_{ox}^2}\right)x^2 \qquad (10\text{-}1b)$$

Trajectories for several specific cases are shown in Fig. 10-2. Trajectory No. 1 represents the path followed by a proton fired from a proton gun operated at 1000 volts into a transverse electric field equal to 8000 N/coul (400 volts across a parallel-plate capacitor whose plates are 0.05 m apart). In traveling forward a distance of 0.10 m, this proton is deflected sidewise a distance of 0.02 m. When the proton emerges from between the plates, it proceeds to travel in a straight line because the electric field intensity suddenly drops to zero.

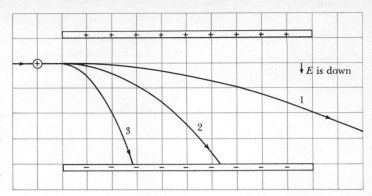

FIG. 10-2

Trajectories of a proton in various uniform
electric fields. Protons fired from a 1000-volt
proton gun enter a uniform transverse electric field. Trajectories 1, 2, and 3 are produced by electric fields of 8000, 40,000, and
200,000 N/coul respectively.

If one increases the transverse force on the proton, it will experience a larger deflection. Trajectories 2 and 3 are the paths followed
when the electric field intensity is increased to 40,000 N/coul and
200,000 N/coul respectively. In the latter two cases, the proton crashes
into the lower deflection plate.

10-2 THE TRAJECTORY OF A CHARGED PARTICLE
IN A UNIFORM MAGNETIC FIELD

Let us next consider the effect of a uniform magnetic field on a moving
charged particle, looking first at the case in which the velocity of the
particle is perpendicular to the field. Such a particle experiences a
force that, by the Q-v-B rule, is perpendicular both to its direction of
motion and to the magnetic field. This force remains constant in
magnitude and perpendicular to the direction of motion of the particle
at all times; therefore, like a stone tied to a string, the particle will
travel with uniform speed in a circular path. Furthermore, unless
it collides with some obstruction or emerges from the field, the particle
will continue to travel indefinitely with uniform circular motion.

The trajectories of a proton injected into various uniform magnetic fields are shown in Fig. 10-3 where the magnetic fields employed

FIG. 10-3

Trajectories of a proton in various uniform
magnetic fields. A 1000-volt proton gun fires
protons into uniform transverse magnetic
fields equal to 0.0183, 0.0915, and 0.4575
wb/m² respectively. The resulting forces were
numerically equal to the electric forces in
the cases described in Fig. 10-2.

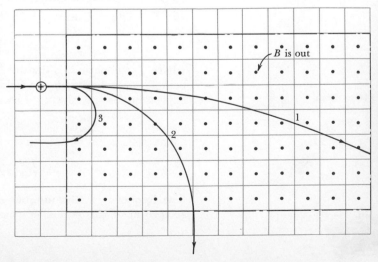

in the three cases are of such strength as to produce forces precisely equal to the electric forces experienced in the three cases shown in Fig. 10-2. The marked differences between the two cases arises, of course, from the difference in the nature of the force—namely, that in the case of the electric field, the force is always constant and in a constant direction (down in the cases shown), whereas in the case of a proton traveling perpendicular to a uniform magnetic field, the force is constant in magnitude but always perpendicular to the instantaneous direction of motion of the proton. For the case of a proton in a uniform electric field, the trajectory is a parabola; in the case of a proton traveling perpendicular to a uniform magnetic field, the trajectory is a circle.

The magnetic interaction force qvB is the centripetal force that produces the centripetal acceleration; that is,

$$qvB = \frac{mv^2}{r} \qquad (10\text{-}2)$$

Two significant conclusions can be reached from this equation. First, upon solving for mv, we secure

$$mv = Bqr \qquad (10\text{-}3)$$

an expression that enables one to infer the momentum mv of a moving charged particle from the radius of its path in a magnetic field.

The second outcome of Eq. (10-2) is a generalization known as the *cyclotron principle*, according to which the frequency (the number of revolutions per second) of a particle in a uniform magnetic field is independent of the velocity of the particle.

To establish this theorem, imagine a charged particle to be injected into a uniform magnetic field. From Eq. (10-2) we see at once that the radius of the circular path of the particle is

$$r = \frac{mv}{Bq} \qquad (10\text{-}4)$$

In one complete revolution the particle will travel a distance equal to $2\pi r$ at a constant speed v. We have, then,

$$2\pi r = vT$$

where T, the period, is the time required for the particle to make one complete revolution in the field. Solving for T and introducing our expression for r, we have

$$T = \frac{2\pi r}{v} = \frac{2\pi}{v}\frac{mv}{Bq}$$

or
$$T = \frac{2\pi m}{Bq} \qquad (10\text{-}5a)$$

The frequency f (the number of revolutions per second) equals the reciprocal of the period T (the number of seconds per revolution).

Hence

$$f = \frac{Bq}{2\pi m} \qquad (10\text{-}5b)$$

This frequency is called the cyclotron frequency of the particle.

Equations (10-5a) and (10-5b) express the period and the frequency, respectively, of a charged particle in terms of B, q, and m, all of which for a given kind of particle in a given experimental arrangement, are, or may be, held constant. In particular, note that these expressions do not involve the magnitude of the velocity v of the particle. This means that a fast particle spends precisely the same amount of time making one complete revolution in the field as a slow particle; a fast particle simply travels in a larger circle than a slow one.

If the vector that represents the velocity of the particle is not perpendicular to the direction of the magnetic field, the trajectory of the particle will be a helix, or "corkscrew." The particle will advance along the field direction with a velocity equal to $v_{\|}$, the component of v parallel to the field. The component of the velocity perpendicular to the field direction, $v_{\perp}$, will play the same role for this motion as was played by v in the previous equations. Hence the frequency of the particle in the field is given by Eq. (10-5b) as before. Thus the frequency of a given charged particle in a magnetic field is independent of both the speed and the direction of the motion of the particle.

10-3 THOMSON'S MEASUREMENT OF e/m

The experimental proof that cathode rays consist of a stream of identical negatively charged particles was carried out by the British scientist Sir J. J. Thomson in 1895. Brief references were made to this important experiment in Secs. 8-7 and 8-9, but because at that point of our discussion we had yet to explain the forces on a moving charge in a magnetic field, the complete discussion was deferred. We now return to this topic, not only because of its importance in the history of physics but also because the apparatus operated on principles that have been utilized in many other experimental arrangements.

The discovery of the electron

Thomson's apparatus was essentially as shown in Fig. 10-4, although it differed somewhat in appearance. It is a cathode-ray tube that combines some of the features of the other cathode-ray tubes described in Sec. 8-7 and shown in Fig. 8-18. Operating on the hypothesis that the cathode rays consist of charged particles, we explain the performance of the device by saying that cathode particles are emitted by the heated cathode and are accelerated toward the positive electrode A. Many of these particles strike the hole in this electrode and make up a very narrow beam of particles, which, in the absence of any other forces, proceed down the tube and strike the face of the

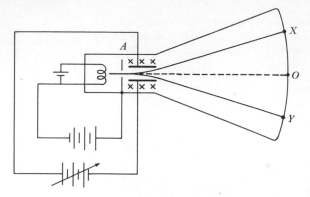

FIG. 10-4

Thomson's *e/m* tube (a modified twentieth-century version). The electric field alone would deflect the moving electrons to *X*; the magnetic field alone would deflect them to *Y*. By adjustment of the relative intensities of these fields, the electrons may travel along a straight path to *O*.

tube at *O*. Fluorescent paint on the inside of the tube may render the point at which they arrive visible.

In a region just beyond the positive electrode, the rays pass through either an electric or a magnetic field or both. The electric field is produced by a battery connected to two parallel metal plates; the magnetic field, into the page in this instance and represented by the crosses, is produced by an electromagnet. The electric field alone would deflect the beam upward to cause the spot to appear on the face of the tube at *X*; the magnetic field alone would produce a downward deflection, perhaps to the point *Y* on the face of the tube. Both fields, acting together, may cancel each other if the magnitudes of the fields are properly chosen.

The first step in Thomson's experiment was a determination of the velocity of the cathode particles. Let us assume, with Thomson, that each particle has the same charge *e* and the same mass *m*. The force on such a particle due to the electric field *E* is given by Eq. (8-2), which in this instance may be written

$$F_E = eE$$

while from Eq. (9-1) the force due to the magnetic field is

$$F_B = Bev$$

The direction and magnitude of both fields are next adjusted to cause the beam to arrive at *O* as though the cathode particles experienced no net force. The electric and magnetic forces therefore are equal but opposite; that is,

$$eE = Bev \qquad (10\text{-}6a)$$

or solving for *v*,

$$v = \frac{E}{B} \frac{\text{m}}{\text{sec}} \qquad (10\text{-}6b)$$

If one substitutes numerical values of the fields into this equation, he secures the velocity of the particles emitted by the cathode.

The fact that various combinations of E and B may be used to direct the beam back to O and that all such combinations yield the same velocity when a given accelerating voltage is employed indicates the validity of the assumption that cathode rays consist of identical charged particles.

The determination of the ratio of charge to mass

The second step in Thomson's experiment is the determination of the ratio e/m. This is achieved by removing the connections to the battery, in which circumstance the particles are deflected by the uniform magnetic field alone. Clearly the path of the particle while in the magnetic field must be circular, for the force due to the magnetic field is at all times perpendicular to its direction of motion.

The force due to the magnetic field constitutes the centripetal force responsible for the departure of the particle from a straight line. We have already seen that, for a charged particle traveling through a magnetic field,

$$Bev = \frac{mv^2}{r}$$

Solving this equation, we find

$$\frac{e}{m} = \frac{v}{Br} \qquad (10\text{-}7)$$

The velocity v of the particles, of course, had been determined in advance, using Eq. (10-6b).

In Fig. 10-5 we show the path followed by the particle. If the magnetic field is uniform over the region indicated and zero elsewhere, the path A-C is part of a circle, whereas the path C-Y is a straight

FIG. 10-5

Magnetic deflection of an electron beam in Thomson's experiment. The path A–C is a circular arc of radius r; C–Y is a straight line. The distance L is known and the distance OY can be measured on the face of the tube. Making a scale diagram, one can determine the radius r.

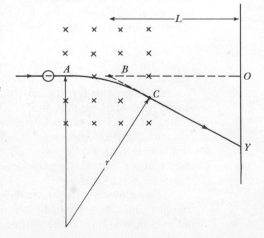

line. Now if one makes a careful drawing to scale, one may determine the radius of the path of the particle while in the magnetic field.

With knowledge of the velocity of the particles, the strength of the magnetic field, and the radius of their path in the field, one may substitute into the preceding equation and compute a value of e/m. Measurements indicate that this ratio has the following value:

$$\frac{e}{m} = 1.7589 \times 10^{11} \text{ coul/kg} \tag{10-8}$$

The particles streaming from the cathode thus all possess the same ratio of charge to mass. We call these particles "electrons" and credit Thomson with their discovery.

10-4 THE ACCURATE DETERMINATION OF ATOMIC MASS

The essential steps in the logic by which nineteenth-century chemists were led to the atomic-molecular view of matter were presented in Chapter 7. The reader was asked to accept the conclusion of these and later observers that the masses and volumes manifested by materials before and after chemical reactions can be explained only by assuming that matter consists of molecules, which in turn consist of simple combinations of a relatively small number of kinds of atoms. These experiments involved observations of matter in bulk, however; in only a few rather isolated occasions did nineteenth-century chemists have the satisfaction of observing an event that involved an individual atom or molecule. The evidence for the existence of atoms and molecules seemed quite indirect at that point in the argument.

The Millikan experiment and the Thomson e/m experiment, on the other hand, both dealt with one kind of submicroscopic particle, the electron, in such a way that its effects were isolated from effects due to other submicroscopic particles. In this section a remarkable instrument will be described that is capable of achieving for atoms the kind of isolation by types that the oil-drop and the e/m experiment achieved for electrons.

The discovery of positive rays

Let us consider the discharge tube shown in Fig. 10-6. The anode is located at one end of the tube, while the cathode, with a small hole at its center, is located near the middle of the tube. A small amount of gas at a pressure of a few thousandths of atmospheric pressure is admitted to the tube, and, under the action of an intermittent high voltage impressed across the electrodes, the tube emits a glow characteristic of this gas.

The glow, for the most part, is confined to the region between the electrodes, but a faintly luminous beam is visible beyond the hole in the cathode. This beam is due to a stream of rapidly moving positive

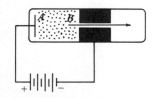

FIG. 10-6

A positive ray tube. A high voltage across the region *AB* causes the residual gas in the region to glow. A tiny streamer may be seen in the space beyond *B*, suggesting that positively charged particles have been projected into that region.

ions—atoms or molecules that have each lost one or more electrons—created by collision in the space between the electrode and shooting through the hole in the cathode. The stream, first observed by the German scientist Goldstein in 1886, has been called a *positive ray*. With a few improvements, this device serves as an ion source. If the gas in the tube is hydrogen, it is a proton gun, to which reference has been made in Sec. 10-1.

The determination of q/M of positive ions

In 1911 Sir J. J. Thomson developed a modified version of the apparatus used by him to measure the ratio of charge to mass of the electron as described in the previous section. The modified version was specifically adapted to measure the ratio of charge to mass (q/M) of positive ions. By assuming that each ion emerging from the ion gun possesses a positive charge that is numerically equal to some easily deduced multiple of the charge on an electron, it became an instrument for determining the mass of the ion itself.

Thomson's first great success with this *mass spectrograph*, as it is now called, was to demonstrate that neon, as found in nature, consists of two isotopes: a very abundant one whose mass number is 20 amu and another, now known to make up 8.82 percent of natural neon, with a mass number of 22 amu.

Many improved mass spectrographs were built in the years immediately following Thomson's landmark discovery. F. W. Aston, working in Thomson's laboratory, designed the first truly precise mass spectrograph in 1927, culminating in an instrument capable of determining the relative masses of ions to 1 part in 100,000.

One of the more straightforward instruments of this kind was built by K. T. Bainbridge in 1937. In this instrument (Fig. 10-7), ions from an ion gun enter a region between S_1 and S_2, where they are exposed to transverse electric and magnetic fields. According to Eq. (10-6b), all ions that manage to travel through this region in a straight line to reach the slit S_2 are assured of having the same velocity, regardless of their masses. In the region beyond S_2, the emergent ions encounter a uniform magnetic field; clearly the heavier ions follow a trajectory of much larger radius in that region than the lighter ions. The record made by the ions when they encounter the photographic plate thus becomes a record of their relative masses.

One can gain an appreciation for the effectiveness of a modern mass spectrograph by examining Fig. 10-8. When this spectrogram was produced, the ion gun was ejecting a variety of ions each with

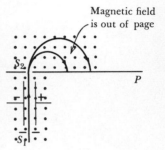

FIG. 10-7

The Bainbridge mass spectrograph. Positive ions enter the slit S_1 but those emerging from S_2 are limited to ions of one velocity ($=E/B$). Beyond S_2, ions travel in circular paths whose radii depend on their masses and produce a "mass spectrum" upon striking the plate P.

FIG. 10-8

A mass spectrogram. This sketch, based on a mass spectrogram by Harry Duckworth, demonstrates the remarkable ability of the mass spectrograph to distinguish between ions having slightly different masses. All of the indicated particles have the same mass number.

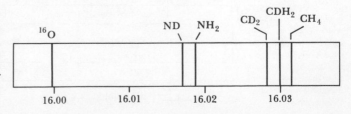

a mass number of 16. One can see, however, that these ions do not all have the same mass. Measurements of such a plate indeed show that the mass of the ND ion equals 16.017176, whereas that of the NH_2 ion equals 16.018724. These two ions have masses that differ by only 1 part in 10,000 and are stated to eight significant figures, implying accuracy to 1 part in 10,000,000.

An abridged list of the masses of the isotopes appears in Appendix 8. As in all such tables, the mass in each case represents the mass of the neutral atom (nucleus plus all of its planetary electrons). Conversion to atomic masses in kilograms, it will be recalled, requires only that one divide the mass in atomic mass units by Avogadro's number— that is, by 6.02252×10^{26} [Eq. (7-1)].

10-5 HIGH-VOLTAGE MACHINES

With the discovery of nuclear transmutation by Ernest Rutherford in 1919 (to be discussed in Sec. 15-4), it became apparent that research in nuclear structure depends on the availability of high-speed nuclear particles, particularly protons, heavy hydrogen nuclei (deuterons), and helium nuclei (alpha particles). Although some nuclear particles are ejected from certain nuclei in radioactive decay, the radiation from such sources travels in all directions and consists of a mixture of particles of different kinds and different energies. The need, then, was for a device that would provide a collimated (unidirectional), monoenergetic, controllable beam of whatever nuclear particles one desires to employ.

Early particle accelerators consisted of a positive ray tube that served as an ion source which ejected ions into an accelerating tube

FIG. 10-9

An early type of accelerating tube. Protons are injected at one end of a series of coaxial cylinders across which a potential difference as large as a million volts has been placed. The cylinders distribute the voltage into a series of steps to prevent sparking.

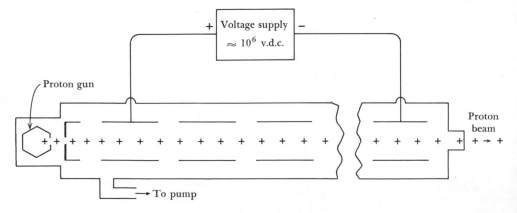

across which a potential difference was established by a battery- or transformer-operated voltage source. The accelerating tube consisted of a series of cylindrical tubes, each of which was maintained at 50,000 to 100,000 volts higher potential than the next in the sequence. This arrangement minimized the tendency of loss of voltage due to charge leakage but, in addition, because of the shape of the electric field, tended to guide the ions down the center of the tube, thus preventing them from getting lost by striking the walls of the cylinder.

One of the more interesting high-voltage sources was an electrostatic generator invented by R. J. Van de Graaff in 1931. In this device electric charges are sprayed on a rapidly moving belt which transports them into a hollow metal sphere. Since the belt operates continuously, the charge on the sphere, and hence its potential relative to other positions in its vicinity, continues to increase until losses due to conduction through the surrounding medium become equal to the rate of transport of charge to the sphere. Modern versions of this machine are capable of producing a potential difference of 10 million volts or more.

The energy and velocity of accelerated particles

When a charged particle falls through a difference of potential, its loss of potential energy is accompanied by an equal gain in its kinetic energy. If the difference in potential is very large, the velocity acquired may approach the velocity of light, in which case the particle will show a "relativistic" increase in its mass. In addition, at such high velocities many of the equations we have become accustomed to using are no longer valid.

Relativistic effects begin to become significant at a velocity of approximately one-tenth the velocity of light. Accordingly, we will classify particles as "nonrelativistic" or "relativistic," depending on whether the velocity is less than or greater than 3×10^7 m/sec. Our calculations in this chapter will be restricted to nonrelativistic particles; relativistic effects will be discussed in Chapter 14.

From the definition of potential difference we know that a particle whose charge equals q coulombs will gain an energy W given by

$$W = qV \qquad (8\text{-}5)$$

in falling through a difference of potential of V volts. For nonrelativistic particles, the gain in kinetic energy equals $\frac{1}{2}Mv^2$. Hence

$$qV = \frac{Mv^2}{2}$$

Solving for v, we find

$$v = \sqrt{\frac{2qV}{M}} \qquad (10\text{-}9)$$

A graph showing the energy and the velocity acquired by a proton upon falling through differences in potential in the range from 0 to 4

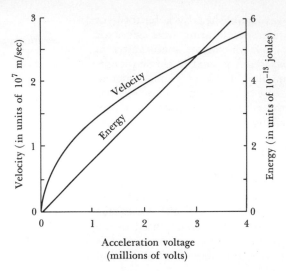

FIG. 10-10

Energy and velocity versus accelerating voltage for protons.

million volts is shown in Fig. 10-10. In order to relate the data presented in this figure to the preceding equations, let us compute the energy and the velocity acquired by a proton ($M = 0.167 \times 10^{-26}$ kg; $q = +1.6 \times 10^{-19}$ coul) in falling through a difference of potential of 500,000 volts. The energy will be

$$\text{Energy} = qV = (1.6 \times 10^{-19}) \times (0.5 \times 10^{6})$$
$$= 0.8 \times 10^{-13} \text{ joules}$$

The energy of the proton is shown on the graph by a colored line; notice that the data point we have calculated lies on this line.

Next we may compute the velocity of the proton that has gained this amount of energy in the form of kinetic energy. From Eq. (10-9) we have

$$v = \sqrt{\frac{2qV}{M}} = \sqrt{\frac{2 \times 8.0 \times 10^{-14}}{0.167 \times 10^{-26}}} = 9.8 \times 10^{6} \text{ m/sec}$$

The velocity of the proton at various voltages is shown in Fig. 10-10 by a black curve; again it should be noted that our computed data point lies on the curve. The curve is terminated at 4 million volts since the proton nearly reaches the relativistic speed of 3×10^{7} m/sec when it falls through this difference of potential.

The cyclotron

The cyclotron, invented by E. O. Lawrence in 1932, is perhaps the most famous of the many high-voltage machines that have resulted from the search for high-energy particles. Although the cyclotron has been superceded by machines that produce particles of higher energy (in fact, the model we will describe is limited to the nonrelativistic range), its performance is based on many of the principles utilized in other, more sophisticated machines.

The essential components of the cyclotron, as shown in Fig. 10-12, are (1) a high-voltage alternating voltage supply, (2) a large electromagnet, (3) an evacuated chamber that contains (4) a pair of "dees," within which the particles from (5) an ion gun are accelerated.

The dees secure their name from the *D*-shape of this accelerating unit in early cyclotrons. To visualize their appearance, one might consider a hollow brass (hence nonmagnetic) cyclinder having a dia-

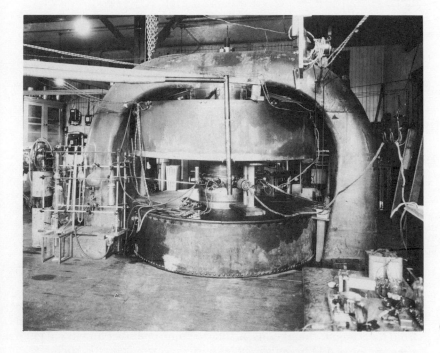

(a)

FIG. 10-11

Early cyclotrons. (a) This model, built in 1932, had dees with a diameter of 27 in. (b) A close-up view of the 60-in. dees of a later model. The deuteron beam shown emerging from a window in the accelerating chamber traveled 5 ft in air.
(*Lawrence Berkeley Laboratory, University of California, Berkeley, Calif.*)

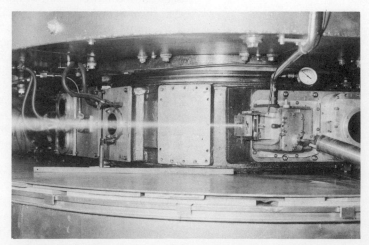

(b)

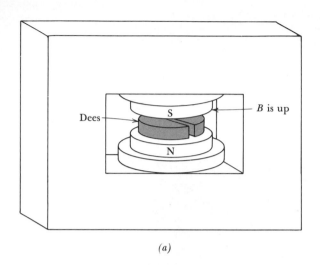

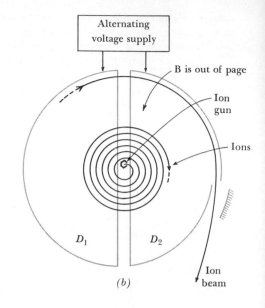

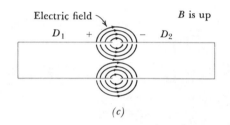

FIG. 10-12

The cyclotron. (a) General plan showing dees in place in the magnetic field.
(b) Top view. (c) Side view showing
the electric field at a given instant.

meter of a few feet and a height of a few inches which has been cut
into two parts along a diameter. These dees are located inside the
evacuated chamber and are electrically insulated from each other and
from the vacuum chamber itself. The vacuum chamber and its enclosed
dees are located between the poles of the electromagnet in such a way
that the magnetic field is perpendicular to the plane surfaces of the dees.
Terminals from the power supply are attached to each of the dees. The
ion source is located near the center of the arrangement in such a way
that it can eject protons (or other charged particles) across the space
between the dees and into the hollow space of the other dee.

The high-voltage supply produces an alternating potential
difference, possibly of 50,000 volts, between dee D_1 and dee D_2 at a
frequency of thousands of alternations per second. The electric field
in the gap between the dees thus changes its direction at this high
frequency, being alternately directed from D_1 to D_2, then from D_2
to D_1, etc.

At a given instant, the electric field is as shown in Fig. 10-12(c).
It is significant that the electric field is concentrated almost entirely in
the space between the dees because this means that a charged particle
inside one of the dees experiences almost no electric force. On the
other hand, should an electrically charged particle arrive at the interface between the dees at the appropriate instant, it would be drawn

into the opposite dee, gaining an energy equal to qV; a proton, for example, would gain an energy of $1.6 \times 10^{-19} \times 50,000$ or 8.0×10^{-15} joules.

Let us consider the "life history" of an individual proton in one of the bursts emitted by the proton gun. This proton is fired from the hollow space of D_1 into the hollow space of D_2, falling through the difference of potential V and gaining an energy qV. Inside D_2, however, it experiences no electric field, but the magnetic field causes it to travel in a circular path of radius r given by

$$r = \frac{mv}{Bq} \qquad (10\text{-}4)$$

This proton will return to the space between the dees after making a half-revolution, having spent a time equal to a half-period Δt, which, from Eq. (10.5a), equals

$$\Delta t = \frac{\pi m}{Bq}$$

If the frequency of alternation of the difference of potential has been adjusted properly, this proton will emerge from D_1 at just the instant the field has reversed direction; hence it will fall through the difference of potential a second time, gaining an additional energy qV. Entering D_1, the proton will travel in a slightly larger orbit, make another half-revolution, and gain energy qV a third time. If this process should continue to occur, the proton should spiral out from the center, gaining energy each time it crosses from one dee to the other.

The cyclotron principle describes the aspect of ion behavior that makes this remarkable instrument possible. Since, according to this principle, the time spent by a charged particle in making a half-revolution is independent of the velocity, a voltage supply whose frequency is adjusted to the cyclotron frequency of the particle being accelerated will always provide an accelerating field at the instant the particle reaches the space between the dees. Although V may be only 50,000 volts, by allowing a proton to pass from one dee to the other many times, it may gain a total energy equivalent to falling through a potential difference of millions of volts. For example, if a proton should make 50 revolutions in the field, it would cross the space between dees 100 times and gain an energy equivalent to 5 million volts. An alpha particle (the nucleus of a helium atom) whose charge equals $+2e$, would gain twice as much energy, other things being equal.

10-6 THE DETECTION OF MOVING CHARGED PARTICLES

The study of nuclear events requires the availability of instruments capable of detecting charged particles and of measuring their energy. In situations in which nuclei or subnuclear particles disintegrate, we must analyze their transmutation into other nuclei and into energy in other forms. As in any situation in which we examine events involving submicroscopic particles, much indirect information will be employed; lacking the means of seeing the particle, we must be content with seeing some effect it is able to produce. Although we will not attempt an exhaustive survey of all the ingenious devices that have contributed to our present understanding, we will examine the performance of several typical particle detectors so that the reader can be aware of the kind of information that has served as the basis for our conclusions. Where necessary in later chapters, we may elaborate on some or all of these instruments as they have been employed in particular experiments.

Most instruments that detect charged particles make use of the fact that a high-speed charged particle traveling through a gas produces a path of ions by collision with the normally uncharged molecules that make up the gas. A typical ionizing collision, however, is not a direct collision between the particle and the molecule that is dissociated in the encounter; ionization occurs, instead, because of the effect of the electric field possessed by the moving particle and/or the magnetic field produced by its motion. The region of influence of the moving particle is therefore hundreds of times larger than the physical size of the particle itself.

The ionization chamber

An ionization chamber is used to measure the total ionization produced by a given radioactive sample and may take the form shown in Fig. 10-13. This model is essentially an electroscope whose discharge rate serves as a measure of the total amount of ionization produced by a radioactive specimen located in the chamber M. At the start of a given observation, the gold-leaf system is charged from a dc voltage source of approximately 400 volts, under which condition the gold

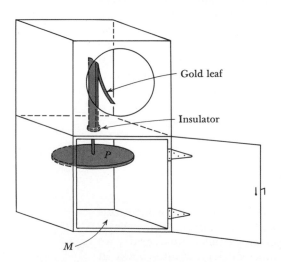

FIG. 10-13

An ionization chamber. Radioactivity in M is measured by the rate of fall of the charged gold leaf.

Gold leaf

Insulator

P

M

leaf may deflect so as to make an angle of approximately 45 degrees with respect to the vertical support bar on which it is mounted. Ions either are drawn toward or repelled by the plate P, depending on whether they are positively or negatively charged; those drawn to the plate collect there at a rate proportional to their abundance in the space M. This plate is in electrical contact with the leaf system; therefore the amount of ionization can be inferred from the rate of fall of the gold leaf as observed by a telemicroscope. Obviously, the ionization chamber measures only the total ionization and cannot distinguish between different kinds of ionizing particles; it is capable, nevertheless, of yielding valuable information regarding the gross behavior of a radioactive specimen. It is also capable of detecting general radiation from exterior sources; one model, for example, is worn by persons whose work requires that they be in the presence of potentially hazardous radioactive sources.

The cloud chamber

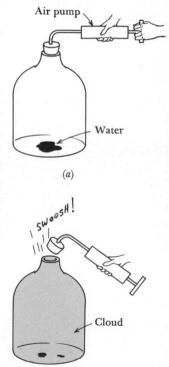

A cloud chamber enables one to see the ion path left by a moving charged particle; in fact, the tracks left by different moving charges differ so greatly that a skilled observer can determine by inspection which type of particle was responsible for a particular track. Photographs of the tracks of certain charged particles are shown in Figs. 15-8 and 15-10.

In order to understand the performance of a cloud chamber, let us digress to describe an experiment that the reader can perform. The experiment utilizes a gallon glass bottle containing a few cubic centimeters of water and closed by a removable rubber stopper with a connection to an ordinary tire pump (Fig. 10-14). The effect that we will describe is enhanced if a small amount of smoke is blown into the bottle before closing it. If one increases the pressure of the air in the bottle with the hand pump, then suddenly removes the stopper, thereby permitting the air to escape, a cloud of water vapor will be seen in the bottle. This cloud will disappear when the pressure of air in the bottle is increased again.

The formation of the cloud in the bottle suggests that the air spontaneously cools when it expands. This change of temperature is explained by breaking the process down into three stages:

1. When air escapes from the bottle, it does work on the surrounding air because it exerts a force through a distance.

2. Because the escaping air has done work, it must have lost an equal amount of energy.

3. Having lost energy, the temperature of the air must decrease.

The density of the cloud depends on the presence of dust or other foreign matter in the expanding air; the smoke particles apparently play the role of providing tiny structures on which water droplets can condense. In the absence of smoke or dust in the bottle, the air may supercool and no water vapor may condense.

FIG. 10-14

A simple cloud chamber experiment. When air that has been pumped into the bottle (*a*) is permitted to escape, a cloud of water vapor is produced (*b*).

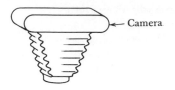

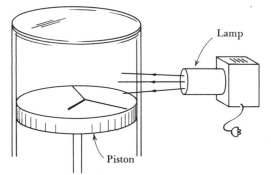

FIG. 10-15

A cloud chamber. High-speed charged particles leave vapor trails that are photographed for later analysis. Examples are shown in Fig. 15-8 and Fig. 15-10.

A cloud chamber, as used in the study of particle tracks, is shown in Fig. 10-15. In contrast to the simplified model described above, the air is confined in a cylindrical chamber by a piston and is caused to expand by suddenly moving the piston downward. The interior of the chamber is as free of dust and smoke as one can manage. However a radioactive source may be in the vicinity with the result that ionizing particles may constantly be passing through the chamber. When the chamber is expanded, water condenses on the ions left in the wake of individual ionizing particles. If the interior is illuminated by a lamp at the side of the chamber, these tracks are clearly visible. A camera mounted as shown in the figure enables one to secure a photograph of any nuclear event that happens to occur in the chamber.

The bubble chamber

The cloud chamber has been discussed in some detail because of its importance in the discovery of many of the particles and phenomena that form the subject matter of this book. Since 1960, however, this historic device has been largely superceded in research by the bubble chamber.

A bubble chamber is a large vessel containing a liquid at its boiling temperature; thus the temperature of a bubble chamber containing hydrogen must be maintained at 21.4°K ($= -252.8$°C). The liquid is enclosed by a piston, which permits one suddenly to change the pressure under which the liquid is confined. If the pressure of the liquid is suddenly decreased, the liquid immediately boils. Boiling, however, preferentially occurs on any ion paths that may be present. Therefore a photograph taken at just the instant of the expansion will reveal the path followed by any ionizing particle that may have passed through the chamber.

The principal advantage of the bubble chamber over the cloud chamber is that there are many more particles per unit volume with which an incoming particle can interact. Thus a high-energy particle that might pass completely through a cloud chamber with very little loss of energy may be brought to rest in a bubble chamber and in doing so reveal much more information about itself. A bubble chamber photograph has been shown in Fig. 7-1.

The Geiger-Muller counter

The Geiger-Muller counter is a device that enables one to count the number of charged particles that enter a small counter tube during a given time interval. The typical arrangement in Fig. 10-16 shows a cylindrical metal tube, along the axis of which is a thin wire held in place by insulating plugs at the two ends of the tube. The tube contains gas, typically argon, at reduced pressure, possible at one-tenth of atmospheric pressure. An electric field is maintained between the wire and the cylindrical tube by means of an external battery.

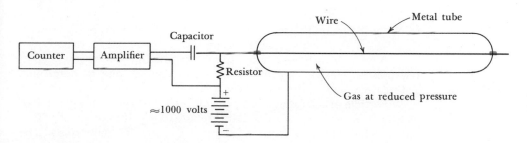

FIG. 10-16

A Geiger-Mueller counter. A high-speed charged particle passing through the tube produces an avalanche of ions. The resulting electric pulse is amplified and registered by the counter.

When an ionizing particle passes through the tube, many ion pairs are formed along its path. The negative ions are drawn to the central wire, whereas the positive ions are swept to the cylindrical sheath by the electric field provided by the battery. As a result, a small pulse of electric current flows from the voltage supply, which, with suitable amplifiers, is used to actuate a counter.

In many counter tubes, the production of even a single ion pair will initiate an avalanche of additional ions; such a tube yields an accurate count of the total number of particles entering the tube irrespective of their ionizing qualities. Another model, the proportional counter, yields a pulse that is proportional to the total number of ions produced by the ionizing particle. Such an instrument can give detailed information about the nature of the radiation to which the counter tube is being subjected.

SUMMARY   The trajectory of an electric charge in a uniform electric field is a parabola.

The trajectory of an electric charge in a uniform magnetic field is a circle if the velocity vector is perpendicular to the magnetic field and a helix otherwise. The radius of curvature and the momentum of the particle may be expressed using the basic condition

$$Bqv = \frac{mv^2}{r} \qquad (10\text{-}2)$$

The *cyclotron principle* states that the frequency of revolution of a charged particle in a uniform magnetic field is independent of the particle's velocity and is given by

$$f = \frac{Bq}{2\pi m} \qquad (10\text{-}5b)$$

Thomson's e/m experiment utilized mutually perpendicular electric and magnetic fields that provided equal but opposite forces to the cathode rays. Thus, assuming the rays to be particles of mass m and charge e, we write

$$eE = Bev \qquad (10\text{-}6a)$$

from which one may determine their velocity. Introducing this velocity into Eq. 10-2, one may solve for the ratio of charge to mass (e/m).

A *mass spectrograph* is a device that, by sending ions through combinations of electric and magnetic fields, enables one to determine the relative mass of the isotopes.

Particle accelerators may provide high energy to charged particles by causing them either to fall through a large difference of potential or to fall through a modest difference of potential many times. The *cyclotron* may provide particles with velocities up to one-tenth the velocity of light.

Three kinds of charge dectectors have been described:

1. Ionization chambers, which measure the total ionization produced by the radiation without regard for the constituency of the radiation.

2. Cloud and bubble chambers, which show individual tracks due to ions produced by moving charged particles.

3. Geiger-Muller counters, which count the particles that produce ionization in a small metal tube.

QUESTIONS AND PROBLEMS

10-A1. What are the most important distinctions between the path traversed by an ion in an electric as opposed to its path in a magnetic field?

10-A2. State the "cyclotron principle."

10-A3. How did the measurement of the ratio e/m of "cathode rays" lead to a determination of the mass of an electron?

10-A4. Your roommate tells you that Thomson's "*e/m*" experiment did not constitute the discovery of an electron. Rather, he says, Thomson only proved that cathode rays consist of clusters having the same "charge concentration," perhaps a bit like showing that 2 kg of water weighs twice as much as 1 kg. What can you say to that?

10-A5. What are "positive rays?"

10-A6. The Bainbridge mass spectrometer utilizes a device called a "velocity selector," which has the property of transmitting ions of only one velocity irrespective of their mass. Make a sketch of this device and tell how it works.

10-A7. Write an expression relating the voltage used to accelerate an ion and the velocity it should attain.

10-A8. To what extent is it correct to say that the vapor trails left by high-flying airplanes correspond to the tracks left by particles in a cloud chamber?

10-B1. Assume that the plates shown in Fig. 10-2 are 0.10 m wide and 0.05 m apart and that trajectory No. 1 was produced when the electric field intensity was 8000 N/coul. (*a*) What was the voltage impressed across the plates? (*b*) What force and acceleration were experienced by the proton while in transit through the field?

10-B2. Given that the protons entering the field in Fig. 10-2 have emerged from a 1000-volt proton gun and that the electric field equals 40,000 N/coul (trajectory No. 2), compute (*a*) their velocity as they enter the field, (*b*) the time required for a proton to travel forward 0.05 m, (*c*) the downward velocity attained in this amount of time, and (*d*) the deflection of a proton in traveling forward 0.05 m.

10-B3. Figure 10-3 depicts protons that have been fired from a 1000-volt proton gun into a transverse magnetic field. (*a*) What is the velocity of the protons as they leave the gun? (*b*) What is the radius of trajectory No. 1, where the magnetic field intensity is 0.0183 wb/m²?

10-B4. Assume that a radioactive atom emits a proton with a velocity of 1×10^7 m/sec in a direction transverse to a uniform magnetic field whose intensity equals 0.80 wb/m². (*a*) What is the radius of curvature of the trajectory of this proton? (*b*) How much time is required for the proton to make one complete circular orbit?

10-B5. Solve the preceding problem, replacing the word "proton" with the word "electron" wherever it appears.

10-B6. Assume that an electric field of 100,000 N/coul deflects a beam of electrons to some point X as in Fig. 10-4 but that this beam can be brought back to O when a transverse magnetic field of 0.01 wb/m² is employed. (*a*) Describe the relative directions of the electric and magnetic fields when this occurs. (*b*) What is the speed of the electrons?

10-B7. The region $S_1 - S_2$ in the Bainbridge mass spectrometer serves to limit the ions that pass through S_2 to a very narrow velocity range. (*a*) If the electric field equals 200,000 N/coul and the magnetic field equals 0.5 wb/m², what will be the velocity of ions that are able to pass straight through from S_1 to S_2? (*b*) Does your answer depend on the mass of the ion? (*c*) Does your answer depend on the charge of the ion?

10-B8. Assume that protons emerging from a cyclotron have a velocity of 2.4×10^7 m/sec. If the radius of the dees equals 0.30 m, what must be the intensity of the magnetic field in the region in which the dees are located?

10-B9. We will see in Chapter 15 that an alpha particle (the nucleus of a helium atom) at one time or another has been regarded as (1) a cluster consisting of four protons plus two electrons and as (2) a cluster of two protons and two "neutrons." Using mass data in Appendix 8, compare the mass of each of these clusters to the mass of an alpha particle (= 4,00150 amu).

The Wave Nature of Light

In the development of theories regarding our physical surroundings, scientists have made extensive use of models, consisting of physical structures from familiar experiences that constitute our image of mechanisms that cannot be seen. Atoms are compared to tiny solar systems, and molecules are alleged to resemble marbles connected to each other by springs.

Historically the development of a model of light has been one of the most difficult problems that physicists have attempted. In retrospect we have learned that the reason is because light is different from other aspects of our physical surroundings; in fact, it is now conceded to be impossible to fabricate a completely satisfactory model of light from facets of our macroscopic physical environment.

Although light is the agency of vision, one cannot "see" light nor any part of the medium that transmits it. We can only see the effects. We look at a lamp—we see a lamp; the nature of the medium that enables us to see the lamp remains as mysterious as ever.

Every optical phenomenon known to scientists prior to about 1890 can be explained by using the wave model. Some additional experimental studies from 1890 to 1914 resulted in extensive modifications of the simple-wave theory and will be discussed in Chapter 12. In this chapter we will limit our attention to the arguments that led to the initial acceptance of the simple-wave theory over the primitive-particle theory of light.

11-1 THE BEHAVIOR OF PARTICLES

Newtonian mechanics deals almost exclusively with the analysis of the behavior of particles under the action of identifiable forces. As we examine the merits of a particle model of light, let us review some of the properties that can be ascribed to particles:

A particle in motion will continue to travel with constant velocity unless acted upon by an external force.

If a force acts upon a particle, the particle will experience an acceleration proportional to that force and in the direction of that force.

If a particle is in the vicinity of another object that possesses mass, the objects will experience equal but opposite gravitational forces of attraction proportional to the product of the masses of the objects and inversely proportional to the square of their distances apart.

If two particles interact with each other in any way whatsoever, the change in the momentum of one is equal to the negative of the change in momentum of the other.

A particle can travel at any speed whatsoever;* when in motion in any inertial frame, a particle possesses an energy relative to that frame equal to $\frac{1}{2}mv^2$.

A primitive-particle model for light proposes that light consists solely of pelletlike objects that conform to all these laws. Should light fail to conform to any of these principles, the model has failed and must be modified or abandoned.

Let us consider a particle in free space traveling with uniform speed in a straight line. At some point in its motion we imagine a force to act on the particle for an instant, after which it continues as a free particle. The velocity of the particle after the force has been applied, however, will be different from the initial velocity.

In order to relate this to a situation close to our experience, we will contrive an arrangement in which a "particle" of significant size experiences a force of short duration. Let us imagine an elevated region, like a low stage, at the center of a gymnasium. Let the edge of this elevated region be skirted by a sloping floor so that if a bowling ball is rolled from any region of the gymnasium, the ball will roll up the skirt and, given sufficient speed, onto the stage. While on the skirted edge, the ball will experience a steady force down the skirt. It should be noted in the discussion that all angles are to be measured relative to a line drawn perpendicular to the interface between the two regions, the regions in this case being the gymnasium floor and the horizontal surface of the platform. We can predict the following behavior:

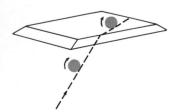

FIG. 11-1

The effect of a short-term force on a moving object. We imagine a bowling ball to be rolled across a smooth floor and to encounter a "skirt" around a raised platform. We will use this situation to develop a particle model of light. "Particles of light" passing from air into glass, for example, are assumed to encounter a zone where they experience a force, hence a change in velocity.

1. If the ball approaches the interface at low speed, it will roll part way up the skirted region and back down again—that is, the ball will be "reflected" back to the gymnasium floor. If the ball encounters the interface obliquely, the angle of incidence (Fig. 11-2) will be equal to the angle of reflection. No kinetic energy will be lost, however, so the speed of the ball will be the same after its encounter with the skirted edge of the platform as it was before.

* However, in Chapter 14 we will show that the theory of relativity places an upper limit of 3×10^8 m/sec on the speed of any particle and introduces a more general expression for kinetic energy.

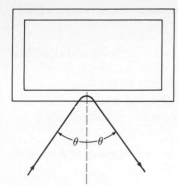 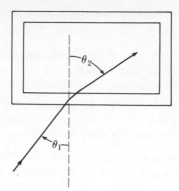 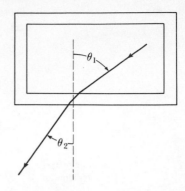

FIG. 11-2

Reflection of a "particle." The bowling ball when rolled at low speed toward the skirted platform is reflected back to the gymnasium floor. Its angle of incidence equals its angle of reflection.

FIG. 11-3

Refraction of a "particle." When the bowling ball is rolled at a high speed, it will roll up the skirt and onto the platform. Its speed will decrease, since it loses energy. The "angle of refraction" (θ_2) is greater than the angle of incidence (θ_1).

FIG. 11-4

Refraction of a "particle." When the bowling ball rolls from the platform to the skirt, it is never reflected; instead its path shows refraction, with the angle of refraction less than the angle of incidence.

2. If the initial speed of the ball is sufficient for it to roll onto the platform, it will experience both a decrease in speed and a change in direction (Fig. 11-3). Because the direction of motion of the bowling ball has changed, we say that it has been "refracted." The angle of refraction will be greater than the angle of incidence—that is, the line of motion is bent away from the normal.

We now reverse the situation. We stand on the platform and roll the ball toward the skirted edge, where it rolls down the skirt to the floor. The following behavior will be observed:

1. Rolled from any point on the platform to the skirt, the ball will always roll down to the floor level and its speed will increase. It will never be reflected back to the platform.

2. When the ball rolls from the platform to the gymnasium floor, it changes its direction in such a way that the angle of refraction is less than the angle of incidence; the line of motion is bent toward the normal (Fig. 11-4).

11-2 THE BEHAVIOR OF WAVES

The behavior of waves has been treated in some detail in Secs. 5.5 to 5.6. In those sections we were dealing with water waves traveling across the surface of a ripple tank. Several characteristics were identified:

Waves of a given wavelength in a given medium travel with a well-defined velocity but may experience a change in velocity and wavelength when they enter a different medium:

The frequency of a wave (the number of waves passing any point in the medium each second) remains unchanged when the wave enters a different medium;

The wavelength of a wave is related to its frequency and velocity in any medium by the equation $v = f\lambda$:

No substance is carried forward with the wave;* only the "waveform" travels. Parts of the medium affected by the wave simply vibrate with the frequency of the source as the wave goes by.

Let us consider a sequence of straight wave fronts traveling through a given medium. From our discussion in Chapter 5, we may recall the following conclusions as to the behavior of these waves:

1. If a sequence of wave fronts encounters a straight obstruction, the waves are reflected without change of wavelength or frequency but in such a way that the angle of incidence equals the angle of reflection (Fig. 5-13).

2. If the waves encounter an "interface" where they experience a change of velocity, they will experience a change in wavelength but no change in frequency (Fig. 5-14).

3. If the wave fronts encounter an interface obliquely, the waves will change the direction in which they are traveling. Figures 5-15 and 11-5 show that when the waves cross an interface at which they undergo a decrease in velocity, their new direction

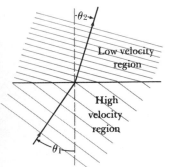

FIG. 11-5

Refraction of waves. When waves pass into a region of decreased velocity, they are refracted; the angle of refraction is less than the angle of incidence. Particles show opposite behavior (see Fig. 11-3).

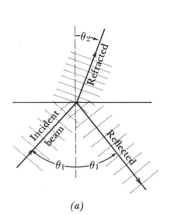

(a)

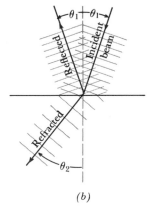

(b)

FIG. 11-6

Waves are always reflected at an interface. Waves are reflected back into the original medium, whether they originate in the medium of high velocity (a) or in the region of low velocity (b).

* We are referring at this point to the simple waves of small amplitude one may observe in a ripple tank. Ocean waves are more complex because of the effect of the wind, the variable depth of the water, and the greater height of the waves.

is such that the angle of refraction is less than the angle of incidence; that is, their line of motion is bent toward the normal.
It should be noted that this is exactly opposite the observed behavior of a
particle.

4. The waves are reflected to some extent at the interface between
two media A and B whether they originate in A or in B (see Fig.
11-6). Again a difference between the behavior of waves and particles may be noted. In our contrived bowling ball experiment in
Sec. 11-1, the bowling ball was reflected back into its original
"medium" only when it encountered the "interface" from the
side on which it had the higher velocity.

11-3 THE BEHAVIOR OF LIGHT: ITS VELOCITY

In order to make a choice between a primitive-particle and a simple-
wave theory of light, we need to examine the properties that light possesses, particularly those properties that might enable us to make a
choice between the two competing models. In this section we will discuss the velocity of light in various media in order to relate its change
of speed to any change of direction that may occur when light passes
from one medium to another.

Measuring the velocity of light
in air and in vacuum

A variety of experimental techniques have been utilized to measure the velocity of light since the first estimate was made by Roemer
in 1676. One of the most straightforward methods was that employed
by A. A. Michelson in 1926, who, improving on an earlier arrangement
by Foucault (1860), utilized a rotating eight-sided mirror in the setup
shown in Fig. 11-7.

First let us consider what happens if the eight-sided mirror is not
rotating and is oriented in such a way that a narrow beam of light is
reflected from "side 4" in the figure to the distant mirror and back to
"side 6." Clearly the beam of light will enter the observer's eye as if it
came directly from the source without reflection. If the mirror is rotated
slowly, the light will flicker but will still appear to come directly from
the source, because light can reach the observer only when the rotating mirror is in an orientation equivalent to that shown in the figure.
If the mirror is rotated at a high velocity, however, the direction of the
beam after its third reflection will change. This change of direction
indicates that although "side 4" or its equivalent must be in exactly the
orientation shown in the figure if light is to reach the distant mirror,
"side 6" will surely rotate through a significant angle as the light
travels the 144 km to the distant mirror and back again. Specifically,
if the mirror is rotated at the rate of 58 rps, the mirror will have
rotated through an angle of 10 degrees while the light is in transit, and
the direction of the reflected beam will shift by 20° away from the

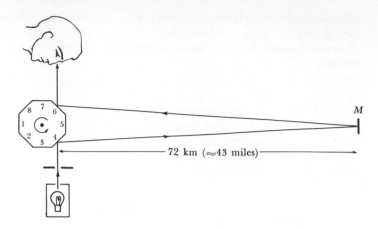

FIG. 11-7

Measuring the velocity of light. Light reflected from side 4 of the rotating mirror will be directed toward the distant mirror M only when it is in the position shown. When light has traveled 144 km to M and back, side 6 will have rotated slightly; hence the emergent beam will be reflected in a slightly different direction.

original direction. The reader who feels so inclined can use this data to show that the light must have traveled at the breakneck speed of 3×10^8 m/sec ($= 186,000$ miles/sec).

In one series of experiments, Michelson measured the velocity of light in a mile-long evacuated tube. These measurements, together with those of other observers using various techniques, show that light of any color travels through free space (a vacuum) at one well-defined velocity c given by

$$c = 2.997925 \times 10^8 \text{ m/sec} \qquad (11\text{-}1)$$

with an accuracy of one part in a million. The velocity of light in air is only slightly less, approximately 2.9970×10^8 m/sec.

The velocity of light in other media

Michelson's method can be adapted to measure the speed of monochromatic (one color) light through any transparent medium. A few representative velocities are shown in Table 11-1. These data are sufficiently typical of how the velocity of light varies from one material to another that one can draw the following generalizations:

1. The velocity of light in free space is the same for all colors.

2. The velocity of light in any medium is always less than its velocity in a vacuum.

3. With notably few exceptions, the velocity of blue light in a given medium is less than the velocity of red light in that medium.

TABLE 11-1		
The velocity of light		
MEDIUM	VELOCITY OF BLUE LIGHT	VELOCITY OF RED LIGHT
Vacuum	2.997925×10^8	2.997925×10^8
Air	2.997034×10^8	2.997116×10^8
Water	2.237×10^8	2.254×10^8
Glass	1.874×10^8	1.897×10^8
Carbon Disulphide	1.795×10^8	1.851×10^8

11-4 THE BEHAVIOR OF LIGHT: REFLECTION AND REFRACTION

In this section we will describe some additional experiments that can be performed with light, experiments that may lead to a definite identification of light to be either a wave motion or a stream of particles. In order to isolate the properties of light that we wish to examine, we will use a light source that produces a narrow beam of high intensity. This beam will be sent through a medium that scatters some of the light, making the path of the light visible (smoke will make a light beam visible in air; a few drops of milk or of dye—for example, flourescein—can be used in water). A photograph will show how the light behaves when it encounters an obstruction or enters a different medium.

Diffuse and specular reflection

We will find it convenient to identify and distinguish between two kinds of reflection called *diffuse* and *specular* reflection. In the case of

FIG. 11-8

Diffuse reflection of a beam of light. Light striking a white card is scattered in all directions.
(*Phillips Hall*)

FIG. 11-9

Specular reflection of a beam of light. A mirror reflects light in a symmetric manner. The angle of incidence equals the angle of reflection.
(*Phillips Hall*)

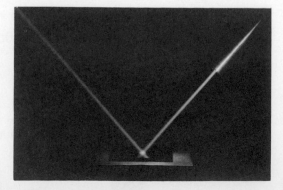

diffuse reflection, light incident upon a surface is scattered in all directions (Fig. 11-8), whereas in the case of specular reflection, as for light incident upon a mirrorlike surface, a beam of light is reflected symmetrically with the angle of incidence equal to the angle of reflection (Fig. 11-9).

Figure 11-10 shows how a single lamp, through the agency of diffuse reflection, can illumine an entire room. The portion of this light that falls on any nonluminous object is scattered, thus causing each point on the object to act like a source of light and making the object visible.

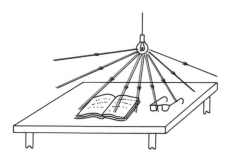

FIG. 11-10

How light makes objects visible. Light from the lamp is scattered from all portions of an object and in all directions. This light leaves the nonluminous object as if that object were the source of the light.

The refraction of light

In Fig. 11-11 a beam of light passes obliquely from air into glass and finally from glass back into air. This photograph shows that when light passes from air into glass, the angle of incidence is greater than the angle of refraction and that when it passes from glass back into air, exactly the opposite behavior is observed. Recalling that light travels less rapidly in glass than it does in air, we conclude that when light enters a medium in which it has a smaller velocity, the angle of refraction is smaller than the angle of incidence; that is, that the light will be bent toward the normal. Thus the behavior of light at an interface between two media more nearly corresponds to the behavior of a wave than of a particle. On this basis alone, *we can reject any further consideration of the primitive-particle model of light.* From this point on we will treat the wave model of light as the active theory, keeping in mind, of course, that one experiment with light that fails to conform to predictions based on the simple-wave theory will force us either to reject the theory or at least to modify it.

11-5　THE BEHAVIOR OF LIGHT: DIFFRACTION

In our discussion of waves in a ripple tank in Secs. 5-5 and 5-6, considerable attention was given to the behavior of straight wave fronts that have encountered an obstruction in which one or more apertures were located. These observations show that (1) if the obstruction has but one aperture that is large compared to the wavelength of the waves,

FIG. 11-11

The refraction of light. A beam of light entering from the lower left encounters a glass plate. At each surface a portion of the beam is reflected; another portion is refracted. (*From PSSC* Physics, *D. C. Heath & Co, Lexington, Mass., 1965*)

the waves passing through have a straight leading edge with curvature on both sides (Fig. 5-12); (2) if the width of the single aperture is comparable to the wavelength, curved wave fronts emerge from the aperture (Fig. 5-11); and (3) if two narrow apertures are present whose separation is only a few times larger than the wavelength of the waves, an interference pattern will be observed (Fig. 5-16*b*).

We wish to set up an arrangement that constitutes the optical equivalent of the interference experiment shown in Fig. 5-16*b*. To do so we need to know the approximate wavelength of light.

Experience tells us that the wavelength of light must be very small because light from the Sun passing through a doorway or even through a ½-in. hole in a window shade does not spread out but travels as a distinct beam with a cross section having the shape of the aperture. These apertures, therefore, must be large compared to the wavelength of the light. Light passing through them corresponds to the situation shown in Fig. 5-12, in which most of the energy remains in the straight portion of the wave front and thus continues to travel in the direction of the incident light. On the other hand, light from a distant point source will not form a sharp beam on passing through a pinhole in a piece of cardboard, presumably because the situation more nearly corresponds to that shown in Fig. 5-11.

Estimating the wavelength of light

In order to secure an estimate of the wavelength of light, we view a distant streetlamp through a pinhole in a piece of aluminum foil. Although the lamp appears to be a point source when viewed with the naked eye, it appears at a blur when seen through the small aperture in the foil. Furthermore, the angular spread of light reaching the retina of the eye appears to be greater as one employs apertures of progressively smaller diameter. If the source is sufficiently intense, alternate light and dark rings will be seen about the central bright region.

Next, if we punch two holes very close together in the aluminum foil and look at the distant streetlamp through the two holes simultaneously, the central bright region that was observed using a single aperture is striated with alternate dark and bright lines. These lines are interpreted to be the equivalent of the alternate nodal and reinforcement lines shown in Fig. 5-16. If one estimates the separation of the holes in the piece of foil and the angular separation of the alternate dark and bright lines, he can even arrive at an estimate that light has a wavelength somewhat less than 10^{-6} m. Light must indeed consist of waves.

The foregoing experiment can be performed with somewhat higher precision if one uses slit-shaped apertures. A convenient method of preparing single and double slits for use in experiments corresponding to the single and double pinhole experiments is to scratch the line or lines on a glass plate that has been coated with lampblack or lacquer. Since the lines must be very narrow, one should use a sharp tool—a razor blade, for example. Two razor blades held tightly together may be used to scribe two lines simultaneously, thereby producing an

FIG. 11-12

Single- and double-slit diffraction of light. (*a*) Lines were scratched on a painted glass plate. The observer looked through these scratches at a distant single-filament lamp. (*b*) The view seen by the observer when he looks through a single slit (above) and a double slit (below).

(*Single-slit pattern, courtesy Brian J. Thompson; double-slit pattern, from Bruno Rossi*, Optics, *1957, Addison-Wesley, Reading, Mass.*)

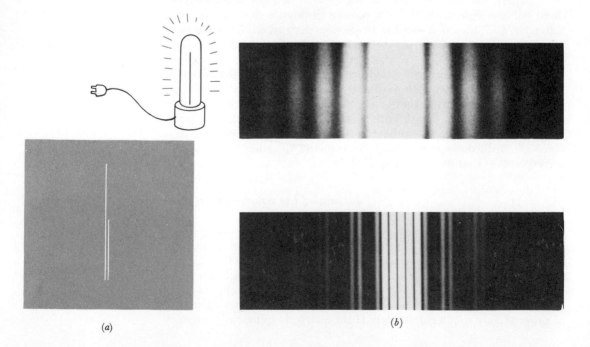

(*a*)   (*b*)

acceptable double slit. Furthermore, the two lines of this double slit will be separated by the readily measured distance equal to the thickness of a razor blade. In use, the plate is held in front of one eye or in front of the lens of a camera. If a camera is used, the pattern can be photographed; one can compute the wavelength of the light from the distance of separation of the dark lines (nodal positions) on this photograph.

11-6 THE DIFFRACTION GRATING

One of the most important contributions of physics has been the establishment of an understanding of the structure of matter by analyzing the light that it either emits or absorbs. Much of the information on which this understanding is based has been secured with apparatus that separates light into its constituent colors or, as we shall see as we elaborate on the wave theory of light, into its constituent wavelengths or frequencies. The most important single optical device for the precision measure of the wavelength of light is the diffraction grating.

A diffraction grating constitutes an extension of the principles involved in the double-slit diffraction device described in Sec. 11-5. Instead of two slits, however, thousands of uniformly spaced, straight, parallel lines are ruled on a plate of glass or metal and play the role of thousands of apertures (for the glass) or line-shaped reflectors (for the metal).

The quality of a diffraction grating depends on the straightness and uniformity of the rulings on the grating surface. In order to achieve quality gratings, elaborate ruling engines have been developed that produce gratings by drawing a fine diamond point repeatedly across a metal plate (to prepare a reflection grating) or across a glass plate (to prepare a transmission grating). Gratings may have as many as 10,000 lines per centimeter.

A schematic sketch of the profile of a diffraction grating is shown in Fig. 11-13. Plane waves of a single wavelength are shown advancing toward the grating from the left. When these plane wave fronts encounter the surface of the grating, much of the energy will be absorbed or reflected. The portion, however, that passes through the grating will emerge as cylindrical wave fronts, which, in the section shown in the figure, appear as circular wave fronts.

At first glance it might seem that only wild confusion could exist in the region beyond the diffraction grating as wave fronts spaced one wavelength apart emerge from each slit and overlap those emerging from all the other slits. However, because of the uniformity of the rulings, a remarkable order is ultimately restored. Consider the line P_0, for example. This line and all the other lines parallel to it one wavelength apart mark the leading front of a plane wave that eventually emerges from the grating. Next consider P_1, which joins a wave front emerging from a given slit to the wave front one wavelength ahead of it at the next slit, two wavelengths ahead from the slit beyond, and so on. Each of these wave fronts eventually will form a plane wave

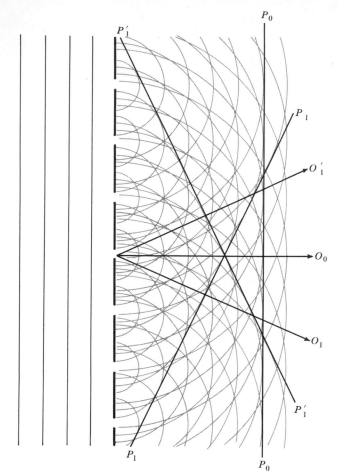

FIG. 11-13

A diffraction grating. Each aperture behaves like a source of circular wave fronts. At a distance from the grating surface, these wave fronts recombine to form plane wave fronts traveling in the directions O_0, O_1, and O'. The recombination of wavelets to form plane waves is more obvious if one holds the page parallel to his line of sight and looks down the lines P_0, P_1, and P'.

front traveling obliquely to the original direction and *at an angle determined by the wavelength of the light*. In a similar way, we can identify wave fronts parallel to P'_1. Hence light having exactly the characteristics of the incident monochromatic light travels away in various directions given by the lines O_0, O_1, and O'_1.

If the incident light consists of a mixture of waves of many wavelengths, light of each of the component wavelengths will leave the grating and travel in a slightly different direction. In the region of O_1, for example, one will see a rainbow of colors with the blue portion on the inside; blue light, with the shortest wavelength, will be bent less than red light, which has the longest wavelength.

A schematic diagram of a diffraction-grating spectrograph is shown in Fig. 11-14. This sketch shows that the light under study is emitted by the source to the left and enters the spectrograph through a narrow aperture or slit at S. This slit is located relative to the lens L_1 in such a way that light leaving L_1 emerges as a parallel beam, and thus it consists of a sequence of plane wavefronts. When these wavefronts encounter the grating, diffraction occurs as described above and lens

L_2 focuses all light of a given color or wavelength at a point on the photographic plate P. A typical spectrogram is shown in Fig. 13-3.

Precision measurements of the wavelength of the light emitted by a source are possible provided that one has accurate data regarding the spacing of the rulings and provided that the spectrograph is equipped to allow precision measurements of the angle θ or the distance x. Examining Fig. 11-14(a) and the enlarged view of the situation in the vicinity of the rulings on the grating [Fig. 11-14(b)], it can be seen that the triangle BCA [Fig. 11-14(b)] is similar to the triangle with two sides x and L in Fig. 11-14(a). Since the sides of similar triangles are proportional, we have

$$\frac{\lambda}{d} = \frac{x}{\sqrt{L^2 + x^2}}$$

where λ is the wavelength of the light. Since d, L, and x are either known or measurable, the wavelength λ is given by

$$\lambda = d\frac{x}{\sqrt{L^2 + x^2}} \qquad (11\text{-}2a)$$

FIG. 11-14

A diffraction grating spectrograph. (a) Circular wave fronts emerging from the rulings of the grating reconstruct plane wave fronts in the region to the right, as shown in Fig. 11-13. The wave fronts travel in the directions indicated by the "rays" in the sketch. (b) A magnified view of the situation near B showing the dependence of the angle of diffraction (θ) on the wavelength of the light.

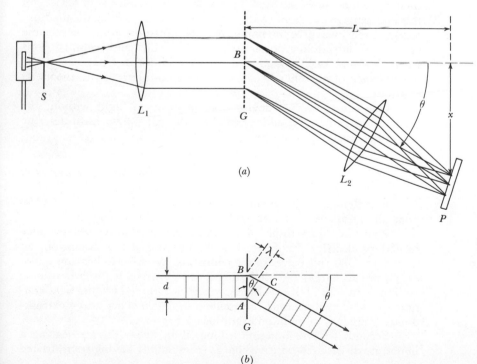

Readers with some knowledge of trigonometry will recognize that this equation may be written in the form

$$\lambda = d \sin \theta \qquad (11\text{-}2b)$$

where θ is the direction in which the reinforced wave proceeds away from the grating.

11-7 THE ELECTROMAGNETIC THEORY OF LIGHT

What is waving?

Whenever, in previous chapters, we discussed waves, we associated the waves with oscillatory displacements in an identifiable medium. Waves traveling over the surface of a ripple tank, for example, were clearly oscillations of portions of the medium (water) which at at any specific point in the medium took place along a line perpendicular to the direction of the motion of the wave. Sound waves were described as oscillations of the air, which, in contrast to water waves, constitute vibrations of the medium along a line parallel to the direction of motion of the waves. In neither case, however, does the medium travel with the wave; only the wave form travels.

These experiences with waves leads one to inquire as to the medium that transports light. Surely one would say, if light is a wave, then something physical must be waving. Any reader who reasons in this manner is in excellent company because the search for the medium (prematurely even give the name *ether*) occupied the attention of some of the most outstanding physicists of the late nineteenth century. This search has now been abandoned; as we will see in Chapter 14, the nonexistence of an ether became the first postulate of the theory of relativity.

One can, however, devise a wave theory of light without the concept of an ether. To understand how this can be, we must describe a very hypothetical experiment.

A thought experiment

Our thought experiment will be carried out in outer space away from all extraneous gravitational, electrical, or magnetic fields.

First, we must think of a simple device that would produce oscillatory electrical impulses. For the purpose of our discussion, let us imagine that this wave source consists of two spheres bearing opposite electrical charges and oscillating up and down with simple harmonic motion as in Fig. 11-15(*b*). Since we are not limited by practical considerations, we will assume the oscillations to take place at the extraordinary frequency of 1 billion vibrations per second.

Second, we will imagine a long, thin, nylon thread bearing a uniformly distributed positive charge (presumably having been stroked

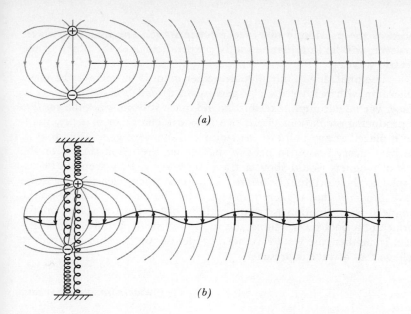

(a)

(b)

FIG. 11-15

Generating an electromagnetic wave. Two oppositely charged objects at rest produce an electric field as in *(a)*, but when vibrating with simple harmonic motion, they generate an electric wave as in *(b)*.

along its length by a silk cloth). This thread leads directly away from the wave generator and, being free from external influences such as gravity and air currents, should remain straight unless acted on by an effect associated with the source.

Our experience with nearly equivalent laboratory devices indicates that, were it practical to carry out the preceding experiment, transverse waves would travel down the thread with a velocity of 3×10^8 m/sec, the positive charges on the thread being alternately pushed up and down very much as leaves floating on a pond are pushed up and down when a water wave goes by. The wave produced by our wave generator is a transverse wave marked by a changing electric field in the space nearby. As in the case of waves discussed earlier, these waves possess a wavelength (λ) that is related to their frequency (f) and their velociy (c) by

$$c = f\lambda$$

The waves generated under the conditions cited would have a wavelength of approximately 1 ft, since $\lambda = 3 \times 10^8/10^9 = 0.30$ m. Such a wave would be called a *polarized* wave because all the electric oscillations lie one plane.

The next step in our thought experiment is to place a long row of magnets along the length of the thread. These magnets must be imagined as so tiny that they are capable of oscillating at a frequency of a billion vibrations per second. With the magnets in place, one will observe a magnetic wave traveling away from the source at the same speed as the previously observed electric wave. The magnetic wave, however, will lie in a plane exactly perpendicular to the plane of the electric wave.

The final step in our thought experiment is to operate the wave generator with neither the charged nylon thread nor the magnets in place. We have complete faith that the space still possesses the *capacity* of producing oscillations of either an electric charge or of a tiny magnet if placed at any point in the vicinity of the wave generator; that is, we have every reason to believe that either the nylon thread or the row of magnets would show a traveling wave if they were to be returned to their original positions. The space, then, clearly possesses a wave property that is merely detected by the electric charges or the magnets. The situation is comparable to observing the waves on the surface of a pool by watching the motion of dust particles on its surface. In neither case would we need to identify the medium in order to believe that a wave is present.

Electromagnetic waves

The waves we have described are called electromagnetic waves because they simultaneously possess both electric and magnetic properties. When produced by a simple pair of oscillating charges (called a dipole), the directions of the electric and the magnetic fields are related to the direction of propagation of the wave as shown in Fig. 11-16. A general rule for identifying the direction of the magnetic field, given the direction of the electric field and the direction of motion of the wave, is as follows:

Place your right hand in the path of the wave with your thumb pointing in the direction of motion of the wave and with your fingers extended in the direction of the instantaneous electric field. Now, upon bending your fingers through 90 degrees, they will be pointing in the direction of the associated magnetic field.

Electromagnetic waves emitted by many common sources (heated filaments or excited gases) originate in a myriad of individual atoms;

FIG. 11-16

An electromagnetic wave. An electric wave in the *E–x* plane is always accompanied by a magnetic wave in the plane perpendicular to the *E–x* plane.

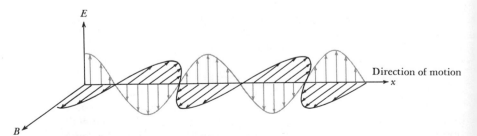

therefore they may be expected to have a very short wavelength because of the small size of the sources. The atoms, in turn, are situated in every conceivable orientation at the instant light is emitted. Thus while the light certainly travels directly from the source to the eye, the light that enters the eye is a wild mixture of waves from different atoms. Upon arriving at the retina, this *unpolarized* light will produce electric impulses in nearly every possible direction perpendicular to the direction of travel of the light. The light will also arrive in all possible relative phases, for there is no reason for the crest of the wave from one atom to be located adjacent to the crest of the wave from another. (Incidentally, in the special case of light from a laser, the emitted light is *coherent*, meaning that the crests are all in phase).

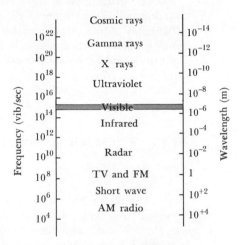

FIG. 11-17

The electromagnetic spectrum. Electromagnetic waves range in wavelength from 10^{-14} m to 10^4 m.

Electromagnetic waves include phenomena as varied as ultraviolet light, visible light, infrared radiation, and radar. Their wavelengths can be as short as 10^{-14} m (x rays) and as long as $\frac{1}{4}$ mile (radio waves). Light that is interpreted by the eye as blue light has a wavelength of approximately 0.40×10^{-6} m, whereas red light has a wavelength of approximately 0.75×10^{-6} m. All electromagnetic waves, however, travel at the same speed in free space (3×10^8 m/sec), differing one from the other only in their frequency and in properties that depend on their frequency (their wavelength and, as we will see in the next chapter, their energy.) The variety of names that have been associated with different kinds of electromagnetic waves has, in fact, much less to do with the waves themselves than with the nature of their sources (see Fig. 11-17). The wavelength ranges of gamma rays and x rays, for example, overlap; their only sharp distinction is in the fact that, for the most part, gamma rays are emitted by radioactive materials while x rays are emitted by x-rays tubes.

Some sources of electromagnetic waves resemble, functionally at least, the wave generator that we have described. In the case of radar or radio waves, an electric circuit alternately charges the two ends of an antenna positively and negatively to produce the equivalent of an oscillating dipole. At the submicroscopic level of atoms, the correlation is less clear, largely, perhaps, because of our faulty knowledge of the structure of atoms themselves.

SUMMARY

In this chapter we have considered the problem of establishing a model that will describe the nature of light by comparing the behavior of light with that of primitive particles and simple waves. We have chosen the wave model over the particle model because (1) light behaves as waves are expected to behave on entering a region in which it experiences a change of velocity and because (2) light can be diffracted.

The diffraction of light enables one to measure its wavelength. Starting with an estimate of the wavelength secured by measuring the diffraction pattern produced by a double slit, one rules a diffraction grating whose grating space is comparable to the wavelength of the light under study. In the first order of the resulting spectrum, the wavelength is given by

$$\lambda = d \frac{x}{\sqrt{L^2 + x^2}} \tag{11-2a}$$

which, for those familiar with trigonometry, becomes

$$\lambda = d \sin \theta \tag{11-2b}$$

Finally, we have attempted to deal with the question, "What is waving?" According to the model we are using, light consists of wavelike electric and magnetic disturbances propagated through space with a speed determined by the medium and the frequency of the disturbance. In its simplest manifestation, the alternating magnetic field is transverse both to the associated electric field and to the direction of propagation of the wave. The relative directions of E, B, and v have been described by a right-hand rule.

QUESTIONS AND PROBLEMS

11-A1. Using a marble, a thin book, and a strip of cardboard to serve as the skirt (Fig. 11-1), experiment with the reflection and refraction of a moving particle as suggested by Figs. 11-2 to 11-4. What is the connection between this setup and a particle model of light?

11-A2. Summarize the experiments that are decisive in making a choice between the wave and the primitive-particle model of light.

11-A3. Compute the frequency of the waves emitted by the following sources: (a) a piece of glass which when held in a flame emits light whose wavelength equals 0.59×10^{-6} m; (b) a radar station which emits waves whose wavelength is 1 cm; (c) a radioactive specimen which emits waves whose wavelength equals 3×10^{-10} m.

11-A4. How would you explain to a doubter our statement that the frequency of a wave is unchanged when it enters a different medium?

11-A5. Blue light ($\lambda_{\text{air}} = 0.4 \times 10^{-6}$ m) passes from air into water with the change in velocity shown in Table 11-1. (*a*) What is the frequency of this light in air? (*b*) What is the frequency of this light in water? (*c*) What is the wavelength of this light in water?

11-A6. Hold a phonograph record horizontal with one edge pressed against the ridge of your nose and look at the reflected image of a distant street lamp. Explain the origin of the rainbowlike spectrum that you see.

11-A7. Section 11-7 started with the rhetoric question, "What is waving?" What was the answer?

11-B1. Assume that a six-sided mirror is used to measure the velocity of light in an arrangement otherwise identical to that shown in Fig. 11-7. At what rotational speed in revolutions per second must the mirror be rotated for the direction of the reflected beam to be shifted 12 degrees away from its original direction?

11-B2. Assume that blue light ($\lambda_{\text{air}} = 0.4 \times 10^{-6}$ m) from a distant source passes from air into glass with an angle of incidence (θ_1 in Fig. 11-6) of 30 degrees. Using ruler and protractor, make a scale diagram of the situation on both sides of the interface. Measure the angle of refraction (θ_2) of the waves in the glass.

11-B3. Two long parallel slits 4 cm wide are cut in a large sheet of aluminum foil with the distance between the centers of the slits equal to 25 cm. A beam of microwaves having a wavelength of 10 cm traveling in a direction perpendicular to the foil produces a double-slit diffraction pattern on passing through these slits. Make a scale diagram of the situation and describe the pattern produced by this arrangement.

11-B4. Assume blue light ($\lambda_{\text{air}} = 0.4 \times 10^{-6}$ m) to advance as in Fig. 11-13 toward a diffraction grating whose rulings are 1.0×10^{-6} m apart. Using ruler, compasses, and protractor, make a scale diagram showing how reinforcement of the waves takes place. Measure the angle between θ_0 and θ_1 in your figure.

MODERN
PERSPECTIVES IN PHYSICS

The world view implicit in the physical theory presented to
this point is one of a gigantic piece of clockwork in which
the positions and velocities of all the atoms and molecules in
nature at one instant ensure the positions and velocities of all
particles at any later instant. According to this view of
nature, every event in the future, while perhaps
indeterminable by man, would be predetermined.

    The clockwork view of nature began to wane early in
the twentieth century. The theory of relativity replaced the
absolutism of classical physics by a revolutionarily different
absolutism; the completeness of this revolution may be
understood by noting that neither mass, nor length, nor time,
basic to all that has been said to this point, remains an absolute
quantity in modern physics. In another development light,
which had been "proven" to consist of waves, was just as
conclusively shown to possess particlelike characteristics;
furthermore, "particles" in motion were found to have a
wavelength. And an uncertainty principle, according to
which man may not simultaneously make precise
measurements of both the position and the velocity of a
particle, became a cornerstone of physical theory.

    With these changes in our view of the physical world
it is no longer possible to describe submicroscopic phenomena
accurately in terms of models based on objects in the
macroscopic world of our daily experience. Light, for
example, is not, strictly speaking, either wave or particle as
we have come to understand waves and particles; indeed, it
may simply be "something else," which outwardly manifests
some of the elements of both. As we abandon mechanical
models, we become more dependent on mathematical
descriptions, and "understanding physics" comes to be
identified with finding a mathematical formulation which
correctly predicts the outcome of a physical experiment.

    Part III, then, is the story of a revolution—not only a
revolution in physics, but, indeed, a revolution in man's
understanding of every aspect of his environment.

Waves and Particles

There is no principle of nature that demands that all things that may be discovered in the future must resemble either waves or particles as we have known them. An electron, for example, is not required to behave exactly like an electrically charged marble nor does light need to have precisely the same properties as water waves. The properties we associate with the waves and particles of the macroscopic world of our everyday experience may be vastly different from their counterparts at the submicroscopic level.

In this chapter we will describe experiments which show that phenomena normally regarded as due to wave motion reveal particlelike characteristics and that phenomena we have associated with particles manifest wavelike properties. This "duality" of waves and particles constitutes a revolutionary change from our conventional view of reality.

12-1 THE PHOTOELECTRIC EFFECT

The photoelectric cell is not one of the greatest commercial devices: many of its most familiar applications, in fact, border on the trivial. A photocell may initiate a mechanism that causes doors to open at the supermarket, may set off a burglar alarm, or may provide the variable electric current needed to actuate the sound system of a motion picture; even these less-than-earth-shaking results can be achieved in other ways. The invention of the photoelectric cell clearly was not a technological breakthrough.

In contrast, the *explanation* of the workings of a photoelectric cell stands as one of the most significant events in the history of science. This explanation reinstated a particle view of the nature of light!

It will be recalled that the wave theory of light obtained favor over a "primitive-particle" theory of light for many seemingly valid reasons. Light does, in fact, behave like waves in all the experiments

12

FIG. 12-1

"There is no principle of nature that demands that all things . . . must resemble either waves or particles. . . . "

that had been conducted prior to approximately 1900. All these experiments, furthermore, still demand explanation, and no particle theory will replace the explanations we have already given. We will amend our wave model, not abolish it.

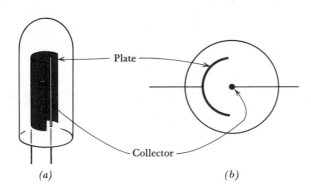

FIG. 12-2

A photoelectric cell. (*a*) This simple model consists of a curved piece of metal (the plate) and a metal pin (the collector) sealed in an evacuated tube. (*b*) A schematic sketch showing the essential parts only.

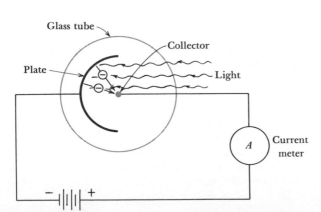

FIG. 12-3

A photoelectric cell as used to measure illumination. Light enters from the right, strikes the plate, and releases electrons, which are attracted to the positively charged collector.

A photoelectric cell consists of a highly evacuated glass tube containing two metal electrodes, of which one is a clean metal *plate* perhaps an inch square curved to form a half-cylinder and the other is a metal wire located at the center of the half-cylinder. Because it receives electrons emitted by the plate, the wire is called the *collector*. A window made of quartz may be provided to permit ultraviolet, as well as visible, light to enter the tube.

In order to actuate the photocell, one must place it in an electric circuit. When the arrangement is intended simply to respond to light (e.g., in measuring the intensity of illumination in some photographic application), the circuit could be as shown in Fig. 12-3. Here the battery provides a negative charge to the plate and a positive charge to the collector. When light falls on the plate, an electric current flows in the circuit in such a direction as to suggest that electrons have been ejected from the plate, go to the positively charged collector, and flow through the wire back to the battery. A sensitive ammeter will detect this flow of electric charge.

The relationship between current and light intensity

Experiments show that the current established by radiation falling on the plate—that is, the number of electrons ejected by the light *per second*—is strictly proportional to the intensity of the light. This observation may not be surprising, but it is significant that an equally "reasonable" alternative possibility is not observed—namely, that an increase in light intensity might have resulted in an increase in the energy of the ejected electrons without an increase in their number. In fact, since the rate of release of electrons is *strictly* proportional to the intensity of light, one must conclude that the energy of a typical electron is precisely the same whether the light is dim or bright.

The relationship between the energy of photoelectrons and frequency of incident light

The crucial experiment dealing with the photoelectric effect involves measuring the energy of the ejected electrons under the action of monochromatic light—that is, light having a specific color or fre-

FIG. 12-4

The photocurrent is proportional to the light intensity. The data shown in this graph were secured using the circuit of Fig. 12-3 and prove that twice as much light ejects twice as many electrons from the plate.

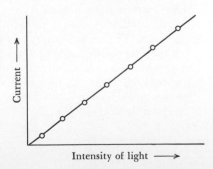

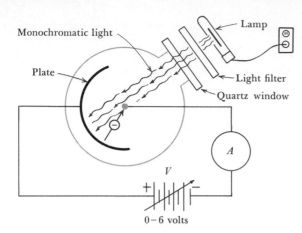

Monochromatic light

Plate

Lamp

Light filter

Quartz window

A

V

+   −

0 − 6 volts

FIG. 12-5

Determining the energy of photoelectrons. Monochromatic (one-frequency) light enters the photocell. The energy of the electrons ejected by light of a given frequency is determined by noting the voltage (V) required to stop all electrons. Their energy equals V multiplied by e, the electronic charge.

quency. To do so one must make a few changes in the electric circuit and utilize a specialized source of light. In the modified circuit shown in Fig. 12-5, the battery has been reversed so that the plate is charged positively and the collector is charged negatively; the voltage provided by the battery, furthermore, may be varied at will in the range from zero to a few volts. In addition, the ammeter used to measure the flow of electrons in the circuit has been replaced by a more sensitive meter, capable of detecting a current as low as 10^{-14} amp (approximately 60,000 electrons per second).

With this setup we perform the following experiment. With the voltage provided by the battery set at zero, a monochromatic light filter is placed between the light source and the photocell, limiting the light that strikes the plate to an ultraviolet beam of one well-defined frequency. Electrons are ejected from the plate and a current is detected. A retarding voltage is then applied and steadily increased to the point that no current is detected by the meter; presumably, no electron is emitted with sufficient energy to overcome the retarding field of the battery. The frequency of the light and the "stopping voltage" are recorded. The filter is successively replaced by filters that transmit light of a different frequency; for each frequency, the stopping voltage is observed and recorded.

Typical data

Let us examine some typical data that may be secured with this arrangement. A photoelectric cell with a plate of clean cesium metal irradiated by light having a frequency of 11.82×10^{14} vib/sec (wavelength = 0.2537×10^{-6} m) required a voltage of 3.05 volts to stop all electrons. We conclude that, using light of this frequency, each electron left the plate with a kinetic energy equal to the work (Ve) required to stop it—that is, with a kinetic energy of 4.88×10^{-19} joules. When the incident light was changed to light having a frequency of 6.88×10^{14} vib/sec (wavelength = 0.4358×10^{-6} m), the ejected electrons had an energy of 1.52×10^{-19} joules; that is, they were stopped by a retarding

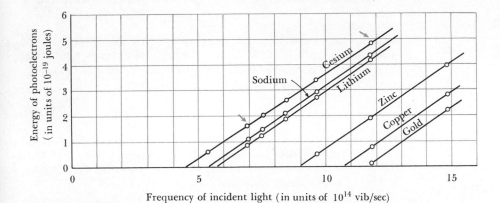

FIG. 12-6

The energy of photoelectrons versus the frequency of incident light. For each frequency of the incident light, one notes the stopping voltage. The graph shows that the energy of the ejected electrons increases linearly with the frequency of the incident light.

voltage of 0.948 volts. These data, as well as other data secured with this apparatus, are shown plotted in Fig. 12-6, where the frequency of the incident light (the independent variable) appears on the horizontal axis and the energy of the ejected electrons (the dependent variable) is recorded on the vertical axis.

The following features of the graph should be noted:

1. For a photoelectric cell with a plate of a given metal, a straight line can be drawn through the data points.

2. The straight lines for different metals never intersect; instead, they are mutually parallel and thus all have the same slope.

3. For each metal there is a frequency below which no electrons are ejected regardless of the intensity of the light. (For example, light having frequency less than 4.5×10^{14} vib/sec will eject no electrons from cesium; this "cut-off" frequency can be read from the graph of Fig. 12-6 by noting the horizontal intercept of the line labeled cesium.)

12-2 THE PHOTON: A PARTICLE OF RADIANT ENERGY

Data essentially like that shown in Fig. 12-6 were first secured by Lenard in 1900 and confirmed by more accurate measurements by Millikan in 1916. We will now consider the interpretation that transformed such seemingly innocuous information into a revolution in our view of light. This interpretation of the photoelectric experiment by Albert Einstein in 1906 may be said to constitute the discovery of the photon, a particle of radiant energy.* His interpretation of the experiment is as follows:

1. Light having a given frequency delivers its energy in the form of individual identical packets, or "quanta," of energy (photons). The energy of a given photon is proportional to the frequency of the light that transports it. This can be written as

$$E = hf \qquad (12\text{-}1)$$

   where E equals the energy of a photon, h is a constant of proportionality called Planck's constant, and f is the frequency of the light. (Note that we are not abandoning the wave theory; the frequency f has meaning only in a wave theory of light.)

2. A definite amount of energy is required to dislodge an electron from the metal of which the plate of a photoelectric cell is constructed. If an incident photon does not possess this much energy, no electron can be ejected. Thus for each metal there is a characteristic frequency f_0 of light whose photons are just adequate to eject electrons. The energy needed to eject an electron from a metal, therefore, equals hf_0. The frequency f_0 corresponds to the horizontal intercept on the graph shown in Fig. 12-6.

3. If the incident photon has a frequency f greater than f_0, electrons will be ejected. Part of the energy of the incident photon (hf_0) will be used in getting the electron out of the metal; the remaining energy will manifest itself as kinetic energy, which, as we have seen, equals Ve, the work to just stop it. Therefore we can write

$$\begin{array}{ccccc} \text{Energy of incident} \\ \text{photon} \end{array} = \begin{array}{c} \text{energy to dislodge} \\ \text{an electron} \end{array} + \begin{array}{c} \text{kinetic energy of} \\ \text{ejected electron} \end{array}$$

$$\text{or} \qquad hf \qquad = \qquad hf_0 \qquad + \qquad Ve \qquad (12\text{-}2)$$

Determining a numerical value for Planck's constant

The experimental data enable us to determine a numerical value for Planck's constant h. Since we have plotted f on the horizontal axis, the most direct way to determine h is to note that h is the slope of any one of the lines in Fig. 12-6. Alternately, we can secure h by substituting the two previously quoted numerical values of the frequency and the associated energy of ejected electrons into Eq. (12-2). We then have two simultaneous equations

$$11.82 \times 10^{14}h = hf_0 + 4.88 \times 10^{-19}$$

$$6.88 \times 10^{14}h = hf_0 + 1.52 \times 10^{-19}$$

* Max Planck had suggested in 1900 that light emitted by certain sources should consist of particles of energy. He associated the property, however, with the nature of the source rather than with the light itself.

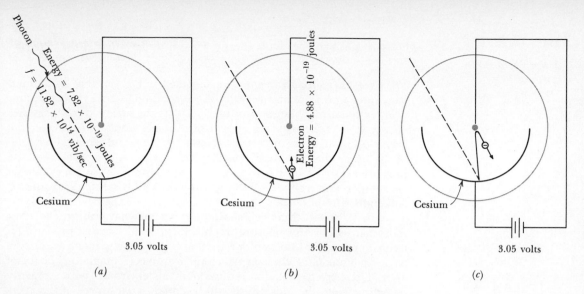

FIG. 12-7

The mechanism of the photoelectric effect. It requires 2.94×10^{-19} joules to release an electron from cesium metal. (a) A photon of ultraviolet light having an energy of 7.82×10^{-19} joules, therefore, (b) is able to give the electron an energy of 4.88×10^{-19} joules. (c) The retarding voltage of 3.05 volts is just sufficient to stop it. This information yields the data point marked by the upper arrow in Fig. 12-6.

Subtracting the first equation from the second and solving for h, we find that $h = 6.8 \times 10^{-34}$ joule sec. An average of results from many (and more accurate) experiments yields

$$h = 6.62559 \times 10^{-34} \text{ joule sec} \qquad (12\text{-}3)$$

Planck's constant is one of the fundamental universal constants of nature and would appear to have far-reaching significance in natural phenomena. As we shall see in subsequent sections, this constant plays a role in physical systems whose relationship with light will appear to be very remote. As a fundamental universal constant of nature, Planck's constant deserves a place on a par with the gravitational constant (G), the general gas constant (R), the velocity of light (c), Avogadro's number (N), the charge on an electron (e), and the thermal equivalent of mechanical energy (J).

12-3 THE MOMENTUM OF A PHOTON

With the discovery of the photon, we have partially reestablished a particle view of light, and one is inclined to speculate at once regarding other particlelike properties a photon may or may not possess. Foremost among the attributes are the properties of *mass* and *momentum*.

The mass of a photon is generally accepted to be equal to zero, and recent theoretical considerations have established that its mass is surely less than 10^{-57} kg. In this text we will regard the mass of a photon to be precisely zero.

We have already become accustomed to the fact that a confined gas exerts a pressure on the walls of the confining vessel. This pressure is an external manifestation of the fact that the enclosed gas possesses energy in that the pressure arises from the accumulative effect of the elastic collisions of gas molecules striking the walls and rebounding. One of the by-products of these considerations was the discovery that the pressure exerted by a gas is equal to two-thirds of the average kinetic energy per unit volume of the gas molecules [see Eq. (7-3) and/or Appendix 6].

In 1871 J. C. Maxwell showed, in a somewhat similar way, that radiation within an enclosure might be expected to exert pressure on its confining walls. Maxwell's prediction of radiation pressure was based on his monumental work in developing an electromagnetic theory of light. He showed that, for radiation in an enclosure with perfectly reflecting walls, the pressure should be equal to two-thirds of the average amount of radiant energy per unit volume—the same as for a molecular gas.

Thirty years passed before Maxwell's prediction was confirmed by laboratory experimentation. This long delay occurred because of the extraordinary small pressure that occurs. For example, if you should hold your hand in the path of sunlight on a summer day, the force exerted by the light against your hand will be considerably less than a millionth of a pound ($\approx 6 \times 10^{-8}$ N). Yet as small as this pressure may seem to be in this situation, the pressure of light is believed to be one of the forces that cause a comet's tail (gases being boiled off the comet due to the heat of the Sun) to be located on the side of the comet away from the Sun.

The important fact in this discussion is that the classical wave theory of Maxwell was adequate to explain the pressure of light; stated alternately, the experimental observation of the pressure of light constituted a forceful confirmation of the wave theory.

FIG. 12-8

The pressure of light on a comet's tail. The pressure of the light from the Sun is partially responsible for the fact that the tail of a comet is always pointed away from the Sun.

A photon interpretation of the pressure of light

With the discovery of the photon, however, an alternate interpretation of the pressure of light becomes possible, requiring only that we ascribe momentum to the photon. If photons possess momentum, radiation within an enclosure should behave very much like a molecular gas. In both cases, the pressure would be due to the momentum imparted to the walls by the many collisions taking place there.

But how much momentum does a photon possess? One procedure for answering this question is to direct a light beam of known intensity (i.e., a known number of photons of known frequency each second) onto a surface and measure the force directly. The force, however, by Newton's Second Law, equals the momentum transmitted to the surface each second. If one divides the momentum delivered each second by the number of photons each second, he finds the amount of

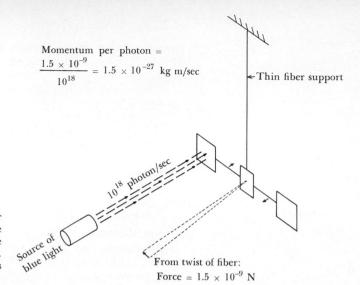

Momentum per photon =
$$\frac{1.5 \times 10^{-9}}{10^{18}} = 1.5 \times 10^{-27} \text{ kg m/sec}$$

←Thin fiber support

10^{18} photon/sec

Source of blue light

From twist of fiber:
Force = 1.5×10^{-9} N

FIG. 12-9

Measuring the average momentum per photon. One arm of the torsion balance absorbs 10^{18} photons each second and the balance registers a force of 1.5×10^{-9} N. On the average, then, each photon possesses a momentum of 1.5×10^{-27} kg m/sec.

momentum delivered by each photon. These experiments confirm that, within 1 percent, the momentum of a photon is given by

$$\text{Momentum of a photon} = \frac{h}{\lambda} = \frac{hf}{c} \qquad (12\text{-}4)$$

where h is Planck's constant and λ, f, and c are the wavelength, frequency, and velocity, respectively, of the light..

Compton's experiment

There is little advantage to associating momentum with a photon unless situations are found that cannot be explained by the wave theory of light. An experiment in which a transfer of momentum can be ascribed to the interaction of a photon with a single fundamental particle would be such an experiment.

The first such investigation was carried out by A. H. Compton in 1923 and involved the collision of an x-ray photon and an electron in a situation that resembles the elastic collision of two particles. For a head-on collision, such an interaction would appear as in Fig. 12-10. Although, for practical reasons, Compton's measurements had to involve glancing collisions, we will limit our attention to head-on collisions as depicted in the figure.

In analyzing this situation, let us accept Eq. (12-4) as the appro-

FIG. 12-10

An elastic collision of a photon with a stationary electron. Both energy and momentum are conserved in the elastic collision of a photon with an electron. In a direct collision, the photon experiences a decrease in frequency and a reversal in direction.

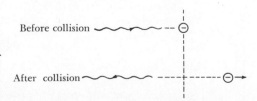

Before collision

After collision

priate expression for the momentum of a photon and see if predictions based on this assumption can be confirmed by experimentation. Since $c = f\lambda$, the energy of a photon will be written as hc/λ. The laws of conservation of energy and momentum, when applied to this interaction, become

$$\frac{hc}{\lambda} = \frac{hc}{\lambda'} + \frac{1}{2}mv^2 \tag{12-5}$$

and

$$\frac{h}{\lambda} = -\frac{h}{\lambda'} + mv \tag{12-6}$$

respectively, where λ and λ' are the wavelengths of the incident and the scattered photons and v is the velocity of the electron after its encounter with the photon.

In solving Eqs. (12-5) and (12-6) for the change in wavelength $(\lambda' - \lambda)$, we will employ one bit of physical insight—namely, that the photon should bounce off the electron with a change in direction but with very little change in energy, something like a piece of buckshot hitting the rear of a freight train. Therefore the photon experiences very little change in wavelength, so that

$$\left(\frac{1}{\lambda} + \frac{1}{\lambda'}\right)^2 = \frac{1}{\lambda^2} + \frac{2}{\lambda\lambda'} + \frac{1}{\lambda'^2} \approx \frac{4}{\lambda\lambda'}$$

Solving Eq. (12-6) for v, substituting this value of v into Eq. (12-5), and using the above approximation, we find

$$\lambda' - \lambda = \frac{2h}{mc} \tag{12-7}$$

FIG. 12-11
Schematic arrangement in the Compton experiment. A narrow beam of photons emitted by the x-ray tube is directed on a small block of carbon. The scattered photons are analyzed with an x-ray spectrometer at various scattering angles.

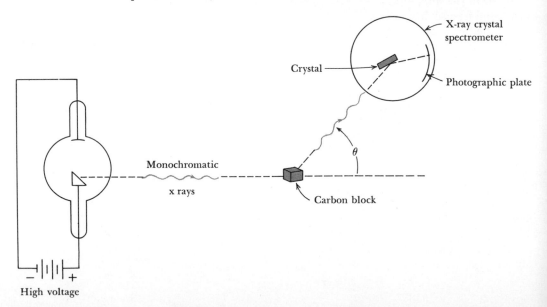

X-ray crystal spectrometer

Crystal

Photographic plate

Monochromatic

x rays

θ

Carbon block

High voltage

This equation predicts that the *change* in wavelength of a photon in a head-on collision with a free electron is always the same regardless of the initial wavelength of the photon. Substituting the known values of h, m, and c into this equation, one finds that

$$\lambda' - \lambda = \frac{2 \times 6.625 \times 10^{-34}}{9.11 \times 10^{-31} \times 3 \times 10^8} = 0.0485 \times 10^{-10} \text{ m}$$

In Compton's experiment x rays were collimated into a narrow beam that struck a carbon block which served as a target containing many electrons. An x-ray spectrometer measured the wavelength of the scattered photons at various angles. Although it was not possible to orient this spectrometer to detect those photons that were scattered directly back along the path of the incident beam, one can extrapolate from measurements carried out for photons scattered in other directions. These measurements confirm that individual photons possess momentum equal to h/λ.

12-4 THE PARADOX OF THE DOUBLE-SLIT EXPERIMENT

The experiments with light described in this and the preceding chapter show that neither a simple-wave nor a primitive-particle model of light will suffice; those of us who would like nature to be simple have not been granted our wish. Light at one and the same time would seem to be both wave and particle, capriciously showing one face in one experiment and the other face in another.

With the information we now possess regarding light, let us reexamine the classic double-slit diffraction experiment, which earlier in this text, as well as in the history of physics, seemed to prove the wave theory of light. How can we reconcile those experiments with the knowledge that energy is delivered by light in the form of the bundles we call photons?

Conducting the double-slit experiment with weak light

The context in which we will reconsider the matter of double-slit diffraction is shown in Fig. 12-12. This figure shows a schematic view from above of the contents of a light-tight box. Light from a weak source at one end of the box is made weaker by a dense filter, and, of the small amount of light that enters the region AB, only the tiniest fraction passes through one or the other of the slits to enter the region BC. At the far end C is a photographic plate that can be removed to reveal a narrow aperture beyond which is located a photoelectric cell. The arrangement thus permits us to detect the light either photographically or by means of the photoelectric cell.

The weakness of the light is important in understanding the significance of this experiment. It is so weak, in fact, that an average time

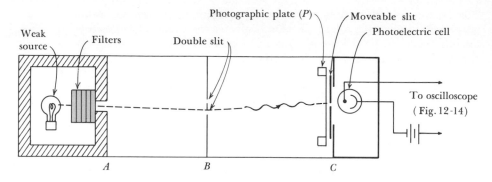

FIG. 12-12

The double-slit experiment revisited. The source is so weak that there is never more than one photon in the box. Even so, individual photons eventually produce a double-slit diffraction pattern on the photographic plate or, if it is removed, on the movable slit in front of the photoelectric cell.

lapse of 4×10^{-8} sec occurs between subsequent photons arriving at the double slit to enter the region *BC*. Since photons travel with the speed of light, on the average, any photon entering one or the other of the slits will travel about 12 m before the next photon comes along. Since the box is less than a meter long, it is very unlikely that two photons will ever be in flight in the box simultaneously. The question is: Can one photon at a time produce a double-slit diffraction pattern?

Statement of the paradox

Let us state the paradox. Consider an individual photon as it approaches the double slit. In order to play any role in the formation of the diffraction pattern, it must pass through the double slit, but since a photon is indivisible, it must pass through one slit or the other. Whichever one it elects to pass through is only a single slit insofar as the photon is concerned. It is difficult to understand how this photon can be guided to the photographic plate by any influence involving the location and size of the slit it did *not* use. On the other hand, if one observes a double-slit diffraction pattern using such weak light, one is forced to believe that the photon passing through one of the slits somehow knows that the other slit is present and responds to it. Stated otherwise, the photon must possess some property that involves both slits.

Experiments show that one does, in fact, observe a double-slit diffraction pattern using light that is so weak that interference cannot possibly occur between two photons. The details of the weak-light experiment, furthermore, provide enormous insight into the phenomenon.

Let us first consider performing the experiment photographically —that is, with a photographic plate in place at *P* in Fig. 12-12. After a very long exposure, upon developing the plate one finds precisely the same diffraction pattern he would see if he were to use a short exposure with an intense source. Under a microscope, however, any photographic plate will appear very grainy. The black portions are not uniformly black; there are simply more grains in that region than in the light portions. In the case of a photographic plate showing double-slit diffraction, one will observe many grains in the region where one would have predicted reinforcement of light and few grains where the diffraction pattern would have resulted in interference (Fig. 12-13). One would be inclined to guess that the grains are initiated by individual photons, that more photons go to positions on the plate where reinforcement of waves is expected than to positions where interference should occur.

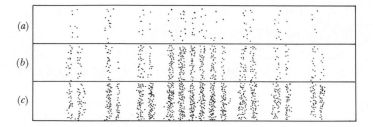

(a)

(b)

(c)

FIG. 12-13

The graininess of a double-slit diffraction pattern. Photons preferentially go to the regions of the photographic plate where reinforcement of waves should occur. A weak exposure may show only a near-random distribution of dots (a). The diffraction pattern becomes distinct as more photons arrive (b and c).

This guess is reinforced by observations made with a photoelectric cell. If the accessory equipment for holding the photographic plate is removed from the arrangement shown in Fig. 12-12, light that passes through the aperture at *C* will be detected by the photoelectric cell. When the very weak light source is used, the photoelectric cell responds by showing individual, sudden surges of current, suggesting action by individual photons (Fig. 12-14). When the aperture is moved slowly sidewise, the number of surges per second alternately increases and decreases, tracing out another grainy double-slit diffraction pattern.

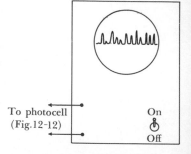

To photocell (Fig.12-12)

On

Off

FIG. 12-14

An oscilloscope trace of the photocell current. The spot on the scope face is swept horizontally while the current is registered by vertical deflection of the spot. The trace shows that the current flows in short bursts as if produced by random arrival of photons.

Resolving the paradox: The wave measures the probability of finding a photon

It would appear that one can make only very general predictions as to where an individual photon will go after passing through the double slit. We can, however, talk about its relative probability of going one place or another, and this relative probability is expressed *exactly* by the laws governing the interference of the waves that we discussed when we described diffraction using the wave theory of light. Photons have a high probability of going to points on the photographic plate where diffraction principles say that light should be intense and a low probability of going to points where these principles say the light should be weak. Our conclusion must be that *when we speak of the intensity of an electromagnetic wave, we are stating the probability of finding photons.*

Light appears, from this analysis, to be both wave and particle, as if the wave structure guides the particle to its destination. The particle accompanies the wave, but one can never be exactly sure where the particle is located in the wave. In the case of the double-slit experiment, the wave clearly is affected by both slits while the photon goes through one or the other with equal probability. The waves that pass through both slits produce alternate regions of reinforcement and interference; the photons (the energy) select regions where the wave is intense. One photon only makes a dot; many photons produce a picture.

12-5 THE WAVE NATURE OF A MOVING PARTICLE

The knowledge that light behaves both as a wave and as a stream of moving particles should prepare us for the next surprise: objects normally regarded as particles also manifest wave properties.

The suggestion that this dual particle-wave character or "duality" should apply to moving objects as well as to light was first made in 1924 by the French scientist Louis de Broglie (rhymes with "doily"). His hypothesis was based on the general observation that nature often reveals physical and/or mathematical symmetry;* no experimental evidence of a particle's wave nature was available to him.

De Broglie's hypothesis involves the following logic:

1. The momentum of a photon is given by

$$\text{Momentum} = \frac{h}{\lambda}$$

2. The momentum of a particle is given by

$$\text{Momentum} = mv$$

3. The wavelength of a particle might then be expected to be

$$\lambda = \frac{h}{\text{Momentum}}$$

or
$$\lambda = \frac{h}{mv} \tag{12-8}$$

Equation (12-8) is the expression for the de Broglie wavelength of a moving particle. That the particle must be moving in order to manifest wave properties becomes evident if one introduces $v = 0$ into the equation; a wave having an infinite wavelength has little meaning.

Let us examine de Broglie's equation to see what wavelength certain specific moving particles should possess. We start with a familar

* For example, a changing magnetic field produces an electric field and a changing electric field produces a magnetic field.

object, a baseball ($m = 0.15$ kg) traveling with a velocity of 17 m/sec; Eq. (12-8) predicts a wavelength λ given by

$$\lambda = \frac{6.6 \times 10^{-34}}{0.15 \times 17} = 2.6 \times 10^{-34} \text{ m.}$$

This wavelength is so small compared with the physical size of the baseball that one cannot imagine an experiment in which it could be detected. Further practice in applying Eq. (12-8) to other objects of macroscopic size will suffice to convince one that the de Broglie wavelength of familiar objects drawn from our day-to-day experience will never concern us.

On the other hand, if we turn to submicroscopic particles, such as electrons and protons, rather different conclusions will be reached. For example, an electron traveling with a velocity of 3×10^{7} m/sec has a wavelength of 0.24×10^{-10} m, and a proton traveling with a velocity of 3000 m/sec has a wavelength of 1.3×10^{-10} m. Since these wavelengths are larger than the presumed size of the particles themselves and, in fact, are comparable to the wavelength of electromagnetic waves (x rays) whose wave character can be verified by diffraction-interference studies, it would appear to be possible to design an experiment that would test the hypothesis.

Experimental verification

De Broglie's hypothesis was first tested in 1927 by G. P. Thomson in Scotland and independently by C. J. Davisson and L. H. Germer in the United States. Their experiments resemble the diffraction experiments in which the wave properties of light and x rays had been established. Electrons were accelerated by an electron gun and caused to strike a crystal that, because of the regular spacing of its atoms, behaves

FIG. 12-15

Verifying the wave properties of electrons. Electrons fired at the thin crystalline film *F* produce diffraction rings on the photographic plate *P*. X rays (photons) having the same wavelength will produce the same ring pattern.

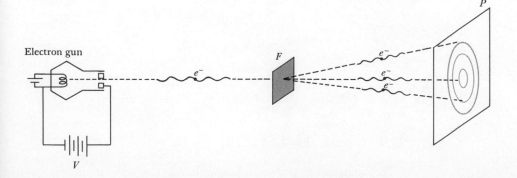

as a diffraction grating. The scattered electrons emerge from the crystal in certain preferred directions (rather than in random directions as primitive particles would be expected to scatter). The direction of scatter permits the calculation of the wavelength of the electrons; these calculated wavelengths agree with de Broglie's predictions and thus confirm the hypothesis. In fact, the wave character of a moving electron is now so well established that every well-equipped hospital possesses one or more electron microscopes whose operation is continual testimony to the wave property of electrons.

More recent experiments have shown that protons, neutrons, alpha particles, and even the more massive atomic nuclei possess a "de Broglie" wavelength as described by Eq. (12-8).

The wave we associate with a moving particle may be called a *matter wave* to emphasize the fact that the kind of wave under discussion is not an electromagnetic wave. It would seem appropriate that this discussion should conclude with a subsection entitled "What is waving?" similar to our earlier conclusion of the discussion of electromagnetic waves. It must be remembered, however, that this subsection was written at the conclusion of our description of the electromagnetic theory of light so that even that discussion may be subject to modification in view of what has been said about the nature of light in this chapter.

The fact is that matter waves are an invention to assist us in describing particles and in predicting what particles will do. The wave property of a particle describes the probability of finding the particle and in diffraction plays the same role in determining the ultimate distribution of particles that the wave property of photons played in comparable experiments. No further qualities of matter waves have ever been detected, nor, for that matter, are they anticipated.

SUMMARY   In this chapter we have extended the wave model of light to provide an explanation of certain aspects of the behavior of light when observed in its encounters with matter at the submicroscopic level. According to this revised version, all the energy and momentum of light is transported as tiny packets or "quanta" called *photons*.

The energy and the momentum of a photon are given by

$$\text{Energy} = hf \tag{12-1}$$

and

$$\text{Momentum} = \frac{h}{\lambda} \tag{12-4}$$

where f and λ are the frequency and the wavelength, respectively, of the light and h is a universal constant of nature known as Planck's constant. The value of h is

$$h = 6.625 \times 10^{-34} \text{ joule sec}$$

The intensity of an electromagnetic wave is now understood to express the probability of finding a photon. Thus in an experiment with a diffraction device such as a double slit, the slits deal with the wave precisely as described in the previous chapter. The detector, however, "sees" the light as a myriad of individual bursts of energy associated with individual photons. The observed diffraction pattern is the cumulative effect of many photons.

This wave-particle "duality" is also observed in connection with particles in motion. A beam of monoenergetic particles can be diffracted, thus it manifests wave properties. The diffraction pattern shows that the wavelength of a particle in motion is

$$\lambda = \frac{h}{mv} \qquad (12\text{-}8)$$

As in the corresponding case of electromagnetic waves, the intensity of a matter wave measures the probability of finding a particle. The diffraction pattern produced by a beam of atoms or electrons is the cumulative effect of the arrival of many particles.

QUESTIONS AND PROBLEMS

12-A1. According to Fig. 12-6, the response of metals to light varies widely. Which of the metals mentioned in the figure are insensitive to visible light?

12-A2. The wavelength of red light is 0.75×10^{-6} m. (*a*) What is the energy of 1 photon of red light? (*b*) What is the momentum of one photon of red light?

12-A3. List all the properties that we have ascribed to a photon. Next list any additional properties that you think we ought to be able to measure if the existence of such a particle is to be believed. Examine your second list carefully; cross out any properties that are not really crucial to the photon model of light.

12-A4. Your roommate says that a photon simply isn't a particle at all and that we would all be better off if no one had ever used the word "particle" in describing it. Prepare either a defense of his position or find an argument to refute it.

12-A5. A 1000-volt power supply is placed across an electron gun. What are the velocity and the wavelength of the matter wave associated with the electrons emitted by the electron gun?

12-A6. The person to whom reference has been made in Prob. 12-A4 says that the wave properties ascribed to moving electrons, protons, etc. are just an invention contrived by some professor to get answers that conform to certain experimental observations. Is there any merit to this contention? What arguments would you present in support of the view that waves associated with moving "particles" have the same kind of reality as those associated with light?

12-B1. According to Fig. 12-6, a photon must have a frequency of 11.7×10^{14} vib/sec or greater in order to eject electrons from gold. (*a*) What is the

wavelength of the light barely able to eject electrons from gold?
(*b*) What will be the energy in joules of electrons ejected from gold
by light having a frequency of 15×10^{14} vib/sec?

12-B2. When blue light ($\lambda = 0.4 \times 10^{-6}$ m) falls on a certain metal plate,
the most energetic electrons emitted by the plate have an energy of
3×10^{-19} joules. (*a*) How much energy is used in simply dislodging
an electron from the metal? (*b*) What retarding voltage is needed
to stop all electrons emitted under these circumstances?

12-B3. By using a piece of smoked glass as a filter, the beam of red light
($\lambda = 0.75 \times 10^{-6}$ m) emitted by a laser can be reduced in intensity
to 10^{-9} joule/sec. This light is then permitted to fall on the plate of
a photoelectric cell as in Fig. 12-3. (*a*) How many photons arrive each
second? (*b*) What current in amperes should the meter read, assuming
that each photon ejects 1 electron?

12-B4. Assume that an x-ray photon ($\lambda = 0.60 \times 10^{-10}$) encounters an
electron at rest in a head-on collision. (*a*) What is the wavelength of
this photon after the collision? (*b*) What is the energy of the electron
after the collision?

12-B5. The electron gun shown in Fig. 12-15 produces a diffraction pattern
that is exactly duplicated by a source of x rays whose wavelength
equals 4×10^{-11} m. What voltage V must have been applied to the
electron gun under these circumstances?

12-B6. Argon gas (mass = 40 amu) in a furnace whose temperature equals
2000°K is permitted to leak out of a small hole in the side. (*a*)
According to Eq. (7-3*b*), what is the average velocity of these atoms
as they emerge from the furnace? (*b*) What is their wavelength?

The
Structure of Atoms

The research of Lavoisier, Avogadro, Mendeleev, and others led to the acceptance of an atomic-molecular model of matter; the prestige of this model has been enhanced by a hundred years of further research and confirmation. This chapter will present evidence for many details of the internal structure of atoms.

13-1 THE SIZE OF AN ATOM

Avogadro's principle (Sec. 7-3) enables one to compute the mass of any atom or molecule in kilograms from its mass in atomic mass units, using the equation

$$m = \frac{M}{N} \qquad (7\text{-}1)$$

where M is the mass of the atom or molecule in atomic mass units and N is Avogadro's number (6.023×10^{26} atoms or molecules per kmol). In this section we wish to consider how much we can infer about the size of atoms and molecules from information presented so far.

The most direct means of estimating the size of an atom is to assume that in solids or liquids atoms are packed tightly together, like oranges in a crate. In the orange crate analogy, we will determine the size of an individual orange in a sealed crate from knowledge of the mass of one orange and from measurements of the volume and mass of the filled crate. The number of oranges present equals the mass of the filled crate divided by the mass of one orange. The volume of one orange is the total volume of the crate divided by the number of oranges. This may not be the most direct way to find the size of an orange!

Let us apply this procedure to the problem of determining the size of one atom of iron. Using a platform balance, we weigh a block of iron to find that it has a density of 7860 kg/m³. From Avogadro's

13

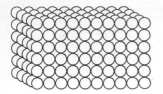

FIG. 13-1

Determining the volume of an atom. If we know the mass of each atom and the mass of the block, we can determine the number of atoms; if, in addition, we know the volume of the block, we can find the volume occupied by one atom.

principle we know that one iron atom has a mass of $55.85/N = 9.27 \times 10^{-26}$ kg; hence a cubic meter of iron consists of $7860/(9.27 \times 10^{-26}) = 8.48 \times 10^{28}$ atoms. Each atom, then, occupies a volume of $1/(8.48 \times 10^{28}) = 11.8 \times 10^{-30}$ m³. If an iron atom is regarded as a sphere sitting in a cubical space of this volume, it must have a diameter of 2.27×10^{-10} m or a radius of 1.14×10^{-10} m.

Next, let us consider a water molecule. As a solid, water (ice) has a density of 917 kg/m³. Assuming that ice consists of a frozen array of water molecules whose masses equal $18/N$ kg each, we find that a water molecule occupies a volume of 32.6×10^{-30} m³. These and other examples are shown in Table 13-1.

The most remarkable feature of these data is the evidence that all atoms are nearly the same size. The most massive atom in the table, uranium, is apparently even slightly smaller than a sodium atom, while its atomic mass is ten times greater.

13-2 EVIDENCE OF ORDER IN ATOMIC STRUCTURE

It is not difficult to find evidence of both order and uniformity in the structure of atoms and molecules. Crystalline materials, for example, show remarkably characteristic shapes whose external order is readily understood to be external evidence of internal order (Fig. 13-2).

Internal order may not always be apparent. Differences, like the difference between yellow and white gold, are ascribable to the presence of an admixture of other metals. Even a handful of mud from a river bank can be separated into elemental pure substances; one specimen of mud differs from another only in the relative abundance of these elemental materials.

TABLE 13-1
The size of typical atoms and molecules

	MASS (amu)	MASS (kg)	DENSITY (kg/m³)	VOLUME (m³)	RADIUS (m)
Beryllium	9.01	1.49×10^{-26}	1,850	8.07×10^{-30}	1.01×10^{-10}
Sodium	22.99	3.82×10^{-26}	971	39.4×10^{-30}	1.70×10^{-10}
Magnesium	24.32	4.03×10^{-26}	1,740	23.2×10^{-30}	1.42×10^{-10}
Iron	55.85	9.27×10^{-26}	7,860	11.8×10^{-30}	1.14×10^{-10}
Gold	196.97	32.70×10^{-26}	19,300	15.9×10^{-30}	1.28×10^{-10}
Mercury	200.61	33.25×10^{-26}	14,193	23.5×10^{-30}	1.43×10^{-10}
Uranium	238.07	39.50×10^{-26}	18,700	21.1×10^{-30}	1.38×10^{-10}
Water	18.016	2.99×10^{-26}	917	32.6×10^{-30}	$(1.59 \times 10^{-10})^*$
Methane	16.032	2.66×10^{-26}	415	64.2×10^{-30}	$(2.00 \times 10^{-10})^*$

* The assumption that these molecules in the solid state behave as neatly stacked spheres may be subject to some question.

FIG. 13-2

A crystal. The external appearance of a crystal is evidence of internal order.
(*Courtesy of Bell Telephone Laboratories*)

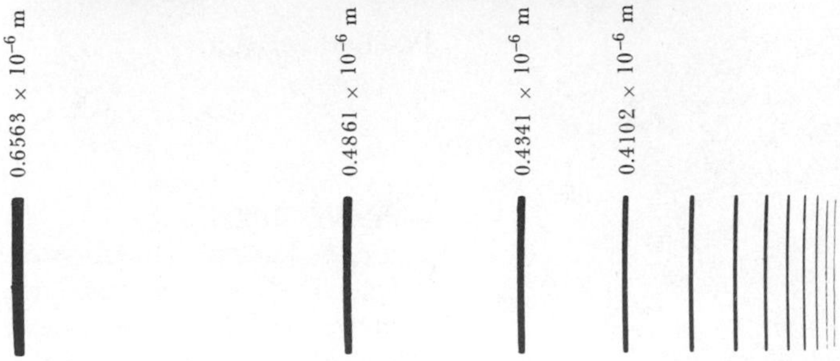

0.6563×10^{-6} m 0.4861×10^{-6} m 0.4341×10^{-6} m 0.4102×10^{-6} m

FIG. 13-3

The Balmer series in hydrogen. The gas that emits light consisting of such an ordered combination of wavelengths must possess mathematically precise order of its own.
(*Courtesy of G. Herzberg*)

The spectrum of hydrogen

More conclusive evidence of order and uniformity in the structure of atoms can be found in the study of the light that is emitted or absorbed by matter. Let us replace the air in the discharge tube shown in Fig. 8-17 with pure hydrogen gas, and, using a spectrograph, let us photograph the spectrum of the light the tube emits in the early stages of evacuation. The resulting spectrum is shown in Fig. 13-3. Hydrogen always produces precisely this spectrum in the visible and near-ultraviolet regions regardless of the origin of the hydrogen gas. Furthermore, even a casual observation of this spectrum is adequate to impress one with the orderly progression of the spectral lines.

The mathematical regularity of the positions of the spectral lines was first expressed in the form of an empirical equation in 1885 by J. J. Balmer, who, by trial and error, arrived at the equation*

$$\frac{1}{\lambda} = 1.0967758 \times 10^7 \left(\frac{1}{2^2} - \frac{1}{n^2} \right) \quad n = 3, 4, 5, \ldots \quad (13\text{-}1b)$$

Balmer's equation not only yields the wavelengths of all the spectral lines in the spectral series known to him, it also correctly predicts many additional lines in the series that have since been observed with improved apparatus. Even more remarkable was the discovery of

* Balmer did not write his equation in precisely this form. Had he done so he would have quoted the value of the constant in the equation as 1.09693×10^7 instead of the more recently determined value of 1.0967758×10^7.

a series of spectral lines in the ultraviolet and of several series of spectral lines in the infrared whose wavelengths may be computed by making minor changes in Balmer's equation. The wavelengths of the lines in three of these series are given by the equations

Lyman Series:

$$\frac{1}{\lambda} = R\left(\frac{1}{1^2} - \frac{1}{n^2}\right), \qquad n = 2, 3, 4, \ldots \qquad (13\text{-}1a)$$

Balmer Series:

$$\frac{1}{\lambda} = R\left(\frac{1}{2^2} - \frac{1}{n^2}\right), \qquad n = 3, 4, 5, \ldots \qquad (13\text{-}1b)$$

Paschen Series:

$$\frac{1}{\lambda} = R\left(\frac{1}{3^2} - \frac{1}{n^2}\right), \qquad n = 4, 5, 6, \ldots \qquad (13\text{-}1c)$$

with $R = 1.0967758 \times 10^7$ in each case. Series of spectral lines in hydrogen have also been observed in which the first term in the bracket is $1/4^2$, $1/5^2$, etc. In fact, with this extension, the group of equations account for *all* spectral lines emitted by atomic hydrogen. The simplicity of these equations indicates the existence of basic simplicity and order in the structure of the hydrogen atom.

13-3 THE RUTHERFORD ATOM

As we begin to formulate models of the structure of atoms, we must draw together certain facts that will play a significant role in determining what models are acceptable. We know, for example, that matter is normally uncharged; hence we may assume that individual atoms in their normal state possess equal amounts of positive and negative charge. Since we have identified a specific particle that possesses negative charge, the electron, and, having learned that electrons can be ejected from a metal by elevating its temperature (the Edison Effect—Sec. 9-3) and by irradiation with light (the photoelectric effect—Sec. 12-1), we will assume that electrons make up the portion of the atom that is negatively charged. Furthermore, since the mass of the electron is very small in comparison with the mass of the atom, we must decide that most of the "stuff" that constitutes an atom must be something other than negative charge.

Thomson's plum pudding model

In 1903 J. J. Thomson proposed a "plum pudding" atomic model according to which electrons are distributed throughout the body of a positively charged sphere. Thomson was unable, however, to devise an arrangement of electrons in a blob of positive charge that would either satisfy certain well-established conditions for electrical stability or explain the spectral lines known to be emitted by the elements. Drastic improvements were clearly needed.

The nuclear atom

The Thomson model was finally discarded when Ernest Rutherford came forward with his interpretation of the now-famous alpha-particle scattering experiment, which had been performed in 1909 by two of this students, H. Geiger and E. Marsden. Rutherford had previously confirmed that the component of the radiation known as alpha particles emitted in the radioactive disintegration of polonium actually consists of positively charged helium atoms ejected at high speed by the radioactive nuclei. In the experiment in question, these high-speed particles were directed at a piece of very thin metal foil. A detecting device was oriented to observe any alpha particles that might be deflected (or "scattered") in passing through the foil. Calculations showed that should atoms consist of a "plum pudding" of electrons in a larger sphere of positive charge, few if any alpha particles should be deflected by more than a few degrees. Instead, Geiger and Marsden reported that some alpha particles are deflected through very large angles and occasionally through an angle of nearly 180 degrees. Such a large deflection is impossible on the Thomson model.

Rutherford devised an alternate model of the atom to explain this wide-angle scattering. According to the revised model, all the positive charge, plus nearly all the mass, is assumed to reside on a tiny nucleus at the center of the atom. The electrons are envisaged to form a cloud in the vicinity of the nucleus; this cloud with very little mass constitutes most of the "volume" of the atom.

Such an atom should have a very open structure, and most alpha particles should be able to plow right through the electron cloud without being appreciably deflected. However, if an alpha particle should pass near a nucleus, a significant force of electrical repulsion would produce a large deflection. Rutherford derived an equation expressing the deflection that individual alpha particles should experience. Although no attempt will be made here to deal directly with Rutherford's equation, it is not difficult to see that a mathematical equation should relate the fraction of incident particles that are deflected through a given angle to the various properties of the impinging alpha particles, and to the properties of the nuclei of the foil. The equation links the following known, measurable, and initially unknown quantities:

1. The velocity of the incident alpha particles (known).
2. The mass of the incident alpha particles (known).
3. The electric charge of the incident alpha particles (known).
4. The fraction of the alpha particles deflected through a given angle (measurable).
5. The nuclear charge of the foil atoms [not known, but with Rutherford's equation, determined by (1), (2), (3), and (4)].

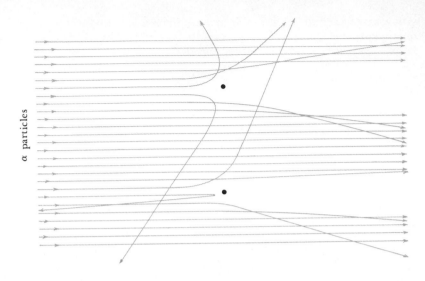

FIG. 13-4

The scattering of alpha particles. Most of the incident α particles pass through the foil without being affected. Those that come close to the nuclei of foil atoms are deflected by the local electric field.

The experimental data of Geiger and Marsden, showing the fractions of the incident alpha particles that are scattered through various angles, agreed with Rutherford's equation and thus confirmed that the alpha particles experience a force described by Coulomb's law even when it has penetrated deep inside the atom itself. Although the initial experiments of Geiger and Marsden were not sufficiently accurate to enable one to compute the nuclear charge, scattering experiments performed by Chadwick in 1920 yielded direct measurments of the nuclear charge of copper (29), silver (47), and platinum (78). This experiment provides the only direct measurement of the charge on a nucleus.

13-4 THE SIZE OF THE NUCLEUS

We may now estimate the sizes of the nuclei of particular atoms, using data that one obtains in scattering experiments.

The alpha particles available to Geiger and Marsden were not energetic by modern standards, and since they were deflected in accordance with equations based on Coulomb's law, we know that no alpha particle at this energy actually makes contact with a nucleus in the foil being bombarded; instead, the alpha particle is simply deflected by the Coulomb field of that nucleus.

An alpha particle that approaches a nucleus along a line directly toward its center should decelerate and come to rest momentarily at some distance r from the center of the nucleus; it should then turn back along its initial path. The initial kinetic energy of such a particle must be precisely equal to its potential energy when at rest in its position of closest approach (Eq. 8-7). Thus

$$\frac{1}{2}mv^2 = \frac{kQq}{r} \tag{13-2}$$

where r is the distance of closest approach, k is the Coulomb constant, Q is the charge on the nucleus, and q, m, and v are the charge, mass, and initial velocity of the alpha particle respectively.

These experiments enable one to estimate the size of the nucleus of a gold atom. Assuming that scattering data have indicated that the nuclear charge for gold equals $79e$, and introducing into Eq. (13-2) the known values of v, m, Q, and q,

$$v = 2 \times 10^7 \text{ m/sec}$$

$$m = \frac{4.00}{6.023 \times 10^{26}} = 0.663 \times 10^{-26} \text{ kg}$$

$$Q = 79 \times e = 126. \times 10^{-19} \text{ coul (gold)}$$

$$q = 2 \times e = 3.2 \times 10^{-19} \text{ coul (helium, the alpha particle)},$$

we find a value for r at the instant both particles are at rest:

$$r = \frac{2kQq}{mv^2} = \frac{2 \times 9 \times 10^9 \times 126. \times 10^{-19} \times 3.2 \times 10^{-19}}{0.663 \times 10^{-26} \times 4 \times 10^{14}}$$

$$= 2.7 \times 10^{-14} \text{ m}$$

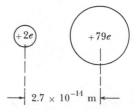

FIG. 13-5

Estimating the size of a nucleus. An α particle has been brought momentarily to rest 2.7×10^{-14} m from the nucleus of a gold atom. This distance is the maximum possible value of the sum of the radii of the two particles.

To interpret this result, visualize the situation at the instant the alpha particle is at rest 2.7×10^{-14} m from the nucleus of the gold atom. At this instant, the centers of the two spheres are at their distances of closest approach. Consequently, this distance represents the largest possible value of the sum of the radius of the alpha particle and the radius of the nucleus of the gold atom. We do not know the size of either particle but neither nucleus can have a radius greater than 2.7×10^{-14} m.

More recent studies, to which reference will be made in Sec. 15-6, show that the radius of the gold atom equals 0.69×10^{-14} m. Rutherford's estimate was approximately four times too large.

These investigations have given us a new picture of matter. Most of the mass of the atom is now seen to be crowded into a nucleus having a radius of less than 0.7×10^{-14} m, while the radius of the atom itself is approximately 1.3×10^{-10} m, 20 thousand times larger than its nucleus. Were one to construct a scale model of a gold atom with a baseball to represent the nucleus, the outermost fringes of the model would extend half a mile in all directions (see Fig. 15-1). So tiny are the nuclei relative to the size of the atom in which they are located that an alpha particle approaching a thin foil "sees" 99.999 percent of the foil as open space even if the foil is 400 atoms thick!

Many perplexing problems, of course, remained unresolved by Rutherford's model of the structure of atoms. The theory on which the model is based says nothing about the structure of the nucleus and gives little information regarding the distribution or motion of the electrons. Why, for example, if electrons are negatively charged, are they not pulled directly into the positively charged nucleus?

An immediate and tempting theory is that the electrons are in orbit about the nucleus, like the planets in their orbits about the Sun. According to electromagnetic theory, however, such a rotating charge should radiate energy, and because of this loss of energy, it should spiral into the nucleus. However, since atoms do not collapse but retain their "size," the electrons, if in orbit about the nucleus, somehow manage to defy conventional electromagnetic theory.

A new era in physics was launched in 1913 when the Danish physicist Niels Bohr proposed his radical "second assumption," which established a planetary model of the hydrogen atom. As in the development of any theory, the assumptions were only guesses whose validity would be accepted only as long as they continue to provide details of nature that are not at variance with experimental observations. The first success of Bohr's model was impressive; the assumptions led directly to Balmer's equation for the lines in the spectrum of hydrogen.

Bohr's first assumption

Bohr's first assumption was that the hydrogen atom consists of a relatively massive, positively charged nucleus, which we will call a *proton*, about which a negatively charged *electron* travels in a circular orbit. Because hydrogen atoms have no net charge, the charge on the proton was assumed to be equal to the charge on the electron but of opposite sign. This Coulomb force holds the electron in a circular orbit where, contrary to electromagnetic theory, it does *not* radiate. The Coulomb force, therefore, provides the centripetal force, so we have

$$\frac{ke^2}{r^2} = m\frac{v^2}{r} \tag{13-3}$$

where e, m, and v are the charge, mass, and speed respectively of the electron and r represents the radius of its orbit. The Coulomb constant k, of course, equals 9×10^9 N m²/coul².

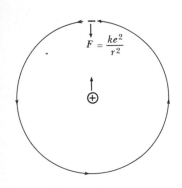

FIG. 13-6

Bohr's first assumption. The electron travels in a circular orbit under the action of a Coulomb force and does NOT radiate energy while in a given orbit.

Bohr's second assumption

In stating Bohr's second assumption regarding the origin of radiation emitted by the hydrogen atom, we will not strictly follow his inspired but complicated logic; rather, we shall make use of a simplifying hypothesis based on the de Broglie matter waves introduced in Sec. 12-5. According to this view of matter, an electron in orbit

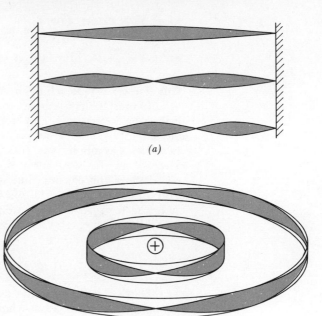

FIG. 13-7

Bohr's second assumption (updated version). (a) A violin string is limited in its vibration by the fact that an integral number of half wavelengths must fit into the space between its supports. (b) The wave associated with the moving electron must fit a whole number of times into the circumference of the orbit.

about a nucleus is accompanied by a matter wave whose wavelength is related to the mass and velocity of the electron by the equation

$$\lambda = \frac{h}{mv} \qquad (12\text{-}8)$$

The ability of a hydrogen atom to hold an electron in a circular orbit without radiating energy, according to this interpretation, results from the formation of "standing waves" by the matter waves accompanying the electron. This means that an electron in orbit in a hydrogen atom has some of the characteristics of a plucked violin string: the wavelength of the wave must fit into the space available to it. In the hydrogen atom, we require that the electron's wavelength fit a whole number of times into the circumference of its orbit; symbolically, this may be written, using Eq. (12-8), as

$$n\frac{h}{mv} = 2\pi r, \qquad n = 1, 2, 3, \ldots \qquad (13\text{-}4)$$

This equation constitutes Bohr's second assumption. Electrons are restricted to orbits whose radii satisfy this equation; all orbits having radii of intermediate magnitudes are ruled out. To determine the "allowed" radii, we solve Eq. (13-4) for v and substitute this value of into Eq. (13-3), obtaining

$$r = \frac{n^2h^2}{4\pi^2kme^2} \qquad (13\text{-}5a)$$

$n = 5$

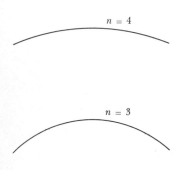

$n = 4$

$n = 3$

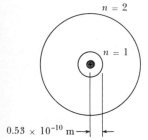

$n = 2$

$n = 1$

0.53×10^{-10} m →

FIG. 13-8

Electron orbits in hydrogen. The circular orbits predicted by Bohr's theory have radii proportional to the square of the number of the orbit.

and using known values of h, m, and e, we find

$$r = 0.529 \times 10^{-10} n^2 \qquad (13\text{-}5b)$$

This enables us to sketch the hydrogen atom as in Fig. 13-8.

It must be emphasized that these radii represent *possible* orbits of the single electron in a hydrogen atom and that (according to Bohr and subject to some reinterpretation) the electron can be in only one of these orbits at any instant. If energy is added to a hydrogen atom, the electron may be lifted out of one orbit into another, but if left to its own devices, this electron, like water running downhill, must seek an orbit of lower energy. Ultimately, unless one continues to supply energy to the system, every atom should be found with its electron in the orbit of least energy, which is the orbit with $n = 1$.

Bohr's third assumption

Bohr's third assumption describes how a photon is created from the energy lost by an atom when an electron makes a transition from one orbit (or "energy state") to another. Letting W_i represent the initial energy of the proton-electron system and W_f represent its final energy, the change in energy of the system equals $W_f - W_i$. Since this loss of energy is precisely the energy emitted, the photon should possess an energy equal to $W_i - W_f$. However the energy of any photon is given by $hf = hc/\lambda$, so we have

$$hf = \frac{hc}{\lambda} = W_i - W_f \qquad (13\text{-}6)$$

The total energy of an electron in any one of its orbits is the sum of its potential plus its kinetic energy:

$$W = -\frac{ke^2}{r} + \frac{1}{2}mv^2$$

which, using $v^2 = ke^2/mr$ from Eq. (13-3), and our expression for r [Eq. (13-5)], becomes

$$W = -\frac{ke^2}{2r} = -\frac{2\pi^2 k^2 me^4}{h^2}\left(\frac{1}{n^2}\right) \qquad (13\text{-}7)$$

Introducing this expression for W into Eq. (13-6) and solving for $1/\lambda$, We have

$$\frac{1}{\lambda} = \frac{2\pi^2 k^2 me^4}{h^3 c}\left(\frac{1}{n_f^2} - \frac{1}{n_i^2}\right) \qquad (13\text{-}8a)$$

Since n_i and n_f are both whole numbers, it is clear that Eq. (13-8), given appropriate choices of n_i and n_f, has precisely the same form as Eqs. (13-1), determined empirically by Balmer and others. Introducing known values of k, m, e, h, and c, we find

$$\frac{1}{\lambda} = 1.0968 \times 10^7 \left(\frac{1}{n_f^2} - \frac{1}{n_i^2} \right) \qquad (13\text{-}8b)$$

Even the constant is the same. Bohr's assumptions must contain a lot of truth.

13-6 THE HYDROGEN ATOM:
A RECAPITULATION OF THE BOHR MODEL

We now have a vastly improved model of the hydrogen atom. Let us summarize the overall process by which light is emitted by gas in a discharge tube, making use of our knowledge of basic principles of chemistry and our knowledge of the nuclear atom given us by Rutherford and Bohr.

Hydrogen gas consists of hydrogen molecules, each of which is made up of two hydrogen atoms held together by interatomic forces. A hydrogen atom consists of an electron and a proton separated from each other by a distance that is enormous when compared to the size of either the proton or the electron.

When hydrogen gas is placed in a discharge tube across which a thousand or more volts are applied, a significant number of molecules are dissociated into individual atoms by collision with other moving

FIG. 13-9
Bohr's third assumption. This diagram differs from Fig. 13-8 in that, by using Eq. (13-7), the hydrogen orbits have been spaced according to the total energy in joules of the system rather than the location of the electron. The energy emitted as a photon equals the loss of energy accompanying the transition of an electron from one orbit to another of lower energy.

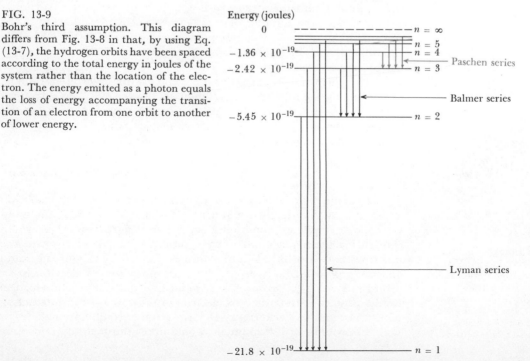

Hydrogen

+1

Helium

+2

2,1

Lithium

+3

2,8

Neon

+10

2,8,1

Sodium

+11

2,8,18,32,18,12,2

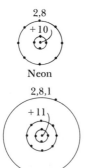

Uranium

+92

FIG. 13-10

The arrangement of
electrons in atoms. A
given atom is most stable
when its electrons are in
nearby orbits. A given
orbit, however, is filled
to capacity when it
possesses $2n^2$ electrons.

particles. The power supply not only dissociates the molecules into individual atoms but gives energy to the atoms as well. The single electron in a given hydrogen atom, therefore, may be in any *one* of a number of possible orbits at a given instant; the possible orbits are those whose circumferences are divisible a whole number of times by the wavelength of the matter wave associated with the moving electron. This condition, together with the fact that the electron in any one of the orbits is held in its path by the Coulomb force of the positively charged proton, limits the energy of the electron-proton system to a relatively small number of values.

A photon is emitted whenever an electron jumps from one orbit to another of lower energy. Because a given specimen of gas consists of a very large number of atoms, photons associated with nearly every possible electron transition are constantly being emitted in the form of billions of individual bursts from billions of individual atoms.

13-7 THE PERIODIC TABLE

The periodic table (see Appendix 4) has been discussed briefly in Sec. 7-5. At that point, since we were speaking from the point of view of nineteenth-century chemistry, this table may have seemed to be but a fortuitous arrangement of the elements. The fact, however, that all elements known to Mendeleev, as well as all elements discovered since, fit into this table would seem to give it special significance.

The significance of the periodic table was greatly enhanced by Rutherford's demonstration that the atomic number of each element (its ordinal number starting with the lightest element, hydrogen) corresponds to the number of positive charges in the nucleus of the corresponding atom. The subsequent x-ray studies of Moseley (discussed in Sec. 13-8) confirmed this interpretation of the atomic number beyond reasonable doubt.

The chemical properties of the elements are determined almost exclusively by the number and arrangement of the electrons surrounding the nucleus, for it surely must be the electrons that serve as the "cement" that holds atoms together as molecules and determines the geometrical arrangement of atoms in molecules.

An "atom-building" procedure

Imagine that we could manipulate the particles that make up an atom in such a detailed manner that we could construct any atom we choose simply by combining some basic ingredients. We would construct a hydrogen atom using a proton with one electron in the innermost $n = 1$ orbit. Helium would be formed by adding mass and another proton to the hydrogen nucleus and a second electron to an orbit in its vicinity. However, we would find that the second electron would insist on joining the first electron in the $n = 1$ orbit, interacting with it in some mysterious way to form a stable combination.

In order to form lithium, we would add a third proton and more mass to the nucleus and another electron to the external region. This third electron, however, would refuse to join the first two electrons in their $n = 1$ orbit but find maximum stability in the next larger orbit, the $n = 2$ orbit. As may be seen in Fig. 13-10 this electron is in a Coulomb field almost identical to the electric field experienced by the lone electron in hydrogen: after all, the nucleus with its three protons combined with the inner "shell" of two electrons produces a core that is electrically the equivalent of one elementary charge. Lithium is accordingly very "hydrogenlike"; it has chemical properties similar to hydrogen and exhibits a similar series of spectral lines, presumably because the lone external electron determines most of its properties.

Proceeding to build up more atoms by adding mass and protons to the nucleus, we would find that eight electrons must be added to the $n = 2$ shell before the kind of stability we see in helium is repeated. This element is neon ($_{10}$Ne). Adding an eleventh proton to the nucleus, we form sodium ($_{11}$Na), another element with many hydrogenlike properties (see Fig. 13-11).

One can continue in this way to form heavier elements. As electrons are added to the $n = 3$ shell, stability is again achieved by adding eight more electrons, now a total of $(2 + 8 + 8) = 18$ electrons; element number 18, argon ($_{18}$A), is very much like helium and neon. The addition of one more proton and an electron forms potassium ($_{19}$K), whose outer electron is in the $n = 4$ shell; potassium resembles the other hydrogenlike elements, lithium and sodium.

Apparently certain combinations of electrons are very stable. Two electrons produce stability; eight electrons may also produce a closed, stable combination; 18 electrons would also be a stable combination and so on. (The series of stable combinations is given by $2n^2$, where n is 1, 2, 3, etc.) An electron added to a stable or "closed" shell produces one of the hydrogenlike elements.

As a final example, consider the elements whose atomic numbers are one less than those with filled shells. These elements ($_9$F, $_{17}$C, $_{35}$Br, and $_{53}$I) each lack one electron in their outer shell; it is not surprising that they all react with other elements in very similar ways.

The principle of atom building just described has a few complications, which we will mention only briefly. Although it is true that

FIG. 13-11

The spectra of hydrogen and hydrogenlike atoms. Transitions into the $n = 3$ orbit in the above three elements result in the series of spectral lines shown in this figure. The yellow light emitted by sodium is due to the first line in the sodium series shown above.

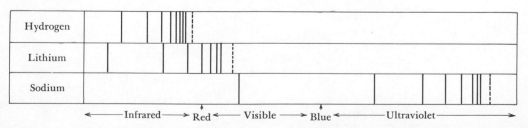

each shell of a given value of n will accept no more than $2n^2$ electrons, one does not find in every case that the "next" electron, as we build up the atoms, falls into the unfilled shell of lowest value of n. An analysis of the spectra of the heavier elements, in fact, shows that many atoms have more than one partly filled shell. On the basis of our simple model, for example, we would expect uranium ($_{92}$U) to have its 92 electrons distributed among five shells with 2, 8, 18, 32, and 32 electrons respectively. One finds, instead, that the electrons are arranged in seven shells with 2, 8, 18, 32, 18, 12, and 2 electrons. Thus although our simplest model has not been followed exactly, it is significant that only one of these numbers is not 2, 8, 18, or 32.

13-8 X RAYS

Although x rays secure their name from the uncertainty that accompanied their early identification ("x" = unknown), it is now well established that they are short wavelength electromagnetic waves. In contrast to visible light, whose wavelength lies in the range 0.4 − 0.75×10^{-6} m, these rays typically have a wavelength shorter than 10^{-9} m. They can be generated by bombarding a metal with high-speed charged particles, or they may be emitted by a radioactive material.

In practice, nearly all x rays are generated by bombarding a metal target with electrons in an x-ray tube similar functionally to the model shown in Fig. 13-12. Such an x-ray tube is, in fact, a discharge tube, differing from the discharge tube shown in Fig. 8-16 in that the x-ray tube possesses a heated cathode and a wedge-shaped anode or "target." The heated cathode, of course, provides more electrons than a cold cathode, while the beveled shape of the target directs the x rays out of the tube.

The accelerating voltage may be in the range 2000 to 200,000

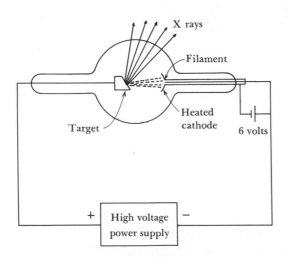

FIG. 13-12

An x-ray tube. The supply of electrons is enhanced by heating the cathode with the 6-volt power supply. X rays are produced at the target by impact of the electrons.

volts for a tube of this design, but by using particle accelerators with internal targets, much higher voltages have been employed. A machine known as a betatron, for example, can produce electrons that have fallen through 100 million volts; an electron accelerator at Cambridge, Massachusetts, has reached 6 billion volts.

X rays are produced in the target of an x-ray tube by the transformation of part or all of the kinetic energy of impinging electrons into photons of electromagnetic radiation. The most energetic photon emitted by a given x-ray tube should therefore be expected to possess an energy equal to the kinetic energy of an individual electron at the instant it strikes the target. Recalling that any charge q falling through a potential difference V acquires an energy equal to Vq [Eq. (8-5b)], we can write

Energy of incident electron = energy of most energetic photon

$$Ve = hf_{max} \tag{13-9}$$

or since $c = f_{max}\lambda_{min}$,

$$Ve = \frac{hc}{\lambda_{min}} \tag{13-10a}$$

where e is the electron charge, c is the velocity of light, and f_{max} and λ_{min} are, respectively, the frequency and the wavelength of the most energetic photon. Introducing known values of h, c, and e into this equation, we find

$$\lambda_{min} = \frac{1.24 \times 10^{-6}}{V} \tag{13-10b}$$

According to this equation, x rays whose wavelength is as short as 6.2×10^{-10} m are produced when one uses a 2000-volt power supply and as short as 2×10^{-16} m at the other extreme of 6 billion volts.

The kinetic energy of an electron may be transformed into electromagnetic radiation by one or both of two processes. Photons produced when an electron simply loses its energy by being slowed down in the target metal is called Bremsstrahlung, a German word meaning "braking radiation" or "energy released when an electron is decelerated." For the purpose of this chapter, however, we are far more interested in the x-ray line spectrum that is superimposed on the continuous Bremsstrahlung background.

A typical x-ray spectrum is shown in Fig. 13-13, where instead of showing the photographic plate on which the spectrum was recorded, we shown a graph of intensity (blackness of the plate) versus wavelength. Several features of this spectrum should be noted:

1. Most of the x-radiation is emitted as a continuous spectrum; notice that the short wavelength limit of this continuum corresponds to that predicted by Eq. (13-10b).

2. Sharp lines appear at certain well-defined wavelengths. The positions of these lines are determined by the metal chosen for the target; hence a line does not change its position when the

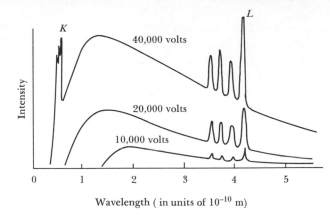

Intensity

K

40,000 volts

L

20,000 volts

10,000 volts

0   1   2   3   4   5

Wavelength (in units of 10^{-10} m)

FIG. 13-13

The x-ray spectrum of silver. This figure shows the spectrum x rays produced at various voltages by an x-ray tube with a silver target. Notice that the K-series is not produced unless 40,000 or more volts are applied.

voltage is changed. As in the case of the continuum, a line will not appear at all if its wavelength lies below the limit for the voltage being employed.

Origin of x-ray line spectra

X-ray line spectra can be explained by using the electron-shell model of the atom that was introduced in the previous section. Each target atom, in its normal or stable condition, consists of a nuclear charge surrounded by shells of electrons with well-defined energies. High-energy electrons bombarding the target may penetrate the target atoms, sometimes losing energy to produce Bremsstrahlung but colliding with shell electrons as well. Such encounters may result in knocking shell electrons out of their stable locations into shells of higher energy.

Figure 13-14 shows a bombarding electron entering from the left and passing close to a bound electron to remove it from the closed shell, where it had formed a stable electronic arrangement. An electron from an outer shell immediately falls into this vacancy, leaving a vacancy for another electron. In each such transition, a photon is emitted whose energy equals the loss in energy of the electron making the transition. Transitions from the outer shells into the "$n = 1$" shell produce a series of spectral lines known as the K-series, transitions into the "$n = 2$" shell produce the L-series, and so forth. These series correspond to the Lyman, Balmer, etc. series in hydrogen but, because the nucleus of a target atom has a much larger electric charge than the hydrogen nucleus, consist of spectral lines with much shorter wavelengths.

Using x rays to determine atomic number

Early studies of x-ray spectra resulted in the conclusive identification of the atomic number of each element as the number of elementary positive charges on its nucleus. Figure 13-15, based on a research paper by A. G. J. Moseley in 1913, shows the positions of the

K-series lines as emitted by x-ray tubes with targets of some of the lighter metals. Prior to this publication, cobalt and nickel, because of their atomic masses of 58.9 and 58.7, respectively, had been assigned atomic numbers of 28 and 27 instead of 27 and 28 as shown here. There can be little doubt regarding which assignment is correct.

SUMMARY

According to the atomic model described in this chapter, an atom consists of a nucleus whose diameter is less than 5×10^{-14} m surrounded by an electron cloud having a diameter of approximately 2×10^{-10} m. The nucleus has a positive charge equal to Ze. Each atom in its stable condition possesses Z electrons arranged in shells whose position relative to the nucleus is identified by a small integer, $n = 1, 2, 3$, etc, with $n = 1$ associated with the shell nearest the nucleus. A given shell can contain as many as $2n^2$ electrons.

According to the Bohr model of the hydrogen atom, a hydrogen atom consists of a proton with an electron held in a circular orbit in its vicinity by a Coulomb force of attraction to the proton. The possible

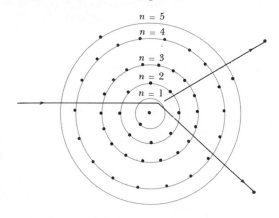

FIG. 13-14

The origin of x-ray line spectra in silver. An electron enters from the left, having fallen through 10,000 or more volts. In collision with one of the inner electrons, a vacancy is produced. Transitions into this or a similar vacancy produce x-ray line spectra.

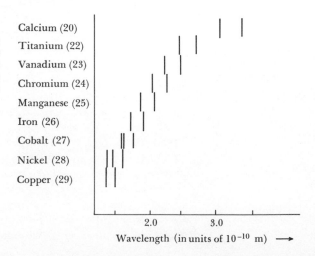

FIG. 13-15

X-ray line spectra of several elements. The positions of the *K*-series spectral lines shift toward shorter wavelengths as one goes to elements of higher atomic numbers. One would have no difficulty telling that scandium ($Z = 21$) is missing.

Calcium (20)

Titanium (22)

Vanadium (23)

Chromium (24)

Manganese (25)

Iron (26)

Cobalt (27)

Nickel (28)

Copper (29)

2.0   3.0

Wavelength (in units of 10^{-10} m) ⟶

orbits are restricted to those for which the wavelength of the matter wave of the electron will fit an integral number of times into the circumference. A definite amount of energy can then be associated with an atom whose electron is momentarily located in a given orbit. Energy is emitted by an atom when an electron makes a transition from one orbit to another of lower energy. These considerations lead to the expression

$$\frac{1}{\lambda} = R\left(\frac{1}{n_f^2} - \frac{1}{n_i^2}\right) \qquad (13\text{-}1)$$

where $R = 1.0968 \times 10^7$ and n_f and n_i are integers with $n_i > n_f$. The fact that this expression, obtained by using the assumptions stated above, yields every line in the spectrum of hydrogen indicates the general validity of the Bohr hydrogen model.

Metals emit x rays when bombarded with electrons. Two separate processes are recognized:

1. A continuous spectrum ("Bremsstrahlung") is emitted when an incident electron passing near the nucleus is accelerated by the intense Coulomb force of attraction.

2. A line spectrum is produced when incident electrons dislodge electrons from inner electron shells. When this occurs, photons are emitted when electrons fall into these "vacancies" from outer shells.

The emission of radiation by an x-ray tube is the response of the target metal to individual electrons. The most energetic photon can, therefore, have no more energy than one of the incident electrons. Consequently, the highest frequency (f_{max}) or the lowest wavelength (λ_{min}) is given by

$$Ve = hf_{max} = \frac{hc}{\lambda_{min}}$$

**QUESTIONS
AND
PROBLEMS**

13-A1. Estimate the number of gold atoms that, laid side by side, would form a single line of atoms around your finger.

13-A2. Cut a slit in a piece of heavy cardboard and place it in front of a fluorescent desk lamp to produce a line source of light. Examine the spectrum, using a phonograph record as a diffraction grating (Prob. 11A6). How does the spectrum of the light from your desk lamp differ from that emitted by a distant streetlamp?

13-A3. You decide to make a scale diagram of a hydrogen atom and start by representing the nucleus as a dot with a radius of 0.5 mm ($= 5 \times 10^{-4}$ m). How far away should the dot representing the electron in the $n = 1$ orbit be located? The electron in the $n = 2$ orbit? NOTE: The radius of a proton equals 0.8×10^{-15} m.

13-A4. In what form is energy emitted by water tumbling over a waterfall?

13-A5. Assuming that electrons in atoms arrange themselves in spherical shells with $2n^2$ electrons per shell, prepare sketches of the assumed distribution of electrons in (a) oxygen ($Z = 8$), (b) sulfur ($Z = 16$), (c) calcium ($Z = 20$), and (d) actinium ($Z = 89$). (Use Fig. 13-10 as a model.)

13-A6. Estimate the wavelengths of the K-series line in the x-ray spectrum of scandium from data you can secure from Fig. 13-15. Using Fig. 13-14 as a model, show what occurs when these two lines are emitted.

13-B1. The mass of a copper atom equals 63.5 amu and the density of copper equals 8,940 kg/m³. If we assume that metallic copper consists of tightly stacked spherical atoms, what is the radius of a copper atom?

13-B2. A proton gun operated at 500 volts fires a proton directly at a distant electrically charged metal sphere. If this proton stops and turns back when it reaches a position 0.1 m from the center of the sphere, what is the charge on the sphere?

13-B3. Using Balmer's equation, compute the wavelength of the first three lines of the Balmer series in hydrogen.

13-B4. What is the frequency of the most energetic photon associated with the hydrogen Lyman series?

13-B5. Equation (13-5b) shows that the radius of the innermost orbit of the hydrogen atom equals 0.529×10^{-10} m. (a) Compute the velocity of an electron in that orbit. (b) Compute the wavelength of the matter wave associated with such an electron.

13-B6. An x-ray tube is operated at a potential difference of 100,000 volts. What is the shortest wavelength of the x rays produced by this tube?

13-B7. The K series in the x-ray spectrum of silver appears at a wavelength of 0.57×10^{-10} m. What is the least voltage that will suffice to excite these lines in an x-ray tube with a silver target?

Relativity

Newton's laws and the relationships stemming from them are accurate descriptions of the behavior of matter when the objects under study are at rest or are traveling with velocities readily attainable in the laboratory. In this chapter we will see that more general equations and some surprising new insights into the nature of space and time are obtained when we consider any system whose velocity relative to us is large, specifically, as large or larger than 1/10 of the velocity of light.

14-1 FRAMES OF REFERENCE

Among the ideas that have been presented so far, it is unlikely that one could identify any concept more basic to our logic than the concept of an inertial frame of reference. It was in such a frame that the laws of physics have been stated to possess authority; whenever we encountered an accelerated frame, the laws either did not seem to apply or, at best, required special treatment. In accelerated frames objects did not even travel in straight lines; fictitious forces had to be invented.

In setting up a definition of an inertial frame of reference, we needed a physical structure that we could depend on to be surely "at rest." In searching for such a structure, our attention was drawn to the "fixed" stars; these stars became basic to the entire fabric of physics when we defined an inertial frame of reference as "any reference system at rest or moving with constant velocity with respect to the fixed stars" (Sec. 2-2).

Having established this definition of an inertial system, we stressed the importance of always thinking about physics from the vantage point of an inertial system. The reader was assured that all the laws of physics are valid in any inertial system. All has seemed well for the past ten or more chapters as we proceeded to act on the assumptions (1) that the fixed stars are a completely valid basis for the definition of the primary inertial frame of reference and (2) that all laws of physics are the same in any inertial frame. Unfortunately, although these as-

14

sumptions have been effective in dealing with the specific situations discussed, some basic inconsistencies are present. Historically these difficulties began to give trouble almost immediately after Maxwell published his famous paper on electromagnetic theory, which describes light as an electromagnetic wave.

Throughout this text we have repeatedly set up models of how we felt our surroundings are constructed. These models have helped us to think about things that we did not really understand very well. We said "Nature behaves as it would, were it to be like this model." Sometimes the model was wrong and had to be abandoned or greatly modified. Sometimes, because of our tendency to retain a model that had lost its usefulness, a model became a hindrance to progress.

Maxwell had said that electromagnetic waves travel with a velocity of 3×10^8 m/sec, and common sense told us (and all nineteenth-century physicists as well) that these waves had to be disturbances in some sort of medium. The fixed stars had been attached to a primary inertial frame. So, obviously (or so it seemed), the waves must travel with the velocity of 3×10^8 m/sec through a medium called the *ether*, and this ether was tacitly assumed to be at rest with respect to the fixed stars.

By the end of the nineteenth century astronomers had shown that the "fixed stars" are not really fixed after all but appear to be in motion relative to each other. Who could say but that perhaps the entire primary reference frame is itself traveling with uniform speed in one direction or another? Standing inside the frame and seeing only the frame itself, one had little chance to make a judgment regarding the absolute motion of anything. When the matter was finally resolved, as we will see, even the *concept* of absolute motion vanished.

14-2 THE MICHELSON-MORLEY EXPERIMENT

A Michelson interferometer is an ingenious combination of mirrors and other optical accessories arranged as in Fig. 14-1. At the heart of this instrument is a half-silvered mirror (O) which, as the name implies, transmits half the incident light and reflects the other half. Fully silvered mirrors are located at A and B. Monochromatic light enters along the path SO; a portion is reflected to A and travels back to the eye at E; a portion is transmitted to be sent back to O and reflected to E. The two portions of the original light beam may arrive at E either in phase or out of phase, depending on the relative lengths of the paths OA and OB. A spot at E will be bright if the waves arrive there in phase (crests from one path coincident with crests from the other path). The spot will be dark if the waves arrive at E out of phase (crests from one path coincident with troughs from the other). If one of the mirrors (B) is moved a half wavelength along the line OB, the spot will go from dark to light to dark again. One such change from dark to light to dark is called a *fringe*.

The interferometer would be famous even without its contribution to our thinking about relativity. Because one sees a pronounced

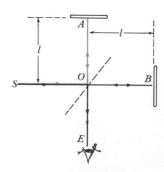

FIG. 14-1

A Michelson interferometer. Light from S is split into two beams that arrive at E either in phase or out of phase, depending on the relative length of the arms OA and OB.

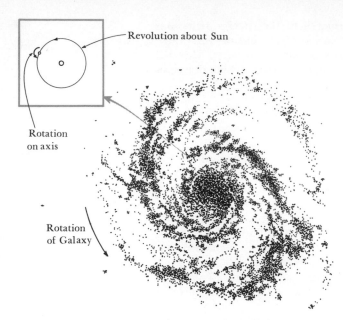

Revolution about Sun

Rotation
on axis

Rotation
of Galaxy

FIG. 14-2

The Earth in perspective. Our galaxy is a
great disk-shaped cluster of stars drifting
and rotating relative to other stars and
galaxies. The Earth rotates on its axis and
revolves about the Sun. The question is:
What is standing still?

change at E when the mirror B is moved a half wavelength, it is used
as a device to measure distances accurately. Using yellow light from
a sodium lamp (wavelength = 0.0000005990 m), one can measure
distances in units of 0.0000002995 m, 3338 fringes are seen as the
mirror B travels 1 mm! In fact it was this instrument that enabled
scientists to define the meter in terms of the wavelength of light (see
footnote, page 3).

If light travels with a well-defined speed relative to a fixed ether
and if the Earth is traveling through this ether, the Michelson interferometer should be able to detect this motion of the Earth. Furthermore, given a sufficiently precise instrument, one should be able, it
was believed, to ascertain the exact velocity of the Earth through the
ether and confirm once and for all whether or not a reference system
based on the fixed stars is at rest in the ether.

In order to understand how the earth's motion should affect the
interferometer, let us reexamine Fig. 14-1. This diagram shows how
the interferometer should respond, if it were at rest with respect to the
alleged ether. We assume that light travels with the same speed in
any direction. Thus if the arms OB and OA are of equal length, we may
be certain to observe reinforcement of the two waves at E.

Suppose, however, that the device is traveling to the right with
a velocity v through an ether as shown in Fig. 14-3. If the light has a
well-defined velocity c relative to the ether, it now has some velocity different from c relative to the apparatus. Light, for example, that is reflected by O travels along the line OA with a velocity c. Its progress toward
the moving mirror is equal to the component of c in the "upward"
direction, which, by the Pythagorean Theorem, is equal to $c' = \sqrt{c^2 - v^2}$. Light traveling from O to B, on the other hand, should be
traveling with a velocity $(c - v)$ relative to the instrument; after reflec-
tion, it should travel relative to the instrument with a velocity $(c + v)$.

Motion of the Michelson interferometer at a velocity v through an
ether should produce a different phase relationship at E than would be

observed if the instrument were at rest in the ether. To verify this, one computes the time required for the light to travel the path OAO' $(= T_A)$ and the time for the light to travel the path OBO' $(= T_B)$. The difference in time $\Delta T = (T_B - T_A)$ is *not* zero. First we find the time T_A.

$$\frac{T_A}{2} = \frac{l}{c'} = \frac{l}{\sqrt{c^2 - v^2}}$$

$$T_A = \frac{2l}{\sqrt{c^2 - v^2}} = \frac{2l}{c\sqrt{1 - v^2/c^2}} \tag{14-1a}$$

One then finds T_B.

$$T_B = T_{B_1} + T_{B_2} = \frac{l}{c - v} + \frac{l}{c + v}$$

$$= \frac{2cl}{c^2 - v^2} = \frac{2l}{c(1 - v^2/c^2)} \tag{14-1b}$$

One can solve for ΔT if he chooses, but the important point is already clear—namely, that T_A is not equal to T_B. If light travels with a well-defined velocity c relative to a fixed ether and if the Earth is traveling with a velocity v relative to that ether, a fringe shift should be observed. In practice, the observer watched the emergent beam while the apparatus was rotated through 90 degrees, since rotation of the apparatus interchanges the roles of the two arms OA and OB and thus doubles the expected phase shift.

The fact is that when the interferometer was rotated through 90 degrees, NO EFFECT WAS OBSERVED. It was as if the Earth were already at rest in the ether! Or that we have to abandon the whole ether concept!

14-3 THE BASIC POSTULATES OF THE THEORY OF RELATIVITY

It is not at all clear that the failure of the Michelson-Morley experiment to detect any evidence of motion of the Earth through space

FIG. 14-3

A Michelson interferometer in motion through an "ether." If light has a definite velocity relative to a fixed "ether," our motion relative to that ether should affect the interference pattern at E because the paths OAO' and OBO' are not identical.

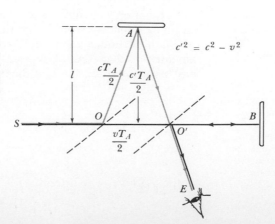

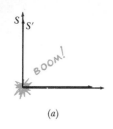

(a)

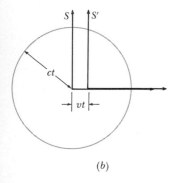

(b)

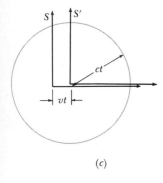

(c)

FIG. 14-4

The light-flash paradox. (a) Frames S and S' move relative to each other at constant velocity v. At the instant the frames coincide, a bomb bursts at their common origin. A spherical light flash spreads out from the source. (b) An observer in the S-frame sees this flash as a spherical wave front with its center at the origin of S. (c) An observer in the S'-frame also sees a spherical wave front but to him it was centered on the origin of the S'-frame.

served as the main stimulant to the development of the theory of relativity; there is little evidence that Einstein regarded it as more significant than a number of other contradictions that existed between Newtonian physics and Maxwell's electromagnetic theory of light. Some historians have even questioned that Einstein had heard of the Michelson-Morley experiment at the time he proposed his solution to the various problems that had arisen. The Michelson-Morley experiment, however, identifies the problem very clearly, and will serve as our starting point.

The basic postulates of the theory are as follows:

1. All observers in inertial frames will secure the same value for the velocity of light independent of their motions relative to other inertial frames.

2. The laws of physics, including the laws governing electromagnetic phenomena, are the same to all observers in inertial frames.

The first postulate is brand new and contradicts seemingly reasonable ideas as to the nature of light and the procedure involved in the vector addition of velocities. The second postulate takes an earlier postulate about the laws of physics, reaffirms it, and gives it added stature.

The paradox of the light flash

The postulates may appear innocent enough until we start analyzing some of the consequences that seem to contradict experience. Let us imagine a frame of reference S' traveling with a velocity v relative to our frame S. Frame S' is in the process of drifting from left to right, but at the very instant the origins coincide, a cherry bomb explodes at that point, emitting a bright flash of light. This bright flash will spread out from the origin in the form of a spherical wave front centered at O and O'; observers in the S-frame located equidistant from the origin all report seeing the flash at precisely the same instant as in Fig. 14-4 (b). However, according to the first postulate, light travels with exactly the same speed relative to frame S'; hence observers in that frame will also see a spherical wave front but centered at O'. Observers traveling in frame S' and located equidistant from O' will all report seeing the flash at the same instant (Fig. 14-4 a).

Later on we will wish to return to this paradox of the light-flash. In preparation, we should note that each set of observers will describe the flash in similar ways, and each observer could, if he chose, write an equation that would tell exactly where the wave front is located at any instant. Letting $t = 0$ at the instant of the flash, he would write

$$x^2 + y^2 + z^2 = c^2 t^2 \qquad (14\text{-}2a)$$

in the S-frame, and

$$x'^2 + y'^2 + z'^2 = c^2 t'^2 \qquad (14\text{-}2b)$$

in the S'-frame. The attentive reader will note and perhaps wonder at the fact that we have introduced a symbol t' for time in the S'-frame, a suggestion that time may proceed at different rates in the two systems.

14-4 PROCEDURES FOR MEASURING TIME AND LENGTH; DEFINITIONS OF SIMULTANEITY AND PROPER VALUES

We have seen that the postulate that light travels with the same speed relative to all observers leads to a seeming paradox in the light-flash experiment. However, the postulate only expresses what Michelson and Morley observed, so we really have only two choices.

1. We can reject the findings of Michelson and Morley and hope that someone will repeat their experiment soon and find a different result.

2. We can accept the findings of Michelson and Morley and the two postulates and see where it takes us. If we elect this course, we must be prepared for some rather different concepts regarding both space and time.

We must keep in mind the fact that physics is a science based on measurement and that some set of rules accompanies the making of any measurement. Two observers cannot get measurements that have meaning to each other unless they agree to follow the same ground rules when they carry out their experiments.

Measuring an interval of time

We start by considering how we carry out a measurement of a time interval. Imagine that a pistol is fired at some position (x'_1) in the S'-frame; later a tape is broken at some other position (x'_2) in the same frame. It is almost irrelevant to the timekeeper that eight sprinters left the position x'_1 and later arrived at x'_2, because the two events under discussion are the firing of a pistol and the breaking of a tape. The procedure for measuring the time interval is simple: Two clocks synchronized with each other will be used, one located at x'_1, the other at x'_2. The clock at x'_1 will be read at the time t'_1 when the pistol is fired; the clock at x'_2 will be read at the time t'_2 when the tape is broken. The time interval will be defined as $(t'_2 - t'_1)$.

If we shift our attention to the sprinter who won the race and utilize the same rules, the procedure goes as follows: When the pistol was fired, the sprinter's wristwatch read t_1; when he broke the tape, his watch read t_2. The watch he was using stayed at the same position in the sprinter's coordinate system throughout the time interval. Therefore, while we utilize the same procedure in measuring $(t'_2 - t'_1)$, we must make a choice in case $(t_2 - t_1)$ differs from $(t'_2 - t'_1)$. The choice, in relativity if not in sports, is the sprinter's watch.

The time interval between two events that take place at a given position in a given reference system as read by a clock located at that position is called the proper time.

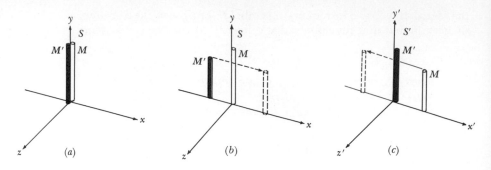

FIG. 14-5

Transverse motion does not affect the length of an object. (*a*) No difference can be seen between rods *M* and *M'* when side by side in one frame or the other. (*b*) *M'* is placed far off to the left in the *S'*-frame whose motion to the right carries *M'* through the origin of the *S*-frame. If the motion of *M'* causes it to shorten, a paintbrush on the top end of *M'* will leave a permanent mark on *M* as it goes by (or if it is lengthened, a paintbrush on *M* would leave a mark on *M'*). (*c*) If all frames are identical, an observer in the *S'*-frame should detect the *same* outcome as the observer in *S*. Thus we have a contradiction unless we assume that transverse motion has no effect on the length of *M* or *M'*.

Measuring the length of an object

Let us next consider the procedure by which we would make a measurement of a length. We have just caught a swordfish. We lay the fish on the deck and bring a tape measure alongside the fish. We read the location of the tip of the sword (x_2) and the tip of the tail (x_1) at the same instant as read by clocks in our frame. The length which we would report would be $(x_2 - x_1)$. Again by implication, observers traveling relative to the fish may secure a different measurement; in any case, the rules of measurement that they would be expected to follow require that they use their own tape measure and make the two observations at the same instant according to *their* clocks. Special significance is given to the measurements made in the frame of the object being measured.

> The length of an object as measured by observers in the frame of reference of the object being measured is called the proper length

The transverse length of a moving object

Having implied that time intervals and lengths may differ, depending on the motion of the observer, the reader may decide that nothing is going to remain unchanged. It can be demonstrated rather easily, however, that the transverse length of a moving object is equal to its proper length. In order to have the satisfaction of knowing that something is not changed by relativity, we will consider this matter briefly.

We start by manufacturing two rods *M* and *M'* and demonstrate to everyone's satisfaction that the rods are identical by placing them side by side as in Fig. 14-5(a). We then place M in a vertical position at the origin of the S-frame and we place M' in a vertical position on a conveyor belt. Just before starting the belt, we place a paintbrush on the upper end of each of the rods M and M', oriented so they will make a mark on anything they pass. We turn on the conveyor belt and M' goes roaring by M at high speed. We stop the belt and examine the two rods. If M' was shortened by its motion, the brush attached to it should leave a mark on M; and, vice versa, if M was shortened by its motion relative to M', its brush should leave a mark on M'. Which will happen?

The answer, of course, is that neither happens. Either frame has exactly the same claim to be called the rest frame, a contention that would be denied if a paint mark should show up on either rod. Stated otherwise, if a paint mark should show up on either M or M', we would have discovered a method of locating a frame of absolute rest, precisely what our postulate has said cannot be done.

Proper mass

Eventually we will discuss the concept of mass as it must be treated in relativity. Again we will find it necessary to distinguish between the *proper* or *rest* mass of an object and its mass as measured by an observer traveling relative to it. The definition of the proper mass will take exactly the same form as our definitions of proper length and proper time. The proper value of the mass of an object, the length of a rod, or time interval between two events is the measurement of the quantity in question as carried out by an observer at rest with respect to the object, the rod, or a clock located where the object or the rod is located or where the events occur.

Synchronizing two clocks in a given reference frame

Finally, we must refine our procedure for measuring the time in a given frame of reference. To be sure, we use a clock that, by definition, ticks off equal increments of time which are registered by the readings on the face on the clock. Time, however, "pervades" the reference system, so our problem becomes a matter of establishing a procedure by which we can all agree that two clocks at different locations in a given frame are synchronized.

The procedure for synchronizing clocks at x_1 and x_2 in the S-frame is shown in Fig. 14-6. An observer stands exactly halfway between the two clocks A and B and, using some mirror device, looks at both clocks simultaneously. One clock or the other is adjusted until the observer declares that they read the same. If the clocks are perfect, they will continue to read the same. In a similar way, clock B can be made synchronous with some third clock C. Experience shows that if A is synchronous with B and B with C, then A will be synchronous with C. In principle, then, one can fill a given reference system with clocks that are each synchronous with all the remaining clocks. The *time*, how-

Two mirrors

A

B

FIG. 14-6

Synchronizing two clocks in a given reference frame. The observer stands halfway between the clocks. The mutually perpendicular mirrors enable him to observe both clocks simultaneously.

ever, is the reading reported by an observer looking at a clock located exactly where he is located.

14-5 THE DILATION OF TIME

We can now prove that an observer in any frame of reference looking at a clock in a frame moving relative to him will find that the moving clock advances more slowly than his own clock. To prove that this happens, we will describe a very special clock, show that it will run slow when in motion relative to us, and, finally, prove that all clocks will behave like our special clock.

The transverse light clock

The clock in question is a "transverse light clock," as shown in Fig. 14-7. Its principle of operation is simple and straightforward. The flashgun *F* gives off a burst of light that travels to the mirror *A* and back to the photoelectric cell *C*. The electric signal generated in the photoelectric cell causes the flashgun to give off another burst of light and, at the same time, advances the hand of the clock by one "tick." The clock will continue to tick indefinitely and can be calibrated in conventional units of seconds if we choose. The time per tick is given by the equation

$$\text{Distance} = \text{velocity} \times \text{time}$$

or
$$T_0 = \text{time per tick} = \frac{2L_0}{c} \text{ sec} \qquad (14\text{-}3)$$

Since this measurement is determined by an observer in the same reference frame as the clock, the time interval T_0 is the proper time between successive bursts of light.

We now place an identical clock in a different reference frame and orient it in such a way that its long dimension *CA* is perpendicular to the direction of its motion relative to us. As we saw in Sec.

14-4, the transverse length L_0 will not be affected by the motion. If we accept the postulate that light travels with a well-defined velocity c relative to us, the time between ticks as measured in the original frame will increase. As shown in Fig. 14-7(b), the flash of light must now travel a significantly greater distance and, to us in the original frame, at the same speed as before. Applying the Pythagorean Theorem and using the same expression to relate distance, speed, and time as before, we have

$$2\sqrt{L_0^2 + \left(v\frac{T}{2}\right)^2} = cT$$

Solving for T, the observed time between ticks, we find

$$T = \frac{2L_0}{c}\frac{1}{\sqrt{1 - v^2/c^2}}$$

or since the proper time (T_0) equals $2L_0/c$, we have

$$T = \frac{T_0}{\sqrt{1 - v^2/c^2}} \tag{14-4}$$

This is the expression for the dilation of time. It states that the time between any two events that occur at a given location in a given frame when measured by a clock at that location in that frame will be less than the lapse of time between the same two events when measured in any frame moving at constant velocity relative to the given frame. The clock in the frame of the events will "run slow" relative to clocks in any other frame.

FIG. 14-7

The transverse light clock. (a) To an observer in the frame of the clock, the light travels a distance $2L_0$ between "ticks." (b) Observed in another frame, the light must travel a much greater distance; hence the moving clock "runs slow."

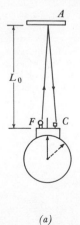

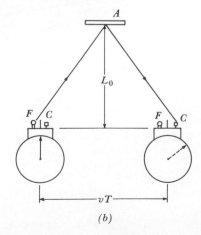

(a)　　　　(b)

Many readers may feel that this proof of dilation of time has meaning only for the transverse light clock. How can my wristwatch, filled with springs, gears, and balance wheels, be affected by motion?

To prove that all clocks are affected in the same manner, we imagine constructing a clear plastic see-through face for our transverse light clock, and we superimpose this face on the face of a perfect mechanical clock operating by any principle whatsoever. We start both clocks and, standing in the same frame as the clocks, we note that they both keep perfect time; the hands of the flashgun clock advance and remain superimposed on the hands of the other clock. We now transfer the combination transverse light clock and superimposed mechanical clock to another frame and watch them in motion. We know that the hands of both clocks will advance together and remain superimposed when observed by someone in the moving frame; all frames are identical. However, the hands cannot appear to be superimposed by one observer and be seen separate and advancing at different rates by another. If they are superimposed, they are superimposed—period. Consequently, although we have not answered the question as to how motion can affect the wheels, gears, and springs of our mechanical clock, we see that we do not need to answer that question. Subtle things take place and the mechanical clock slows down.

14-6 *THE CONTRACTION OF LENGTHS*

We next show that any object in motion in a direction parallel to one of its dimensions will appear shortened to an observer in any frame other than the frame of the object. The measured length of any moving object will be less than its proper length.

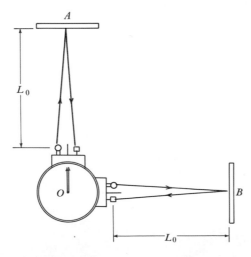

FIG. 14-8

The superimposed longitudinal and transverse light clocks. In their own frame of reference they are identical; hence the two sets of hands stay together. These hands will stay together when observed in any other frame; hence if the transverse light clock "runs slow," the longitudinal light clock also "runs slow."

In deriving an expression for the length of a moving object, we will superimpose the faces of a transverse light clock and a "longitudinal" light clock as shown in Fig. 14-8. Standing in the frame of these two clocks, the transverse and the longitudinal clocks are identical, because in that frame "longitudinal" and "transverse" have no meaning since the clocks are not in motion. To an observer in the frame of these superimposed clocks, light travels from the flashgun along the dimension L_0 and directly back to the appropriate photoelectric cell. Since the clocks are identical, they are certain to keep the same time; the hands are sure to remain superimposed.

Next we place the combination of two light clocks in motion along the x axis—that is, in a direction parallel to OB. One clock now is clearly our transverse light clock and, relative to an observer in the laboratory, it must slow down. However, since the hands of the two clocks remain superimposed, we know that the longitudinal light clock has also slowed down. How can we explain the fact that the longitudinal light clock slows down?

The path followed by light in the longitudinal light clock is easy enough to follow. Light leaves the flashgun and, traveling with a velocity c relative to *us*, reaches the mirror B, is reflected, and, still traveling with a velocity c relative to us, returns to the photoelectric cell. It does this in a time equal to

$$\frac{2L_0}{c} \frac{1}{\sqrt{1 - \dfrac{v^2}{c^2}}}$$

as read by the transverse clock. The only way the light can travel this distance in this amount of time is for the distance OB to change from its proper value L_0 to some other value L. Our question is: What must be the new length L that will enable the beam to get back to the photocell in a time equal to

$$\frac{2L_0}{c} \frac{1}{\sqrt{1 - \dfrac{v^2}{c^2}}}?$$

FIG. 14-9

The longitudinal clock in motion. The mirror B travels a distance equal to vt_1 while the burst of light is in transit from O. After being reflected by B, the light encounters the clock at O'. We can explain that the longitudinal clock "runs slow" only by saying that its length changes.

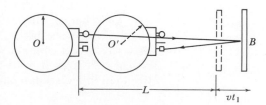

To answer this question, we divide the time interval into two parts, the time (t_1) for the light to travel to the mirror B, and the time (t_2) for the light to travel back to the photoelectric cell. Hence, referring to Fig. 14-9, we see that the light flash must travel a distance equal to

$(L + vt_1)$ to get to mirror B and a distance $(L - vt_2)$ in going back to the photoelectric cell. Since the light travels with a velocity c, we have

$$L + vt_1 = ct_1$$

and

$$L - vt_2 = ct_2$$

Solving for t_1 and t_2 and equating the total time $(t_1 + t_2)$ to the time that has been registered by the transverse light clock, we have

$$\frac{L}{c - v} + \frac{L}{c + v} = \frac{2L_0}{c} \frac{1}{\sqrt{1 - \dfrac{v^2}{c^2}}}$$

Solving for L, we find,

$$L = L_0 \sqrt{1 - \frac{v^2}{c^2}} \tag{14-5}$$

The bar holding the mirror B has been shortened.

Let us summarize the conclusions we have reached so far.

1. Measurements show that one cannot detect absolute motion of the Earth by any experiment whatsoever, including experiments involving measurements of the velocity of light. Thus we conclude that the velocity of light is the same for all observers.

2. We analyze the means by which a transverse light clock would measure time and conclude that a moving transverse light clock will advance more slowly than one at rest in our frame of reference. Superimposing the face of this clock on the face of any clock whatsoever and making use of the postulate that light has one well-defined velocity relative to all observers, we conclude that any clock manufactured and tested in our frame of reference will be seen by us to run more slowly when moving at constant velocity relative to us.

3. The length of an object is not affected by motion perpendicular to that length. However, superimposing the faces of a transverse and a longitudinal light clock, and with the certainty that their hands will remain superimposed, we find that the longitudinal clock has been shortened. We conclude, therefore, that any object manufactured and measured in our reference system will appear to us to be shortened if set into motion relative to us in a direction parallel to the length of the object.

14-7 THE LORENTZ TRANSFORMATIONS

It is possible to arrive at the consequences of the two postulates of relativity in a more formal manner than the procedure we have followed. The more formal, and hence more mathematical, procedure requires that we secure transformations (a set of equations) that express positions and clock readings observed in one frame of reference in terms of those observed in another. The transformations must be unique; that is, one must prove that there is only one set of equations that will work. Finally, the resulting equations must be applicable to situations in which the velocity of a frame of reference is very great yet yield the familiar results discussed in previous chapters if the velocity of the frame of reference is small.

The equations that satisfy all these conditions are known as the Lorentz transformations. In writing these equations, it must be remembered that it is incorrect for us to say that one frame is at rest and the other in motion because it is equally correct to assume either frame to be at rest. However, we will refer to a "laboratory frame" as the S-frame or unprimed frame and a "moving frame" as the S'-frame or primed frame. The S'-frame is represented as moving to the right with a velocity v relative to the S-frame. Therefore the person in the laboratory will introduce a positive value for the velocity v if the S'-frame is moving to the right relative to the S-frame, and a negative value for the velocity if the motion is to the left.

The Lorentz transformations are as follows:

$$x' = \frac{x - vt}{\sqrt{1 - v^2/c^2}} \quad (14\text{-}6a); \qquad x = \frac{x' + vt'}{\sqrt{1 - v^2/c^2}} \quad (14\text{-}7a)$$

$$y' = y \quad (14\text{-}6b); \qquad y = y' \quad (14\text{-}7b)$$

$$z' = z \quad (14\text{-}6c); \qquad z = z' \quad (14\text{-}7c)$$

$$t' = \frac{t - vx/c^2}{\sqrt{1 - v^2/c^2}} \quad (14\text{-}6d); \qquad t = \frac{t' + vx'/c^2}{\sqrt{1 - v^2/c^2}} \quad (14\text{-}7d)$$

In these equations x, y, z, and t refer to space and time coordinates in the S-frame; x', y', z', and t' are corresponding coordinates in the S'-frame.

One can make several very simple tests to verify that these equations do in fact achieve their purpose. However, the algebraic manipulations involved in making these tests will not be carried out here but will be left as exercises for the zealous reader (Prob. 14-B1, 14-B2, and 14-B3). The tests are as follows:

1. One can solve Eqs. (14-6) for x, y, z, and t; one must secure Eqs. (14-7).

2. One can substitute Eqs. (14-6) into Eq. (14-2b); he must secure Eq. (14-2a). Also one can substitute Eqs. (14-7) into Eq. (14-2a); he must secure Eq. (14-2b).

3. Reexamining the procedure given in Sec. 14-4 for determining a length, one can define the length of a moving object, introduce the appropriate equations for position and time as expressed by the Lorentz transformation, and secure the expression for contraction of a moving object [Eq. (14-5)].

4. Reexamining the procedure for determining a time interval and

applying the Lorentz transformations, one can secure the expression for dilation of time [Eq. (14-4)].

5. Finally, we can test to see what form the Lorentz transformations take at low velocity. Consider a velocity of 1860 mi/sec—that is, 3,000,000 m/sec—a low velocity compared to the velocity of light. For such a speed

$$\frac{v}{c} = 0.01 \quad \text{and} \quad \frac{1}{\sqrt{1 - (v^2/c^2)}} = \frac{1}{\sqrt{0.99990}} = 1.0005$$

a number so close to unity that it can be set equal to 1. Consequently, the Lorentz transformations [Eqs. (14-6)] at low velocities become

$$x' = x - vt, \quad y' = y, \quad z' = z, \quad t' = t \quad (14\text{-}8)$$

These equations are known as the Galilean transformations and are just the common-sense transformations that we have been led to expect in our discussions of vector addition (Fig. 14-10).

The Lorentz transformations have survived all the tests. As a result, they may be judged to be the mathematical equivalent of all that has been done to this point.

14-8 *THE TRANSFORMATION OF VELOCITIES*

We now return to Fig. 14-10 and ask a different question: Suppose that a train is traveling with a velocity u, with respect to the ground,

FIG. 14-10

The Galilean transformations. (*a*) The man on the flatcar is at rest at position x' with respect to his frame of reference. His position (x) with respect to the station is continually changing. The Galilean transformation is the "obvious" one: $x' = x - vt$. The Lorentz transformations must agree with this at low velocities. (*b*) A simplified representation of the same situation.

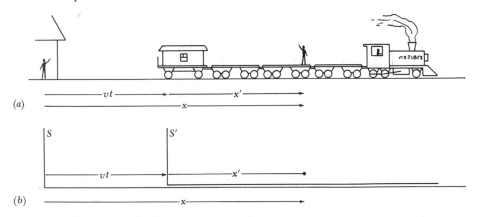

and that a man is traveling with a velocity v' with respect to the train. What is his velocity (v) relative to the ground?

The answer to this question in Galilean physics is straightforward; if a train is traveling with a speed of 20 m/sec and a man walks in the same direction with a speed of 5 m/sec relative to the train, his speed relative to the ground will surely be $(20 + 5) = 25$ m/sec. In general, then, in Galilean physics the velocity transformation equation will be

$$v = v' + u \qquad (14\text{-}9)$$

Everyone is now prepared to find that this transformation will not hold at relativistic speeds; on the other hand, we know that whatever transformation is found must reduce to the Galilean velocity transformation at low speeds.

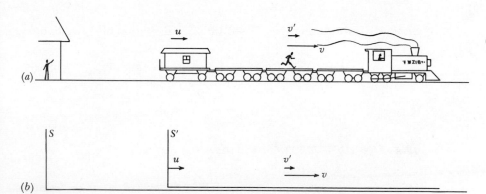

FIG. 14-11

The transformation of velocities. (*a*) The velocity of the man on the flatcar with respect to the flatcar is different from his velocity with respect to the station. The Galilean transformation is the "obvious" one: $v' = v - u$. The transformation used in the special theory of relativity must reduce to this at low velocity. (*b*) A simplified representation of the same situation.

Figure 14-11 describes the relative velocities of the frames and objects in question. The laboratory frame is the S-frame. Another frame, the S'-frame, travels relative to the S-frame with a velocity u. Some object travels relative to both of these frames, with a velocity v relative to the S-frame and a velocity v' relative to the S'-frame.

Equations (14-7) are the Lorentz transformations in the form in which we need them. At time t_1 as measured by clocks in the S-frame, the object is located at x_1; at that instant the object is adjacent to a position x_1' in the S'- frame, and a clock at x_1' reads t_1'. Later the object is at x_2 at time t_2 in the S-frame but at x_2' at time t_2' as observed in the S'-frame. By definition,

$$v = \frac{x_2 - x_1}{t_2 - t_1} \quad \text{and} \quad v' = \frac{x_2' - x_1'}{t_2' - t_1'}$$

Substituting Eqs. (14-7) into the definition of v, we obtain

$$v = \frac{\gamma(x_2' + ut_2') - \gamma(x_1' + ut_1')}{\gamma(t_2' + ux_2'/c^2) - \gamma(t_1' + ux_1'/c^2)} = \frac{(x_2' - x_1') + u(t_2' - t_1')}{(t_2' - t_1') + \dfrac{u}{c^2}(x_2' - x_1')}$$

$$= \frac{\dfrac{x_2' - x_1'}{t_2' - t_1'} + u}{1 + \dfrac{u}{c^2}\left(\dfrac{x_2' - x_1'}{t_2' - t_1'}\right)}$$

where

$$\gamma = \frac{1}{\sqrt{1 - u^2/c^2}}$$

This becomes

$$v = \frac{v' + u}{1 + \dfrac{v'u}{c^2}} \tag{14-10a}$$

Similarly, substituting Eqs. (14-6) into our definition of v', we find

$$v' = \frac{v - u}{1 - \dfrac{vu}{c^2}} \tag{14-10b}$$

14-9 THE DEPENDENCE OF MASS ON VELOCITY

It is very common to refer to a particle as "nonrelativistic" or "relativistic," depending on whether one can secure accurate numerical results using the equations of Newtonian mechanics or finds that he must use the somewhat more formidable equations of the theory of relativity. Strictly speaking, of course, all particles are relativistic particles because the equations of relativity are equally valid at low and high velocities. A decision, therefore, as to the use of a Newtonian versus a relativistic expression is only a judgment regarding the accuracy one demands in his calculations.

Nearly every equation in relativistic mechanics differs from the corresponding Newtonian equation by the presence of the term $\sqrt{1 - v^2/c^2}$; hence this quantity may serve as the criterion for making a choice. Its numerical value is less than unity for all nonzero values of v but approaches unity as v approaches zero; when $v/c = 0.1$ (that is, when $v = 3 \times 10^7$ m/sec), it differs from unity by only 0.5 percent. For most practical purposes, then, one can regard a particle having a velocity greater than 3×10^7 m/sec as "relativistic."

Although talk of relativistic space ships is still a topic only for science fiction, it is common for scientists to conduct experiments with relativistic electrons, protons, and other elementary particles. A relativistic electron, in fact, can be produced by a voltage source of only 3000 volts; even the electrons in most home television sets are "relativistic."

Experiments with relativistic electrons have been conducted for over a half century. As early as 1908 Bucherer reported that the ratio e/m for electron is not constant but decreases with their velocity. The change in mass of a moving particle as reported by Bucherer, and confirmed in hundreds of later experiments, supports the expression

$$m = \frac{m_0}{\sqrt{1 - v^2/c^2}} \qquad (14\text{-}11)$$

where m represents the mass of the moving particle, m_0 represents its rest or proper mass, and v and c represent the velocity of the particle and the velocity of light respectively.

The observation of a relativistic change of mass has become a commonplace to scientists involved in research employing high-energy particles. Since the mystique has largely been removed from this phenomenon by these day-to-day experiences in the laboratory, and in the interest of brevity, we will not present a formal proof of Eq. (14-11); instead let us regard it as an experimental law. Formal proofs may be found in the references.

14-10 MASS, ENERGY, AND MASS-ENERGY

The fact that mass is a form of energy and that under certain circumstances a given amount of mass can be converted into energy in another form was probably well known to all readers before they opened this book. Our proof that

$$\Delta E = \Delta m\, c^2 \qquad (14\text{-}12)$$

where Δm is the change of mass and ΔE is the energy released or absorbed is, then, anticlimatic. We end our discourse on relativity proving what many readers already knew. Hopefully our discussion, however, will have made a contribution to the reader's understanding of the basis for our knowledge of the equivalence of mass and energy.

The proof of Eq. (14-12) that we will present involves only a bit of algebraic manipulation followed by an interpretation. We start with our definition of momentum, namely,

$$p = mv$$

where m, the mass of the object, is given by Eq. (14-11). Squaring both sides and making several straightforward manipulations of symbols on the right side of the expression, we have

$$p^2 = m^2v^2 = m^2c^2\frac{v^2}{c^2} = m^2c^2\left[1 - \left(1 - \frac{v^2}{c^2}\right)\right]$$

$$= m^2c^2 - m^2\left(1 - \frac{v^2}{c^2}\right)c^2$$

From Eq. (14-11) we see that

$$m^2\left(1 - \frac{v^2}{c^2}\right) = m_0^2$$

so we have

$$p^2 = m^2c^2 - m_0^2c^2$$

or

$$p^2 = (m + m_0)(mc^2 - m_0c^2) \qquad (14\text{-}13)$$

We are now in a position for the interpretation phase of our derivation.

Our first step is to recall that all relativistic equations must reduce to an appropriate Newtonian equation at low velocity. For all non-relativistic velocities, we use the familiar equations from earlier chapters. Kinetic energy, for example, can be written as

$$(\text{KE})_{\text{low}} = \frac{1}{2}m_0v^2 = \frac{m_0^2v^2}{2m_0} = \frac{p_{\text{low}}^2}{2m_0}$$

Therefore

$$p_{\text{low}}^2 = 2m_0(\text{KE})_{\text{low}} \qquad (14\text{-}14)$$

where p_{low} represents the momentum and $(\text{KE})_{\text{low}}$ represents the kinetic energy at low velocity.

Our second step is to notice that term by term Eq. (14-13) becomes Eq. (14-14) if we replace each relativistic velocity in (14-13) by a nonrelativistic velocity—that is, p^2 becomes p_{low}^2, and $(m + m_0)$ becomes $2m_0$. Clearly $(mc^2 - m_0c^2)$ and $(\text{KE})_{\text{low}}$ approach zero as v approaches zero. Hence $(mc^2 - m_0c^2)$ must be the relativistic form of the equation for kinetic energy, which, with Eq. (14-11), becomes

$$\text{Kinetic energy} = m_0c^2\left(\frac{1}{\sqrt{1 - v^2/c^2}} - 1\right) \qquad (14\text{-}15)$$

Our third step is to make an additional interpretation of Eq. (14-13). Since $(mc^2 - m_0c^2)$ represents the gain in energy when the object is put into motion, it must follow that m_0c^2 represents energy that the object possessed before it is was put into motion; it is the energy one can associate with its rest mass. Furthermore, we can associate the energy that is put into the object (which, in this case, all shows up as kinetic energy) with the change of mass by writing

$$\Delta E = \Delta m\, c^2 \qquad (14\text{-}12)$$

where $\Delta m = (m - m_0)$, the change of mass.

SUMMARY   The laws of physics have equal validity in any inertial frame of reference; even the velocity of light is the same to all observers. These conclusions mean that there is no preferred reference frame in nature.

A measurement of a length, a mass, or a time interval carried out by an observer in the same reference frame as the object or clock under observation is called the *proper length* (L_0), *the proper mass* (m_0), or the *proper time* (T_0). Any measurement of length, mass, or time $(L, m, \text{or } T)$ carried out by an observer in a frame other the frame of that object or clock will differ from the proper value as follows:

1. A moving clock will appear to run slow when observed by an observer in another frame:

$$T = \frac{T_0}{\sqrt{1 - v^2/c^2}} \qquad (14\text{-}4)$$

2. The longitudinal length of a moving object (L) will be shortened in relation to its length (L_0) measured by an observer in the frame of the object:

$$L = L_0 \sqrt{1 - \frac{v^2}{c^2}} \qquad (14\text{-}5)$$

3. The mass of a moving object (m) will be greater than its mass (m_0) measured by an observer in its own frame:

$$m = \frac{m_0}{\sqrt{1 - v^2/c^2}} \qquad (14\text{-}11)$$

An object with a velocity v' relative to a frame whose velocity relative to the laboratory is u has a velocity relative to the laboratory equal to

$$v = \frac{v' + u}{1 + v'u/c^2} \qquad (14\text{-}10a)$$

Mass is a form of energy. Whenever mass is converted into energy in another form, the conversion occurs according to the equation

$$\Delta E = \Delta m \, c^2 \qquad (14\text{-}12)$$

The kinetic energy of an object is given by

$$\text{KE} = m_0 c^2 \left(\frac{1}{\sqrt{1 - v^2/c^2}} - 1 \right) \qquad (14\text{-}15)$$

At low speeds ($v < c/10$), this expression reduces to the familiar

$$\text{KE} = \tfrac{1}{2} m v^2$$

QUESTIONS AND PROBLEMS

14-A1. How is an inertial frame of reference to be defined now that it has been established that the "fixed stars" are in motion relative to us and to each other?

14-A2. Write the equation of the sphere formed by the light-flash emitted by a cherry bomb 10^{-6} sec after it explodes.

14-A3. A space ship whose velocity relative to the laboratory equals $c/3$ carries a transverse flashgun clock. If the distance from flashgun to mirror L_0 equals 5 m, how far will light travel between ticks of the clock as observed by an observer in the laboratory frame?

14-A4. A *light year* is the distance light travels in one year. (*a*) What is the distance in meters to the nearest star *Alpha Centauri,* known to be located at a distance of 4.3 light years from the Earth? (*b*) What will this distance be if measured by an observer traveling toward that star with a velocity equal to half the velocity of light?

14-A5. An object passes by with a speed equal to half the speed of light and appears to be 1 m long. How long is this object as measured by an observer in its own frame?

14-A6. What is the mass of an electron traveling with a velocity of 2×10^8 m/sec?

14-A7. A proton is given such a velocity that its mass as measured by an observer in the laboratory frame is increased tenfold. What is the velocity of this proton?

14-A8. Your roommate (skeptical as ever) says that he knows that the theory of relativity is wrong. As evidence, he points out that when he glances from one star to another, his line of sight travels much faster than c. How do you counter this argument?

14-B1. Equations (14-6) are a group of simultaneous linear equations expressing certain "unprimed" quantities in terms of certain "primed" quantities. Solve these equations for $x, y, z,$ and t and show that you secure Eqs. (14-7).

14-B2. Show that upon substituting Eqs. (14-6) into Eq. (14-2b), one secures Eq. (14-2a).

14-B3. Let us assume that while standing in the laboratory (unprimed) frame, we observe two events that take place at the same location in another (primed) frame—that is, $x'_2 = x'_1$. We define the elapsed time between these events, as seen by us, as

$$\Delta t = (t_2 - t_1)$$

Introduce the Lorentz transformations into this expression and show that you secure Eq. (14-4).

14-B4. Assume that a star is at a distance of 100 light years ($= 9.48 \times 10^{17}$ m) from the Earth and that a spaceship travels toward this star with a velocity of $2c/\sqrt{5}$. (*a*) What is the Earth-star distance as observed by a passenger in the ship? (*b*) How much will the passenger age in going from the Earth to the star? (*c*) How much will people on Earth age in the interval between the departure of the spaceship and its arrival at the star?

14-B5. A muon is a particle with a mass of approximately 200 electron masses and with the remarkable property of disintegrating into energy in other forms with an average lifetime of about 2×10^{-6} sec as judged by clocks in the muon's frame of reference. What will be the average lifetime of such a muon as measured by an observer relative to whom the muon is traveling with a velocity of $0.99c$?

14-B6. A certain nuclear reaction is known to occur in which the net mass of all the ingredients decreases by 0.2 percent with the emission of an equivalent amount of energy in other forms. What is the fuel value of this material in joules per kilogram? Compare the result you get with the fuel value of dynamite (5.4×10^6 joules/kg).

14-B7. The sketch (Fig. 14-12) depicts a photon that has just been emitted by a high-speed radioactive nucleus. At the instant shown, the nucleus is traveling to the right with a velocity of 6×10^7 m/sec and the photon is traveling to the right (relative to the laboratory) with a velocity of 3×10^8 m/sec. What is the velocity of the photon relative to the nucleus?

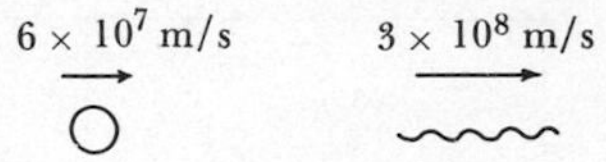

14-B8. An electron falls through a difference of potential of 10 million volts. (*a*) Compute the anticipated velocity on the basis of classical physics [Eq. (10-9)]. (*b*) Compute the velocity anticipated on the basis of the theory of relativity. (*c*) What mass will the electron have?

6×10^7 m/s 3×10^8 m/s

FIG. 14-12

Radioactivity
and
Nuclear Transmutation

The discovery of nuclear energy was an event that undoubtedly will rank in history on a par in significance with the discovery of fire. Whereas fire, a chemical reaction, involves a release of energy accompanying the reorganization of electrons in the proximity of nuclei, nuclear reactions release or absorb energy when structural changes take place in the nucleus itself. Our final chapters will summarize this remarkable part of our story.

15-1 THE SEARCH FOR A NUCLEAR MODEL BASED ON ELEMENTARY PARTICLES

We have seen that all atoms, when judged by the size of their charge clouds, are about the same size, ranging in diameter from approximately 2×10^{-10} m to 3×10^{-10} m (Table 13-1). The size of nuclei, however, increases steadily with nuclear mass,* but even in one of the more massive atoms, the nucleus is less than 2×10^{-14} m in diameter. The electron cloud, then, is tens of thousands of times larger than the nucleus.

When it was found that all nuclei possess a positive charge that is numerically equal to an integral multiple of the charge on a proton, nuclear models were developed in which the nucleus was treated as a clustering of protons. This hypothesis was strengthened by the observation that all nuclear masses are also nearly (but, alas, not exactly) equal to an integral multiple of the proton mass. It was also supported by the discovery by Sir Ernest Rutherford in 1919 that a nitrogen nucleus can be induced to emit a proton, a part of the story to which we will return in Sec. 15-4.

The only particles known at that time which could serve as the "building stones" of a nuclear model were protons and electrons. One of the nuclear models that was used for a time is shown in Fig. 15-2(a),

---

* An empirical equation relating nuclear mass and nuclear radius is given in Sec. 15-6 [Eq. (15-10)].

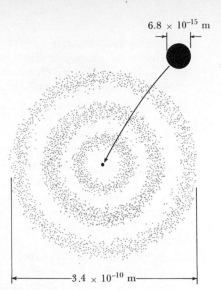

6.8 × 10⁻¹⁵ m

FIG. 15-1

The relative size of the nucleus and its electron cloud in a typical atom (sodium). Had the nucleus been drawn the size of a baseball, the charge cloud would have a diameter of 2 miles.

3.4×10^{-10} m

where it is applied to one of the isotopes of tin. The isotope in question has an atomic number Z equal to 50 and a mass number A equal to 120 and thus may be designated as $^{120}_{50}\text{Sn}$. The mass of the atom was accounted for by imagining a nucleus of 120 protons, while the nuclear charge of 50 elementary charges was explained as the net effect of 120 protons and 70 electrons. In general, then, according to this now out-of-date model, the nucleus was pictured as consisting of A protons and $(A - Z)$ electrons, which, because of their small mass, made little contribution to the total mass of the atom.

This model, however, raised as many questions as it answered. How could such an electric structure be stable? Why didn't the total mass of the ingredients turn out to be exactly equal, instead of only approximately equal, to their combined mass?

In 1932 the neutron, an uncharged particle with a mass very nearly equal to that of the proton, was discovered and provided the substance for an alternate nuclear model, as in Fig. 15-2(b). According to this model, the nucleus of the $^{120}_{50}\text{Sn}$ atom consists of an aggregate of $Z = 50$ protons combined with $(A - Z) = 70$ neutrons. Superficially this may not seem to be much of an improvement because neither the problem of the electrical stability of the nucleus nor the question of the inequality of the mass of the nucleus and the mass of its constituent parts was immediately resolved by this changeover. However, for reasons associated with the total angular momentum (spin) possessed by nuclei, this model was immediately favored over the now-abandoned earlier one.

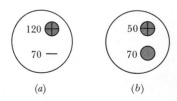

(a)   (b)

FIG. 15-2

Early nuclear models of tin isotope ($^{120}_{50}\text{Sn}$). (a) An early model showing 120 protons and 70 electrons: $Z = 50$, $A = 120$. (b) A later model showing 50 protons and 70 neutrons; again $Z = 50$ and $A = 120$.

15-2 NATURAL RADIOACTIVITY

The nucleus of the atom became the subject of research and investigation with the discovery by Henri Becquerel in 1896 that certain uranium salts possess the property of darkening photographic plates even though the plates may remain carefully wrapped with dark paper. Shortly thereafter, Marie Curie discovered the radioactive element thorium and, in collaboration with her husband, Pierre Curie, the radioactive elements polonium and radium.

Alpha, beta, and gamma rays

Three types of rays were found to be emitted by radioactive materials, distinguished by their different abilities to penetrate materials normally opaque to light. These rays were called alpha rays, beta rays, and gamma rays in the relative order of their ability to penetrate through increasing thicknesses of matter. Furthermore, when radiation from a uranium sample was directed into either an electric or a magnetic field, the beam was split into three beams. The beam of alpha rays was deflected as if it were a stream of positive charges; the beam of beta rays was deflected as if it were a stream of negative charges; and the gamma rays proceeded straight ahead without out detectable change in direction.

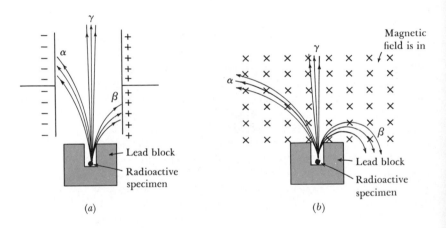

FIG. 15-3

Separation of radiation into three components. Whether sent through an electric (*a*) or a magnetic (*b*) field, the radiation from a naturally radioactive material reveals three components, called α, β, and γ rays, bearing $+$, $-$, and no charge respectively.

Within a few years the three kinds of radiation from naturally radioactive materials were identified as follows:

1. Alpha rays, or alpha particles, are helium nuclei. As a representative nuclear reaction in which alpha particles are emitted, we may consider the radioactive isotope uranium-238 ($^{238}_{92}$U). The disintegration of this isotope may be described by the reaction equation

$$^{238}_{92}\text{U} \Longrightarrow {}^{234}_{90}\text{Th} + {}^{4}_{2}\text{He} + Q \qquad (15\text{-}1)$$

where Q is the energy released or, if negative, absorbed. An atom of thorium-234 is produced by the reaction through the loss by the nucleus of four mass units and two charge units. The fact that there is no net change in mass number ($238 = 234 + 4$) nor in charge number ($92 = 90 + 2$) is a characteristic of nuclear reactions that is never violated.

2. Beta rays, or beta particles, are electrons. Although electrons are surely emitted by nuclei, it is believed that nuclei do not normally contain any electrons, rather that an electron is created in the nucleus and ejected (along with a chargeless and massless particle called an "antineutrino") at the instant disintegration occurs. It is as if a neutron in the nucleus changes into a proton, an electron, and an antineutrino. Beta particles are emitted by thorium-234 ($^{234}_{90}\text{Th}$), the very isotope that was produced in the emission of an alpha particle in (1) above. The disintegration of $^{234}_{90}\text{Th}$ may be written as follows:

$$^{234}_{90}\text{Th} \Longrightarrow {}^{234}_{91}\text{Pa} + {}^{0}_{-1}e + {}^{0}_{0}\bar{\nu}_e + Q \qquad (15\text{-}2)$$

where Pa designates the element protoactinium formed in the reaction, ${}^{0}_{0}\bar{\nu}_e$ represents the antineutrino, and Q, as before, is the energy released. Again there is no net change in mass number ($234 = 234 + 0 + 0$) or in charge number ($90 = 91 - 1 + 0$). Little will be said here about the neutrinos (${}^{0}_{0}\nu_e$) or the antineutrinos (${}^{0}_{0}\bar{\nu}_e$) except that they are remarkable particles possessing energy and angular momentum (or spin) but no mass and no charge. They travel with the speed of light, but since they are not electromagnetic in nature, their interaction with matter is extremely weak. Calculations show that either neutrinos or antineutrinos can travel through 10^{14} miles of solid rock without interacting with a single atom.

3. Gamma rays are electromagnetic waves (photons) differing from x rays only in that they may originate in the nucleus rather than in the electron cloud surrounding the nucleus. Since photons possess neither charge nor mass, neither the atomic (charge) number nor the mass number of the nucleus changes when a gamma-ray photon is emitted. Some change in nuclear arrangement must occur, however, since energy is lost by the nucleus in the process.

The series of naturally radioactive isotopes

There are approximately 50 naturally radioactive isotopes, all but a very few of which can be classified as belonging to one of three radioactive families or series. All these isotopes were present when the Earth's crust was formed; aside from a relatively small contribution from radiation reaching the Earth from outer space (cosmic rays), there appears to be no process for the regeneration of radioactive materials in the Earth's crust.

Each of the three series of naturally radioactive isotopes owes its continued presence in the Earth's crust to the existence of one member that disintegrates very slowly. Consequently, that isotope has a long "half-life," meaning, as will be explained in more detail in Sec. 15-3, the time required for half of a given specimen of the isotope to disintegrate. For example, the first member of the uranium series is $^{238}_{92}$U with a half-life of 4.5×10^9 years. Since this time interval is comparable with the age of the Earth, one may conclude that approximately half of the uranium-238 that was present when the Earth was formed is still present.

The long-lived member of a given radioactive series is the "parent" of other members of the series. For example, we have already seen that $^{238}_{92}$U disintegrates into $^{234}_{90}$Th, which disintegrates into $^{234}_{91}$Pa. These processes constitute the first two of 15 steps, each involving the emission of either an alpha particle or an electron. At the end of the sequence of disintegrations, the nucleus is finally reduced to an

FIG. 15-4

The uranium series. Uranium-234 disintegrates into lead-206 in 14 steps. When an α particle ($^4_2$He) is emitted, the mass number decreases by 4 and the atomic number by 2; when a β particle ($_{-1}^{0}$e) is emitted, the mass number is unchanged and the atomic number increases by 1. This table omits reference to gamma rays, which are emitted by most of the isotopes listed here. [1 MeV equals 1.6×10^{-13} joules (see page 314).]

PARENT ISOTOPE	PARTICLE EMITTED (ENERGY IN MeV)	HALF-LIFE
$^{238}_{92}$U	$\alpha(4.18)$	4.50×10^9 yr
$^{234}_{90}$Th	$\beta(0.19)$	24.1 day
$^{234}_{91}$Pa	$\beta(2.32)$	1.14 min
$^{234}_{92}$U	$\alpha(4.76)$	2.5×10^5 yr
$^{230}_{90}$Th	$\alpha(4.68)$	8.0×10^4 yr
$^{226}_{88}$Ra	$\alpha(4.79)$	1620 yr
$^{222}_{86}$Rn	$\alpha(5.49)$	3.82 day
$^{218}_{84}$Po	$\alpha(6.00)$	3.05 min
$^{214}_{82}$Pb	$\beta(0.65)$	26.8 min
$^{214}_{83}$Bi	$\beta(3.15)$	19.7 min
$^{214}_{84}$Po	$\alpha(7.68)$	1.6×10^{-4} sec
$^{210}_{82}$Pb	$\beta(0.025)$	25 yr
$^{210}_{83}$Bi	$\beta(1.65)$	4.8 day
$^{210}_{84}$Po	$\alpha(5.30)$	140 day
$^{206}_{82}$Pb	Stable	∞

isotope of lead ($^{206}_{82}$Pb), which is stable. The details of the steps by which uranium-238 finally disintegrates into lead-206 are shown in Fig. 15-4. It will be noted that the half-lives of the various isotopes formed along the way vary from 1.6×10^{-4} sec ($^{214}_{84}$Po) to 2.5×10^{5} years. Clearly, none of these intermediate isotopes would still be present were it not for the long half-life of the parent, $^{238}_{92}$U.

The other two naturally radioactive series are known as the thorium series, which initiates with $^{232}_{90}$Th, whose half-life is 1.39×10^{10} years, and the actinium series, which initiates with $^{235}_{92}$U, having a half-life of 7.1×10^{8} years. A fourth series, the neptunium series, whose first member is $^{237}_{93}$Np with a half-life of 2.2×10^{6} years, has not been present in detectable amounts for millions of years because the half-life of the parent isotope is short when compared to the age of the Earth.

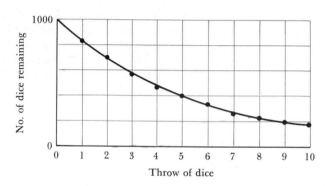

FIG. 15-5

The "disintegration" of 1000 dice. The actual numbers remaining after each throw are represented by the data points. The predicted behavior, that one-sixth will disintegrate in any given throw of the dice, is shown by the curve.

15-3 RADIOACTIVE DECAY

Probability in throwing dice

Consider a box containing 1000 six-sided dice, each of which is painted red on one side. We shake the box vigorously and throw the dice onto a level floor, removing each die that shows its red side up, and declare that all such dice have "disintegrated." On each throw we can expect to remove approximately one-sixth of the remaining dice.

Let us say that on the first throw we remove 171 dice, thus leaving 829. We shake up the 829 remaining dice and throw them, this time removing 137. On the third throw we remove 115 of the remaining 692 dice, leaving 577. In the next throw we remove 95, leaving 482, and so on.

This game could go on for a very long time because the number that disintegrate per throw decreases as the number of dice decreases. However, in four throws the number of dice has diminished to slightly less than half of the original number. Furthermore, in another four throws the number of remaining dice will diminish by half again—that is, to one-fourth of the original number of dice. A graphical display of the law obeyed by the dice in this dice game is shown in Fig. 15-5.

Radioactive atoms obey a similar law except that, in the case of radioactive decay, "time" plays the role that was played by a "throw" in the dice game. Starting from a given number of radioactive atoms, some definite fraction will disintegrate in one second; the same fraction of those remaining will disintegrate during the second second, and so on. Therefore, of the atoms originally present, a definite fraction will disintegrate in a specific number of seconds, a fact that is utilized to describe the decay of a radioactive material. We define the *half-life* of the material as follows:

The half-life of a radioactive sample is the time required for half of the nuclei in the sample to disintegrate.

The numerical values of the half-lives of radioactive materials vary from times shorter than a billionth of a second to billions of years. As examples, the half-life of $^{238}_{92}$U equals 4.5×10^9 years, whereas the half-life of $^{234}_{90}$Th into which it decays is only 24 days. The decay curves of these naturally radioactive isotopes are shown in Fig. 15-6.

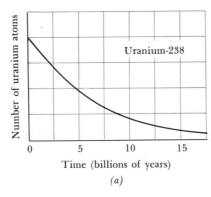

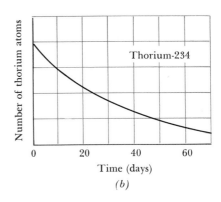

FIG. 15-6

The disintegration of uranium-238 and thorium-234. (*a*) U-238, an emitter of alpha particles, has a half-life of 4.5 billion years, while (*b*) Th-234, which emits electrons, has a half-life of 24 days.

The disintegration of a specific nucleus of a radioactive sample, like the "disintegration" of a specific die in the dice game, is to be regarded as a matter of chance. Were one to focus his attention on one particular $^{238}_{92}$U nucleus, assuming that to be possible, it might disintegrate at once or it might remain inert for millions of years. The only thing that we can say with certainty about it is that we have a 50-50 chance of seeing it disintegrate in 4.5×10^9 years. On the other hand, if we were dealing with 1 gram of uranium (25×10^{20} atoms), it is certain that nuclei in the sample will disintegrate at the rate of approximately a thousand atoms each second.

These considerations show the remarkable certainty that may govern a phenomenon involving a myriad of individual events, each of which is governed by the same laws of chance. This aspect of nature is observed in various contexts in physics; it is frequently known as the "orderliness of chance."

We can express the law that governs the disintegration of a radioactive material by saying that the fraction of the material which disintegrates in a given time period is proportional to that period of time; that is,

$$\frac{\Delta N}{N} = -k\,\Delta t \qquad\qquad (15\text{-}3)$$

where N is the number of atoms of a given kind present at any given instant, Δt represents a short lapse of time, and $\Delta N/N$ is the fraction of the atoms that disintegrate in that amount of time. The constant k is a constant of proportionality and differs from one radioactive material to another. The negative sign signifies that the amount of material (the number of atoms N) decreases due to the disintegration.

15-4 NUCLEAR TRANSMUTATION

All the nuclear processes we have described so far are ones that take place spontaneously. All have involved a net decrease in mass, a change that always accompanies the transformation of some of the internal energy of the nucleus into some other form of energy—for example, into kinetic energy of the emitted particles or into the creation of photons or neutrinos. These nuclear events have something in common with the case of water running downhill, where some of the original potential energy of the water is transformed into kinetic energy or is radiated away as heat (the temperature of the water and, consequently, the temperature of the surrounding air increase) and as sound (a brook babbles.)

In this section we shall examine nuclear processes that are stimulated by interaction with some external particle. The atom that results may be an isotope of the same or of another element; it may be stable or it may be radioactive. The reaction may release energy, in which case it will be referred to as *exothermic*, or it may absorb energy—that is, it may require more energy to produce the reaction than the reaction yields (*endothermic*).

The discovery of nuclear transmutation

As an example of such a nuclear reaction, let us look briefly at the experiment conducted by Ernest Rutherford in 1919, in which a nuclear transmutation was first observed. In this experiment a probe containing a radioactive material was placed inside a vessel filled with nitrogen gas. The alpha particles emitted by the radioactive material were expected to collide with the nearly stationary nuclei of nitrogen atoms. Were this encounter to be an elastic collision, the nitrogen nucleus should have attained a velocity only one-ninth of that of the impinging alpha particle. Instead, particles able to travel up to 40 cm through the nitrogen gas were observed, a distance in excess of the range even of the incident alpha particles. Rutherford's explanation of this observation was that the nitrogen nucleus and the alpha particle must have fused momentarily into a composite but unstable nucleus that exploded by throwing off a proton. This event can be represented as follows:

$$\ce{^4_2He + ^{14}_7N \longrightarrow ^{18}_9F \longrightarrow ^{17}_8O + ^1_1H} \tag{15-4}$$

Rutherford detected the protons by observing tiny flashes of light emitted when they struck a fluorescent screen at a distance from the source.

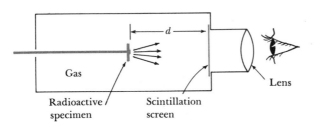

FIG. 15-7

The discovery of nuclear transmutation. Sir Ernest Rutherford found that particles of unanticipated long range are produced by alpha particles in nitrogen gas. These particles were identified to be protons produced in nuclear transmutation of $^{14}_7$N.

This reaction has also been observed in a cloud chamber [Fig. 15-8(c)]. For comparison, Fig. 15-8(a) and (b) shows elastic collisions between alpha particles and nuclei that make up the gas of the chamber.

The discovery of nuclear transmutation opened a floodgate of research and discovery. These developments have been accelerated by the invention of high-voltage machines capable of providing a well-collimated beam of particles that had fallen through a difference of potential of billions of volts. By 1970 the distinguishable nuclear reactions that had been confirmed numbered in the thousands. Hundreds of new isotopes had been studied; many had been isolated in quantity for use or possible use in industrial, military, or medical applications.

The discovery of the neutron

In 1930 the German scientists Bothe and Becker conducted an experiment in which a sheet of beryllium was bombarded by alpha particles. They found that a very penetrating radiation was emitted from the beryllium, capable of passing through a significant thickness of lead. The radiation was assumed to be gamma rays.

Shortly afterward, Frederic Joliot and Irene Curie-Joliot found that this radiation caused protons to be ejected from a block of paraffin

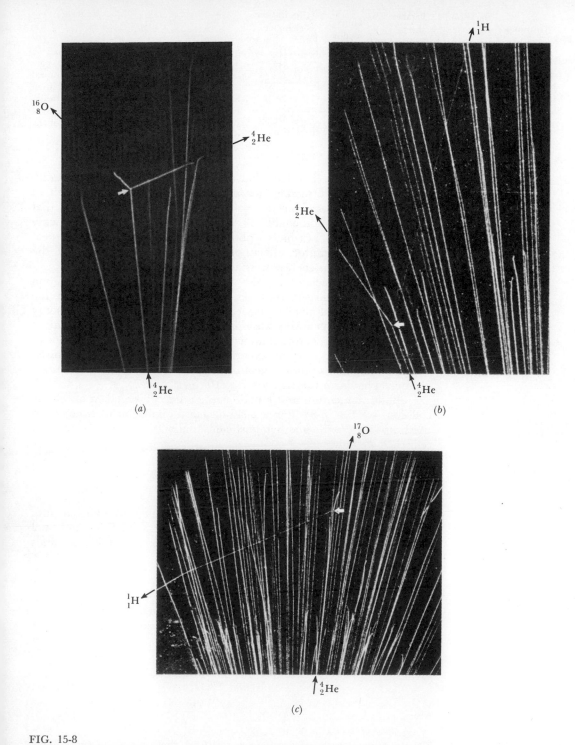

FIG. 15-8

Cloud chamber photographs of alpha particle collisions. (*a*) Elastic collision with oxygen nucleus. (*b*) Elastic collision with a proton. (*c*) Inelastic collision with nitrogen nucleus.

(*By permission of the Royal Society*)

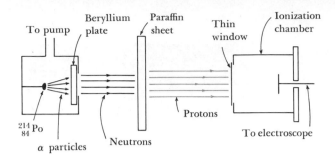

FIG. 15-9

The discovery of the neutron. Alpha particles striking the beryllium plate produced a very penetrating radiation, capable of ejecting protons from paraffin. This radiation was proven to consist of particles possessing no charge and a mass near that of the proton.

placed in its path. The fact that protons emerged from the paraffin was in itself not surprising, for paraffin, a hydrocarbon, is rich in hydrogen. In England Sir James Chadwick immediately argued that the initiating radiation could not possibly be gamma rays, for the only known mechanism by which a photon could transfer energy to a proton (a proton Compton effect, Sec. 12-3) would demand that the gamma-ray photons each possess an unreasonably high energy.

Chadwick decided to test the hypothesis that the radiation consisted of uncharged particles. Such particles would have the necessary quality of penetrating lead sheets and could transfer energy to other particles in elastic collisions. He found that when the radiation encountered a hydrogen-rich material, protons emerged with a velocity of 33×10^6 m/sec, whereas experiments by Norman Feather showed that the same radiation, in encountering nitrogen gas, produced nitrogen nuclei having a velocity of 4.7×10^6 m/sec.

Both momentum and kinetic energy are conserved in an elastic collision (see Sec. 3-6). For a head-on elastic collision of a non-relativistic particle with another particle initially at rest in our frame of reference we have

$$mv = mv' + MV'$$

and

$$\tfrac{1}{2}mv^2 = \tfrac{1}{2}mv'^2 + \tfrac{1}{2}MV'^2$$

where m, v, and v' represent the mass and the initial and final velocities of the impinging particle and M and V' represent the mass and final velocity of the struck particle. Solving these equations for V', one finds

$$V' = \frac{2m}{m + M}v \qquad (15\text{-}5)$$

The mass m and the velocity v of the incident particle are the same in the experiment with protons ($M = 1$) as in the experiment with nitrogen nuclei ($M = 14$). Therefore we have

$$33 \times 10^6 = \frac{2m}{m + 1}v$$

and

$$4.7 \times 10^6 = \frac{2m}{m + 14}v$$

Eliminating v and solving for m, we find that $m = 1.14$ amu. This analysis by Sir James Chadwick in 1932 constituted the discovery of the neutron (see also Fig. 5-10).

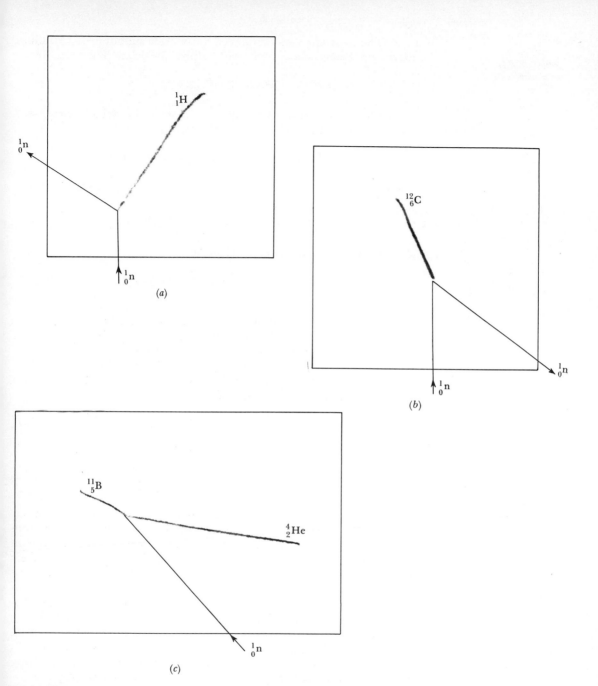

FIG. 15-10

Cloud-chamber photographs (negatives) of neutron collisions. Since the neutron leaves no track, the indicated neutron tracks at best only represent possible collisions in which momentum is conserved. Shown are elastic collisions with a proton (a) and with a carbon-12 nucleus (b) and an inelastic collision with a nitrogen-14 nucleus (c). In the latter case the compound nucleus disintegrated into a boron-11 nucleus and an α particle.

(*Courtesy of Norman Feather and the Royal Society*)

The neutrons were produced by a transmutation of beryllium atoms on being bombarded with alpha particles. The reaction is

$$\ce{^4_2He} + \ce{^9_4Be} = \ce{^1_0}n + \ce{^{12}_6C} \tag{15.6}$$

More recent measurements of the mass of the neutron are based on rather different experiments and show

Mass of the neutron = 1.008665 amu

and clearly differs significantly from the mass of the proton, now being quoted as

Mass of the proton = 1.007277 amu

By comparison, the mass of an electron is

Mass of the electron = 0.0005485 amu

Several events involving neutrons are shown in the cloud chamber photographs of Fig. 15-10. The first two photographs show collisions of neutrons with a hydrogen nucleus and a carbon nucleus. The neutrons, of course, leave no track, for they are uncharged particles. In Fig. 15-10(c) a neutron stimulates a nuclear reaction upon collision with a nitrogen nucleus. The reaction is

$$\ce{^1_0}n + \ce{^{14}_7N} = \ce{^{11}_5B} + \ce{^4_2He} \tag{15-7}$$

15-5 THE POSITRON

The Earth is being bombarded continually by radiation from outer space, which we call *cosmic rays*. Although the existence of this radiation has been known since 1912 when Victor F. Hess carried out the balloon flights that resulted in their discovery, a full understanding of their nature and origin has been difficult to achieve. The *primary rays*, the radiation that actually encounters our upper atmosphere, are now known to consist largely of very energetic protons. When they encounter our atmosphere at an altitude of 20 or more miles, the primary rays produce cascades of complex nuclear transmutations. Lacking until recently the means of getting beyond our atmosphere, scientists have had to content themselves with the difficult task of inferring the nature of the primary rays by studying the secondary rays.

In 1932 C. D. Anderson of the California Institute of Technology was much involved in cosmic ray studies using cloud chambers. Photographs of his cloud chamber showed some tracks that resembled those normally left by electrons except that they appeared to be deflected in the wrong direction in a magnetic field. In order to resolve the problem Anderson designed a cloud chamber that could be located in a magnetic field. The interior of the chamber was divided into two com-

FIG. 15-11

The discovery of the positron. We know that the particle originated at the bottom of the photograph because it clearly lost energy in passing through the lead plate. Since it was deflected to the left by a magnetic field directed into the page, by the Q-v-B rule, the particle must have had a positive charge. (*Courtesy of C. D. Anderson*)

partments by a lead plate placed along the diameter of the cylinder which constituted its active volume. The photograph shown in Fig. 15-11 was secured with this arrangement.

The track shown in the photograph was identified by its density and its curvature to be that of a particle having the mass and charge of an electron. Its direction of motion can be ascertained by noticing that the curvature of its path is different on one side of the lead plate than it is on the other. Because the particle must lose energy in passing through the lead plate, it must have originated on the side where it had the least curvature—that is, on the side where the track was more nearly a straight line. Knowing that the magnetic field was into the plane of the photograph, one has only to apply the Q-v-B Rule (Sec. 9-3) to prove that the particle possessed a positive charge. This interpretation of Fig. 15-11 constituted the discovery of the *positron*, a positively charged electron.

Annihilation of electrons and positrons

Soon after the discovery of the positron it became apparent that a sophisticated theory of the electron proposed earlier by P. A. M. Dirac predicted the existence of just such a particle. The Dirac theory even went so far as to endow the positron with the "antiparticle" characteristic that it should annihilate and be annihilated by any electron it should encounter. The mass energy of both particles should be released in the form of two or, on occasion, three photons.

The mutual annihilation of an electron and a positron is represented schematically in Fig. 15-12. In the specific situation depicted here, the positron drifted in from the left and encountered an electron coming in from the right; both particles are assumed to be traveling at a nonrelativistic velocity before the encounter. Two photons were produced in the small-scale explosion that followed, and, to conserve momentum, these photons left the scene in nearly opposite directions.

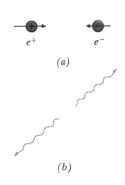

e^+ e^-

(a)

(b)

FIG. 15-12

The annihilation of a positron. This schematic diagram depicts a positron encountering an electron. Both are annihilated; two photons are created and carry away the energy of the electron-positron pair.

It can be shown, in fact, that the process must generate two (or more) photons in order to conserve both energy and momentum.

An application of the Law of Conservation of Energy enables us to determine the nature of the photons produced in electron-positron annihilation. Let us consider the situation in which two identical photons emerge from the encounter of an electron and a positron. Clearly the total energy of the two photons $(2hf)$ was secured at the expense of the mass-energy $(2mc^2)$ of the two particles that were annihilated. Since $f = c/\lambda$, we have

$$\frac{2hc}{\lambda} = 2mc^2 \qquad (15\text{-}8)$$

Solving for λ, we obtain

$$\lambda = \frac{h}{mc} = 0.244 \times 10^{-11} \text{ m}$$

a photon that lies in the wavelength range of gamma rays. The energy of each photon equals the mass-energy of an electron, namely, mc^2 or 8.2×10^{-14} joules.

The fact that any positron is doomed to be annihilated whenever it encounters an electron means that a positron is a very transitory particle in our electron-dominated environment. Although capable of lasting indefinitely in outer space, its lifetime near the surface of the Earth where electrons abound is less than a billionth of a second.

Origin of positrons

Positrons are emitted in the radioactive decay of many short-lived isotopes. A positron is also created simultaneously with the creation of an electron from the energy of a gamma-ray photon in a process called *pair production*.

The creation of an electron-positron pair can take place with the loss of a single photon provided that the photon possesses sufficient energy and provided that matter is present in the vicinity to assist in the conservation of momentum. The most likely way for these conditions to be satisfied is for an energetic photon to pass near an atom where, interacting with the intense electric field of the nucleus, it can transfer momentum to the atom while conserving both total momentum and total energy. In this case, a single photon produces two electrons ($+$ and $-$) each of mass m; such a photon must have a wavelength as small as or smaller than λ_0 in the equation

$$\frac{hc}{\lambda_0} = 2mc^2 \qquad (15\text{-}9)$$

or $$\lambda_0 = 0.122 \times 10^{-11} \text{ m}$$

This photon has a wavelength precisely half, and a frequency precisely twice, that of one of the photons resulting from the annihilation of a positron and an electron. If the photon that produces an electron-

FIG. 15-13
Several pair production events. Gamma rays entered from the left and produced several sets of electron-positron pairs.
(*Lawrence Berkeley Laboratory, University of California, Berkeley, Calif.*)

positron pair has a wavelength less than λ_0, it will possess excess energy that will manifest itself as kinetic energy of the electron-positron pair.

15-6 THE NUCLEUS AS A LIQUID DROP

Any satisfactory model of the nucleus must be endowed with properties that are compatible with the known behavior of nuclei. In this section we shall discuss a model commonly known as the "liquid-drop" model, examining the anticipated properties of a nucleus that conforms to this model and showing that the model does, in fact, possess many of the desired qualities.

Behavior of a liquid drop

A drop of water consists of an enormous number of molecules held together by intermolecular forces: for example, a drop of water only 2 mm in diameter contains approximately 10^{20} molecules. Large drops, however, cannot be formed; under the best of conditions, a drop as large as a pea is about the maximum.

The density of water in a drop of water is independent of the size

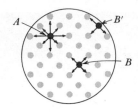

FIG. 15-14

Intermolecular forces in a drop of water. The net force on an interior molecule A equals zero. A surface molecule B or B' is tied to its nearest neighbors and experiences a net inward force.

of the drop; that is, if the volume of one drop is twice the volume of another, its mass is twice as large also. This constancy of the density of a liquid suggests that the molecules that constitute a liquid are in contact with each other and that external forces make little change in the distance of separation of their centers. Experience, indeed, shows that water is almost incompressible.

If intermolecular forces were to act between every pair of molecules in a drop of water, one should be able to form as large a drop as he chooses. The fact that only small liquid drops are possible infers that the intermolecular forces act primarily between nearest neighbors. The effect of short-range forces in holding a drop together will be described by making reference to Fig. 15-14.

Consider molecule A in the figure. This molecule is located in the interior of the drop and is subjected to forces of attraction toward neighbors on all sides. Because these forces act on the molecule in every conceivable direction, the net force on the molecule equals zero. Molecule B, on the other hand, is located on the surface. Forces of attraction tie this molecule to all of its nearest neighbors on the surface and, in addition, produce a net force toward the center of the sphere. Because this is true of every molecule on the surface, the drop acts as if it were enclosed in an elastic membrane which causes it to assume a spherical shape. If one attempts to form a large drop, these "surface-tension" forces continue to operate, of course, but the ratio of the number of surface molecules to the number enclosed becomes smaller, with the result that only a small amount of vibration of the drop will suffice to cause it to "fission"—that is, to break into two or more smaller drops.

Liquid-drop properties of nuclei

We now wish to examine the characteristics of nuclei, noting the characteristics they share with liquid drops. We start by considering the density of nuclear matter.

Several experimental techniques have been developed through the use of which quite accurate measurements of the size, and thus of the density, of various atomic nuclei have been secured. We have already discussed one of these techniques, namely, that employed in Rutherford's early experiments, through the scattering of alpha particles. Recently, the precision of the measurements has been improved significantly through the use of neutrons and high-energy electrons. Although these studies show that nuclei do not manifest sharp well-defined surfaces, and thus cannot be described as simple hard spheres, their characteristics are such that they can be treated as hard spheres to a first approximation. Data on the radii of nuclei over a wide range of mass numbers can be reproduced quite accurately by the equation

$$R = 1.20 \times 10^{-15} A^{1/3} \text{ m} \qquad (15\text{-}10)$$

where R represents the radius of the nucleus and A represents its mass number. According to this expression, the nucleus of one of the lighter atoms with $A = 4$ ($^4_2$He, for example) should have a radius of $1.90 \times$

10^{-15} m, while one with $A = 208$ ($^{208}_{82}$Pb, 52 times more massive) should have a radius of 7.12×10^{-15} m. (The measured values of the radii of $^4_2$He and $^{208}_{82}$Pb are 2.07×10^{-15} m and 7.00×10^{-15} m respectively.)

We are now in a position to compute the density of nuclear matter. Since

$$\text{Density} = \frac{\text{mass of nucleus}}{\text{volume}}$$

and the volume of a sphere is given by

$$\text{Volume} = \tfrac{4}{3}\pi R^3$$

the density of a nucleus of mass number A is

$$\text{Density} = \frac{A/(6.02 \times 10^{26})}{\tfrac{4}{3}\pi(1.2 \times 10^{-15}A^{1/3})^3} = 0.23 \times 10^{18} \text{ kg/m}^3*$$

Since this result does not depend on A, all nuclei must have the same density, at least within the accuracy of Eq. (15-10).

Nuclei also share with liquid drops the characteristic of having a maximum size: nuclei having radii greater than about 7.7×10^{-15} m ($A = 260$) simply cannot be formed. Even at a mass number considerably below 260, nuclei are unstable and tend to disintegrate into smaller nuclei under minor or no external provocation. There are no stable nuclei having mass numbers of 210 or greater. By analogy with liquid drops, one can infer that the binding forces between nuclear particles act only between nearest neighbors. These short-range, attractive, internuclear forces must be large compared to the electrostatic forces of repulsion between protons in the nucleus.

SUMMARY

Most naturally radioactive isotopes are found in one of three major radioactive series, presumably formed with the galaxies and still existent because of the slow decay of their senior parent isotope. In decaying, they emit alpha particles (helium nuclei), beta particles (electrons), and gamma rays (energetic photons).

All radioactive materials disintegrate at a rate proportional to the amount present at any given instant. Thus

$$\frac{\Delta N}{N} = -k\Delta t \tag{15-3}$$

The time required for half of the nuclei in a given sample to decay is called the *half-life* of that isotope.

A nucleus may be transmuted into an isotope of the same or of

* The reader is encouraged to contemplate the enormous density of nuclear matter represented by this figure. According to this calculation, were the Earth to be squeezed down until all of its nuclei were in contact with each other, the radius would be about 180 m (600 ft).

a different element by bombardment with another nuclear particle. The discovery of nuclear transmutation was made by Rutherford through his identification of the nuclear transmutation of nitrogen into oxygen according to

$$\ce{^4_2He + ^{14}_7N = ^{17}_8O + ^1_1H} \qquad (15\text{-}4)$$

The neutron was discovered by Chadwick in experiments involving the bombardment of beryllium with alpha particles:

$$\ce{^4_2He + ^9_4Be = ^1_0\mathit{n} + ^{12}_6C} \qquad (15\text{-}6)$$

The mass of the neutron is nearly but not identically equal to that of a proton.

Positrons possess a mass equal to that of an electron and an equal but positive charge. An electron-positron pair may be created from a sufficiently energetic photon. A positron and an electron may annihilate each other with the emission of two or more photons.

Measurements show that the radii of nuclei can be expressed rather accurately by the equation

$$R = (1.2 \times 10^{-15})A^{1/3} \text{ m} \qquad (15\text{-}10)$$

from which it can be seen that the densities of all nuclei are the same and equal 0.23×10^{18} kg/m^3. Thus nuclei appear to be made up of protons and neutrons held together by short-range nuclear forces. These forces, like intermolecular forces in a drop of water, act only between nearest neighbors. Hence a large nucleus, like a large drop of water, is not stable.

**QUESTIONS
AND
PROBLEMS**

15-A1. A student performs a "dice experiment" starting with 1000 pennies. At each throw he removes all pennies that fell with the head up. (*a*) How many throws constitute a "half-life"? (*b*) In how many throws should the student find that he is dealing with less than 10 pennies?

15-A2. A student performed an experiment using 500 eight-sided dice. One face of each die bore a distinguishing mark, and in his experiment a die was declared to have decayed if that face came up on top in a given throw of the dice; all such dice were removed from the set before the next throw. In how many throws should the number of remaining dice have decreased to 250 dice?

15-A3. Prepare the graph of a decay of a specimen of $\ce{^{214}_{83}Bi}$, assuming that one starts with a 1-gram specimen and that bismuth-214 has a half-life of 19 min.

15-A4. The hydrogen isotope tritium ($\ce{^3_1H}$) disintegrates with a half-life of 12.26 yr by emitting electrons. (*a*) What isotope is formed as a result? (*b*) What fraction of a given sample will remain after five half-lives (approximately 60 yr)?

15-A5. Prepare a list of the fundamental particles mentioned to this point and list the physical properties that you know each possesses.

15-A6. Compute the radii of the nuclei having mass numbers of 1, 8, 27, 64, 125, and 216. Prepare a graph showing how the radius (y axis) depends on the mass number (x axis).

15-B1. The *thorium series* of naturally radioactive elements initiates with $^{232}_{90}\text{Th}$, having a half-life of 1.4×10^{10} yr. In this series of nuclear decays, particles are emitted in the following sequence: $\alpha, \beta, \beta,$ $\alpha, \alpha, \alpha, \alpha, \beta, \beta, \alpha$, where α represents a helium nucleus and β represents an electron. Determine the mass number and atomic number of each nuclide formed in this series.

15-B2. Carbon-11 ($^{11}_{6}\text{C}$), an emitter of positrons, has a half-life of 21 min. How many positrons should be emitted by a 1-mg ($= 10^{-6}$ kg) specimen in 21 min?

15-B3. Compute the diameter of the Moon (mass $= 7.3 \times 10^{22}$ kg) if it were to be compressed until all its nuclei were in contact. Neglect any consideration as to what would happen to the electrons.

15-B4. Complete the following nuclear reactions:
  (*a*) $^{3}_{1}\text{H} + ^{2}_{1}\text{H} = ? + ^{1}_{0}n$
  (*b*) $^{55}_{25}\text{Mn} + ^{1}_{0}n = ? + \gamma$ (gamma-ray photon)
  (*c*) $^{10}_{5}\text{B} + ^{4}_{2}\text{He} = ^{13}_{7}\text{N} + ?$

15-B5. Complete the following nuclear reactions:
  (*a*) $^{3}_{1}\text{H} = ^{3}_{2}\text{He} + ?$
  (*b*) $^{43}_{20}\text{Ca} + ^{4}_{2}\text{He} = ? + ^{45}_{21}\text{Sc}$
  (*c*) $^{9}_{4}\text{Be} + ^{1}_{1}\text{H} = ^{4}_{2}\text{He} + ?$

15-B6. Confirm that a gamma-ray photon must have an associated wavelength as short as or shorter than 0.122×10^{-11} m in order to create an electron-positron pair.

15-B7. Confirm that when a positron and an electron mutually annihilate each other with the emission of two photons, these photons will have a wavelength no greater than 0.244×10^{-11} m.

Nuclear Energy

In the years since Sir Ernest Rutherford discovered nuclear transmutation, thousands of nuclear reactions have been observed and studied. Some release energy; others require the input of an enormous amount of energy. Some produce useful and interesting radioisotopes; others produce radioisotopes having no known application.

The significance of many promising applications of nuclear reactions has been eclipsed by public awareness of the promise and the threat of nuclear reactions capable of sustaining themselves, the so-called *chain reactions* involving uranium or hydrogen. These reactions have made nuclear bombs possible; they also provide the basis for hope that a nearly endless supply of energy may become available. That the Sun is itself a self-sustaining chain reaction demonstrates that we are already much in debt to one specific nuclear energy source. Whether or not man will control nuclear processes as a source of energy and materials for the betterment of life depends largely on whether or not man can learn to control himself.

16-1 NUCLEAR VERSUS ELECTRIC FORCES

Research in nuclear physics has shown that while the properties of free neutrons are very different from those of free protons, neutrons and protons become almost indistinguishable when joined to form an atomic nucleus. For this reason, we will refer to the particles that constitute nuclei of atoms as *nucleons*, making no distinction between nuclear protons and neutrons.

A nucleus, then, consists of a cluster of nucleons held together by short-range forces of attraction that are stronger than the electric forces of repulsion. The nuclear forces operate over such short distances that, as a general guide, we can say that a nucleon exerts no nuclear force on another nucleon unless they are literally in contact with each other.

16

The observations we have made regarding the short range of nuclear forces also has relevance to nuclei in their interaction with each other; one nucleus exerts no nuclear force on any other nucleus unless the nuclei can be brought into the close range of nuclear forces.

Electric forces, however, are not short-range forces. Because they are forces of repulsion, they act in such a direction as to tend to cause nuclei to explode or *disintegrate* and, for the same reason, they act to prevent one nucleus from approaching another nucleus. Although Coulomb's law shows that these forces of repulsion are inversely proportional to the square of the distance between any two positive charges, and thus decrease rather rapidly with distance of separation, when compared to nuclear forces, the electric forces act over a relatively long range.

The reason charged nuclear particles must have high energy in order to stimulate a nuclear reaction is now clear: one nucleus can react with another only if they are brought close enough together that nuclear forces can become of significance. Because of the shielding effect of the electric field in the vicinity of a nucleus, this condition will occur only if the particles approach each other at high speed.

Let us compute the energy that must be possessed by a proton for it to overcome the electric field and reach the nucleus of a gold atom. The radius of a proton is approximately 0.8×10^{-15} m,* whereas, by Eq. (15-10), the radius of the gold atom ($A = 197$) is approximately 7.0×10^{-15} m. At the instant the proton touches the gold nucleus, the distance between their centers equals 7.8×10^{-15} m.

We will assume that the proton shown in Fig. 16-1 has been accelerated by some high-voltage machine to such an energy that, at the instant shown, it is momentarily at rest but in contact with the nucleus of a gold atom. The energy given the proton by the high-voltage machine Ve is equal to the electrical potential energy of these charges at this distance of separation. From Eq. (8-7) we have

$$Ve = \frac{kQe}{r} \qquad (16\text{-}1)$$

where V is the potential difference in volts provided by the high-voltage machine, e is the charge on the proton, Q is the charge on the gold nucleus ($= 79e$), and r is the distance between the centers of the proton and the gold nucleus. Solving for V, we find that, for the case cited, the high-voltage machine must be operated at 14.6 million volts or higher if one is to expect a nuclear transmutation to occur. This calculation, for the case in which the impinging particle is an alpha particle, shows that 12.8 million volts will be needed.

Experience shows that both protons and alpha particles produce nuclear reactions at voltages somewhat lower than what one finds in using Eq. (16-1). Therefore it is apparent that our requirement that nuclei must touch in order to react is somewhat too stringent. However,

* Equation (15-10) fails for $A = 1$.

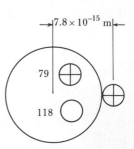

FIG. 16-1

A proton in contact with a gold nucleus. A proton must approach a nucleus with a very high velocity if it is to overcome the electric field and come within range of nuclear forces.

the general principle is valid: for charged particles to undergo a nuclear reaction, they must be brought close enough so that nuclear forces become operative.

It must be emphasized that one cannot be assured that a reaction will occur simply because a nuclear particle is brought near another nuclear particle. Whether or not a nuclear reaction does occur in a specific instance depends on intricate details of nuclear structure; the act of providing sufficient energy to bring nuclei into the range of nuclear forces is only the first condition that must be satisfied.

High-voltage machines

Prior to the invention of high-voltage machines (Sec. 10-5), scientists were limited in their choice of bombarding particles to those alpha particles, electrons, and photons that are emitted by naturally radioactive materials. In addition, they had little control over the direction of motion or energy of the bombarding particles.

With the invention of the Van de Graaff accelerator in 1931 and

FIG. 16-2

The Brookhaven accelerator. Protons from a 50 million volt preaccelerator enter along the 4-in. pipe in the foreground and are injected into a circular "race track" whose diameter is 843 ft, where a changing magnetic field accelerates the protons until they have an energy equivalent to falling through 33 billion volts. A proton travels 170,000 miles in the process.
(*Brookhaven National Laboratory.*)

of the Lawrence cyclotron in 1932, and with the development of vastly more sophisticated machines in the years that followed, scientists became able to choose from an extensive list of seven or more bombarding particles. As one specific example, an "alternating gradient synchrontron" at the Brookhaven National Laboratory (Fig. 16-2) produces pulses of 0.3×10^{12} protons per pulse at an energy equivalent to falling through a potential difference of 33 billion volts.

16-2 AN OVERVIEW OF NUCLEAR REACTIONS

In this text we will mention briefly only a specimen few of the thousands of nuclear reactions that may be induced by charged particle bombardment. We will, in fact, limit ourselves to certain reactions that have played a particularly significant role in the historical development of the subject. We will rule out any discussion here of cases in which the bombarding particle has fallen through a potential difference in excess of 140 million volts under which conditions a complicated splash of very strange particles may be produced.

Our list, plus a brief explanation, is as follows:

1. $$^{14}_{7}\text{N} + {}^{4}_{2}\text{He} \Longrightarrow {}^{1}_{1}\text{H} + {}^{17}_{8}\text{O} + Q \qquad (16\text{-}2a)$$

The observation of this reaction by Ernest Rutherford in 1919 constituted the discovery of nuclear transmutation (Sec. 15-4).

2. $$^{7}_{3}\text{Li} + {}^{1}_{1}\text{H} \Longrightarrow {}^{4}_{2}\text{He} + {}^{4}_{2}\text{He} + Q \qquad (16\text{-}2b)$$

This nuclear reaction, first observed by Cockcroft and Walton in 1932, was the first occasion in which a nuclear reaction was induced by a machine-accelerated particle.

3. $$^{9}_{4}\text{Be} + {}^{4}_{2}\text{He} \Longrightarrow {}^{1}_{0}n + {}^{12}_{6}\text{C} + Q \qquad (16\text{-}2c)$$

This reaction, first observed by Bothe and Becker in 1930 and interpreted by James Chadwick in 1932, constituted the discovery of the neutron (Sec. 15-4).

4. $$^{27}_{13}\text{Al} + {}^{4}_{2}\text{He} \Longrightarrow [{}^{30}_{15}\text{P}] + {}^{1}_{0}n + Q$$
$$[{}^{30}_{15}\text{P}] \Longrightarrow {}^{30}_{14}\text{Si} + {}^{0}_{+1}e \qquad (16\text{-}2d)$$

This nuclear reaction was observed by Irene Curie-Joliot and Frederic Joliot in 1934 and was the first occasion in which a material was made radioactive by nuclear bombardment; hence this reaction constituted the discovery of induced radioactivity.

5. $$^{14}_{7}\text{N} + {}^{1}_{0}n \Longrightarrow {}^{1}_{1}\text{H} + {}^{14}_{6}\text{C} + Q \qquad (16\text{-}2e)$$

In 1946, W. F. Libby showed that this reaction could be initi-

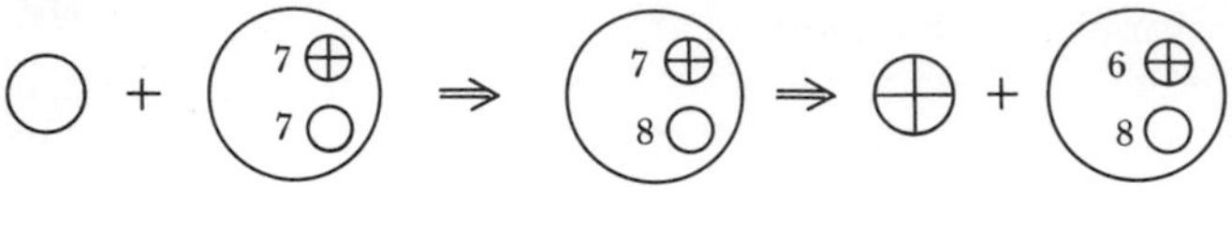

$$\begin{array}{ccccccc}
{}^1_0 n & + & {}^{14}_7 N & \Longrightarrow & {}^{15}_7 N & \Longrightarrow {}^1_1 H & + & {}^{14}_6 C
\end{array}$$

FIG. 16-3

A nuclear reaction that produces carbon-14. The first observed reaction triggered by neutrons resulted in the production of carbon-14.

ated in the atmosphere by the neutron component of secondary cosmic rays and that the ${}^{14}_6 C$ isotope with its long half-life of 5770 years can serve to establish the age of archeological finds. The technique is applicable to any specimen 500 to 25,000 years old provided it was fabricated from growing plants.

6. $$\qquad {}^9_4 Be + {}^2_1 H \Longrightarrow {}^1_0 n + {}^{10}_5 B + Q \qquad (16\text{-}2f)$$

This reaction, produced by deuterons from a Van de Graaff generator or a cyclotron served as the primary source of neutrons from 1934 to 1942 prior to the invention of the nuclear reactor.

7. $$\qquad {}^2_1 H + {}^2_1 H \Longrightarrow {}^1_1 H + {}^3_1 H \qquad (16\text{-}2g)$$

This reaction, first produced by M. L. E. Oliphant in 1934, constituted the discovery of hydrogen-3, the extra-massive isotope called *tritium*. With a half-life of 12.26 years, it is probably one of the constitutent materials in the hydrogen bomb.

16-3 ENERGY BOOKKEEPING IN NUCLEAR REACTIONS

Isotopes that are radioactive as found in nature are said to be "naturally" radioactive; stable isotopes, on the other hand, can be transmuted into "artificial" radioisotopes by bombardment by charged particles or by irradiation with neutrons or gamma rays. The naturally radioactive materials emit alpha particles, electrons, and photons; artifically produced radioactive isotopes may emit any of dozens of types of particles, depending in part on the means by which the isotope was made radioactive. Positrons and electron emitters, however, predominate among the artificial radioactive materials, at least among those that, on being produced, disintegrate slowly enough to permit them to be separated for study from the other materials with which they are mixed after the irradiation.

Mass as a form of energy

One of the most important consequences of the theory of relativity was the proof of the equivalence of mass and energy (Sec. 14-10). The clearest demonstration of the validity of this conclusion is found in nuclear disintegrations, where the amounts of energy emitted or absorbed are so large that the mass changes accompanying the processes can be measured. The mass-to-energy transformation that takes place is given by

$$\Delta E = \Delta m c^2 \qquad (14\text{-}12)$$

in which Δm represents the overall change in mass, c equals the velocity of light, and ΔE is the energy that is either released or absorbed. The amount of energy released or absorbed in any nuclear process can be calculated if one knows the change in nuclear mass that takes place. Because of the enormous "conversion factor" of 9×10^{16} joules/kg, a minute change in mass may involve a very large change of energy.

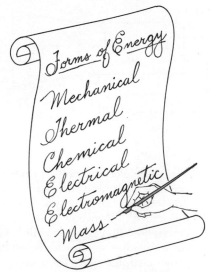

FIG. 16-4

Mass-energy is another form of energy. From Einstein's $E = mc^2$ equation we learn that 1 kg equals 9×10^{16} joules.

Energy units in high-energy physics

Nuclear reactions occur as individual nuclear events. Therefore we will wish to express the amount of energy absorbed or released per reaction.

Even for a nuclear reaction, the energy per reaction is a very small number when expressed in joules. For this and reasons associated with procedures used in calculating the energy involved in nuclear reactions, we may wish to express the energy in other units, specifically in *atomic mass units* (amu) and in *electron volts* (eV).

The energy released can be measured simply by the change in mass, since, according to Eq. (14-12), mass and energy are strictly proportional to each other. In order to find the relationship between energy in atomic mass units and energy in joules, we note first that a nucleus having a mass of 1 amu has a mass of $1/(6.0225 \times 10^{26})$ kg. The energy equivalent of 1 amu can then be calculated from the mass-energy equation to yield

$$1 \text{ amu} = \left(\frac{1}{6.0225 \times 10^{26}} \right) \times 9 \times 10^{16} = 1.493 \times 10^{-10} \text{ joules}$$

Energy is also commonly expressed in electron volts (eV) [or in millions of electron volts (MeV)], where the electron volt is defined as the amount of energy acquired by an electron in falling through a difference of potential of 1 volt. Applying the definition of potential difference [Eq. (8-4)], we find that

$$1 \text{ eV} = 1.6 \times 10^{-19} \text{ joules}$$

and one million electron volts equals

$$1 \text{ MeV} = 1.6 \times 10^{-13} \text{ joules}$$

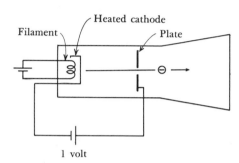

FIG. 16-5

A particle with an energy of 1 eV. An electron volt of energy is attained by an elementary charge in falling through a difference of potential of 1 volt.

The electron volt is a particularly convenient unit to use whenever one is dealing with particles that have attained their energy in an accelerator. For example, an accelerator that generates a potential difference of 8 million volts will surely provide protons whose energy equals 8 million electron volts because the charge on a proton is numerically the same as the charge on an electron. The same accelerator will provide alpha particles with an energy of 16 million electron volts (16 MeV), for the charge on an alpha particle equals twice the electronic charge.

The disintegration of uranium-238

Let us demonstrate the use of the mass-energy equation in computing the energy released in a typical nuclear reaction, specifically, the disintegration of uranium-238. As we have seen, this reaction takes place with the emission of alpha particles according to the equation

$$^{238}_{92}\text{U} \Longrightarrow ^{234}_{90}\text{Th} + ^{4}_{2}\text{He} + Q \tag{16-3a}$$

Up to now we have regarded "equations" like (16-3a) simply as descriptions of the nuclear reaction, but because of the presence of Q, which expresses the energy released or, if negative, absorbed, we can also regard it as an expression for the Law of Conservation of Energy in the process. By introducing the known values of the masses of the nuclei as measured by the mass spectrograph (see Appendix 8), the equation becomes

$$238.0508 = 234.0436 + 4.0026 + Q \tag{16-3b}$$

or upon solving for Q,

$$Q = 0.0046 \text{ amu}$$

or in other units,

$$Q = 6.9 \times 10^{-13} \text{ joules}$$
$$= 4.3 \text{ MeV}$$

Most of the energy is carried off as kinetic energy by the smaller of the two fragments, the alpha particle. (As when a rifle is fired, most of the energy is given to the bullet.) Measurements show that the actual energy of the emitted alpha particles equals 4.18 MeV.

The disintegration of the neutron

In Sec. 16-1 we noted that neutrons become almost indistinguishable from protons when combined with other neutrons and protons to form a nucleus but that neutrons and protons have many very distinguishing characteristics when "free"—that is, when no longer under the influence of short-range nuclear force from other neutrons or protons. A free neutron, in fact, is radioactive, disintegrating into a proton, an electron, and an antineutrino with a half-life of 12.8 min according to the reaction equation

$$_{0}^{1}n \Longrightarrow {}_{1}^{1}\text{H} + {}_{-1}^{0}e + {}_{0}^{0}\bar{\nu} + Q \qquad (16\text{-}4a)$$

Introducing known values of the masses of the neutron, the proton, and the electron into the preceding reaction equation, we have

$$1.008665 = 1.007277 + 0.000549 + {}_{0}^{0}\bar{\nu} + Q \qquad (16\text{-}4b)$$

Solving, we have

$$_{0}^{0}\bar{\nu} + Q = 0.000839 \text{ amu}$$
$$= 12.5 \times 10^{-14} \text{ joules}$$
$$= 0.78 \text{ MeV}$$

FIG. 16-6

Energy distribution of electrons emitted in neutron decay. In each disintegration, 0.78 MeV of energy is released, part of which is carried off by an antineutrino.

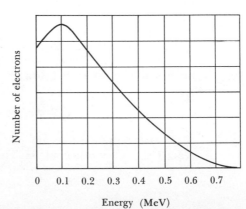

Experimental studies of the disintegration of neutrons show that the ejected electrons emerge with energies that range from zero to a maximum of 0.78 MeV distributed according to the curve shown in Fig. 16-6. Presumably, in every reaction, the total energy given off equals 0.78 MeV, but in nearly every disintegration a portion of this energy is carried off by the antineutrino.

16-4 THE BEHAVIOR OF NEUTRONS

With the discovery of the neutron by James Chadwick in 1932, scientists turned their attention to experiments in which neutrons were used to initiate nuclear reactions. Clearly, most of what has been said in Sec. 16-1 regarding the need for high energy does not apply here, because neutrons have no electric charge and thus can pass freely through matter without regard for the electric field due to nuclei and their associated charge clouds. Even at low speed, a neutron can encounter or even enter the nucleus of any atom that happens to lie in its path.

Carbon as a moderator of neutrons

A neutron, however, may or may not stimulate a nuclear reaction when it encounters a nucleus. Many nuclei, the nucleus of $^{12}_{6}C$ for example, constitute a very stable combination of nucleons that will reject any neutron which happens to come by. An encounter of a neutron with such a nucleus, therefore, is an elastic collision in which, it will be recalled, both momentum and kinetic energy are conserved.

Let us assume that by some as-yet-unidentified process we are able to release 1000 neutrons near the center of a large block of carbon, each neutron possessing at the outset an average velocity of 2×10^7 m/sec (KE = 2 MeV). We will assume that these neutrons behave like a neutron "gas" under these conditions, making repeated elastic collisions with other neutrons and with the carbon nuclei. In a collision with a carbon nucleus, like a golf ball striking a baseball, the neutron transmits a portion (actually about 16 percent) of its energy in each encounter. By making repeated collisions in the carbon "moderator," the neutrons gradually lose energy. If the moderator is at room temperature (300°K), their average energy [$= \frac{3}{2}(R/N)T$ by Eq. 7-3(a)] finally will equal 0.04 eV. Each neutron achieves this energy in less than 100 collisions; therefore, since a neutron has millions of collisions per second in the carbon block, it will spend most of its lifetime (half-life = 12.8 min) traveling at the low velocity of approximately 2800 m/sec. Such neutrons are called slow or "thermal" neutrons.

Cadmium as a controller of neutron population

Although neutrons do not react with carbon-12, they readily react with many other nuclei. They stimulate a reaction of particular

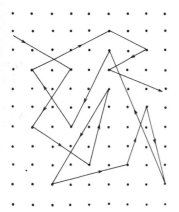

FIG. 16-7

The loss of energy of a neutron in a nuclear reactor. In each collision of a neutron with a carbon nucleus in a nuclear reactor, the neutron loses 16 percent of its energy. In 100 collisions a 2 MeV neutron achieves equilibrium with the carbon nuclei at an average energy of 0.04 eV.

importance to our story when they encounter a nucleus of cadmium-113. The reaction in question is

$$_{48}^{113}\text{Cd} + {}_0^1 n \Longrightarrow {}_{48}^{114}\text{Cd} + \gamma \qquad (16\text{-}5)$$

where γ represents a gamma-ray photon that is emitted. Because this particular reaction has a high probability of occurring in the presence of thermal neutrons, a cadmium rod inside a block of carbon moderator literally drinks up the neutrons; hence cadmium rods thrust in a region where neutrons abound may serve as a means of controlling their population and retarding, if desired, other reactions that may be produced by them. More attention will be given to this use of cadmium in our discussion of nuclear reactors.

16-5 NUCLEAR FISSION

In the early 1930s a group of scientists at the University of Rome under the leadership of Enrico Fermi were investigating the reactions produced by irradiating various materials with neutrons. One of the early findings was that, in many cases, the end result turned out to be an isotope of the element of next-higher atomic number. Quite probably they anticipated that by irradiating uranium, the last element in the periodic table, they might produce the *trans*-uranic element No. 93, not present as a natural element in the Earth's crust. Fermi's results (1934) were inconclusive but intriguing enough to stimulate a great deal of contagious interest in other laboratories.

Repeating Fermi's experiment in Germany, Otto Hahn and co-workers performed some painstaking chemical separations of irradiated uranium samples and identified a trace of barium (1938). This conclusion was verified independently in 1939 by Otto Frisch and Lisa Meitner in Denmark. Surely, they said, the energy added to the uranium nucleus when it captures a neutron has caused it to split into two nearly equal parts.

Only a very few isotopes undergo fission when irradiated by thermal neutrons; for all practical purposes, there are only three: uranium-235, thorium-233, and plutonium-239. Of these, only uranium-235 is present in the Earth's crust; the others must be produced by nuclear transmutation from other materials.

The 17 known isotopes of uranium have half-lives ranging from a few minutes to billions of years, but only two, $_{92}^{238}\text{U}$ and $_{92}^{235}\text{U}$, have half-lives of sufficient length (4.9×10^9 yr and 7.1×10^8 yr respectively) that they remain as a significant natural resource. The most abundant isotope of uranium ($_{92}^{238}\text{U}$) fissions, but only for incident fast neutrons (energy > 1 MeV). In contrast, the other isotope ($_{92}^{235}\text{U}$) is prone to fission by capture of either a thermal or a fast neutron, although, of the two possibilities, fission by thermal neutrons is the more probable event, given equal numbers of thermal and fast neutrons.

Only 0.71 percent of uranium as found in nature is the fissionable

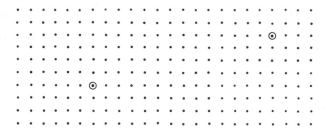

FIG. 16-8

The relative abundance of U-235 in natural uranium. Only one atom in 141 in natural uranium is U-235; the rest are U-238 and a trace of U-234.

isotope uranium-235. Had nature failed to supply the small amount of uranium-235 that is present, the discovery and use of nuclear fuels would have been delayed for many years.

Description of the fission process

When uranium-235 fissions, it divides into two fragments and releases several additional neutrons, which are immediately available to produce fission in other nuclei in the vicinity. Thus, given the proper amount of uranium in the appropriate physical circumstance, the fission of one nucleus may trigger a chain reaction of subsequent fissions, which may go out of control (a nuclear bomb) or may be controlled by the use of cadmium rods and serve as a steady source of energy (a nuclear reactor). The difference between these two levels of activity may be compared to the difference between igniting a stick of dynamite and burning a piece of coal.

The capture of a neutron by a $^{235}_{92}U$ nucleus produces a $^{236}_{92}U$ nucleus that is very unstable; thus it is said to be in an "excited" state. The excess energy is released as a gamma-ray photon by approximately 15 percent of such nuclei to produce a radioactive form of $^{236}_{92}U$ that disintegrates by emitting an alpha particle. In 85 percent of the occasions, however, the $^{236}_{92}U$ nucleus does not decay by emitting a gamma ray but, instead, splits into two nuclear fragments, one with a mass in the range $A = 80 - 108$ amu, the other in the range $A = 126 - 153$ aum. The distribution of fragment sizes is shown in Fig. 16-9. As one

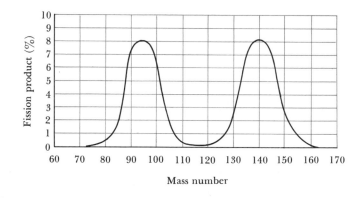

FIG. 16-9

Fragment mass distribution in nuclear fission. A wide range of masses is found in the fragments that result from the fission of U-235. A large and a small fragment are produced in nearly every fission; it is a rare event for the atom to split into two exactly equal parts.

may see from this graph, it is a rare event for a nucleus to produce two fragments of equal size.

Although nuclear fission in a nuclear reactor or in a nuclear bomb is apparently a chaotic event even when viewed at a submicroscopic level, a certain kind of order governs this chaos. The fragment sizes observed in many events distribute themselves on a smooth curve. In addition, smaller particles and gamma rays are emitted during fission. The number of such smaller particles and their energy distributions are also governed by some internal law of chance, as seen by the fact that a study of many events enables us to describe an average behavior.

Let us describe this average behavior briefly. The energy released during nuclear fission is carried off in the form of kinetic energy of the fission fragments (167 MeV), kinetic energy of 2 to 4 neutrons (5 MeV), electrons (5 MeV), gamma rays (10 MeV), and neutrinos (11 MeV). The neutron component is of particular significance; an average of 2.3 neutrons are emitted in each fission event. The energies of these neutrons range from zero to 10 MeV with a distribution that reaches a maximum at 0.7 MeV. Nearly all the neutrons are emitted at the instant fission occurs; however, a small but significant number of "delayed" neutrons (less than 1 percent) are emitted as much as a minute later.

The energy released in nuclear fission

When one considers the wide variety of nuclear events that may occur in nuclear fission, it becomes apparent that it is impossible to write specific nuclear equations that describe the phenomenon in detail. However, perhaps by some stretch of the imagination, the following reaction could represent at least one of the possible fissions that may occur.

$$^{235}_{92}\text{U} + ^{1}_{0}n \Longrightarrow ^{95}_{38}\text{Sr} + ^{138}_{54}\text{Xe} + 3^{1}_{0}n + Q \qquad (16\text{-}6a)$$

To determine the energy Q released in this event, we reinterpret Eq. (16-6a) as a statement of the Law of Conservation of Energy, expressing it as follows:

$$235.043915 + 1.008665 = 94.91424^* + 137.91381$$
$$+ 3 \times 1.008665 + Q \qquad (16\text{-}6b)$$

Solving for Q, we have

$$Q = 0.19853 \text{ amu}$$
$$= 185,000,000 \text{ eV}$$

Measured values of the energy released in fission are, on the average, somewhat larger than our calculated value, because the total reaction is not yet complete when expressed as in Eq. (16-6a). The reason is that

---

* The mass of this isotope of strontium was determined by extrapolation from the known masses of neighboring isotopes in the periodic table.

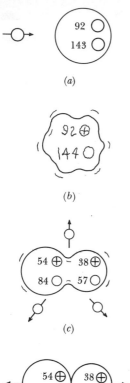

FIG. 16-10

Fission according to the liquid-drop nuclear model. (a) A thermal neutron approaches a nucleus of U-235 and is captured. (b) The nucleus is put into violent oscillation. (c) It splits, emitting three neutrons. (d) The nuclei each experience the electric force of the other and move rapidly apart.

the fragments ($^{95}_{38}$Sr and $^{138}_{54}$Xe in this case) are highly radioactive and should emit many millions of electron volts of energy in the form of gamma rays, electrons, and neutrinos to contribute to the total energy released. Some of the radioisotopes formed along the way may have long half-lives (hours or even months) and thus constitute, in nuclear bombs, a dangerous by-product known as "fallout."

Fission viewed as the splitting of a liquid drop

In order to gain some insight into the source of the energy released by nuclear fission, let us return to our liquid-drop model of the nucleus. We will think of the fission of a uranium nucleus as a sequence of nuclear events (Fig. 16-10) as follows:

1. A nucleus with 235 nucleons becomes a nucleus with 236 nucleons upon capturing a neutron [Fig. 16-10(a)].

2. This composite nucleus is unstable and is set into violent vibration [Fig. 16-10(b)].

3. The nucleus divides into two smaller nuclei with 138 and 95 nucleons respectively. Three neutrons escape [Fig. 16-10(c)].

4. The two fragments are now out of range of nuclear forces but are still under the influence of very large electric forces. These electric forces result in the fragments moving violently apart [Fig. 16-10(d)].

The energy involved in (4) can be computed using Eq. (8-7) for the electric energy of the two charges. Their distance apart at the instant the fragments are formed is given by the sum of the radii of the fragments, which, by Eq. (15-10), equals 5.5×10^{-15} m and 6.2×10^{-15} m respectively; hence r in Eq. (8-7) equals 11.7×10^{-15} m. Letting $Q_1 = 38e$ and $Q_2 = 54e$, we have

$$\text{Energy} = \frac{kQ_1Q_2}{r} = \frac{9 \times 10^9 \times 38 \times 54 \times (1.60 \times 10^{-19})^2}{11.7 \times 10^{-16}}$$

$$= 40.3 \times 10^{-12} \text{ joules}$$

$$= 253 \text{ MeV}$$

While this result does not agree precisely with the experimental observation that the average energy released per fission equals 198 MeV, it is sufficiently close to indicate the correctness of the basic idea in this simplified model of the fission process.

The nuclear reactor

A nuclear reactor is a device that plays the role for uranium that a furnace plays for coal in that it contains the fuel, controls the rate at which it is used, and transmits the released energy to the location desired by the designer. Just as a furnace can either heat a house or melt an ingot of iron, so a nuclear reactor can heat a boiler or make a cube of metal radioactive.

The first nuclear reactor, an artist's view of which is shown in Fig. 16-12, was put into operation on Dec. 2, 1942, at the University of Chicago under the leadership of Enrico Fermi. Much progress in reactor design has occurred since then, but because the basic principle of operation has not changed, we will describe a reactor very much like the original model.

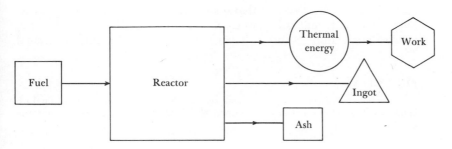

FIG. 16-11

A furnace and a reactor compared. Both a furnace and a nuclear reactor require a fuel, produce thermal energy that can do work, and can change a material located near it, either melting it (in the case of a furnace) or making it radioactive (in the case of the reactor).

FIG. 16-12

The first nuclear reactor. The first self-sustaining fission reactor was demonstrated by Enrico Fermi at the University of Chicago on December 2, 1942. It consisted of a pile of 385 tons of graphite laced with uranium and cadmium.
(*Argonne National Laboratory.*)

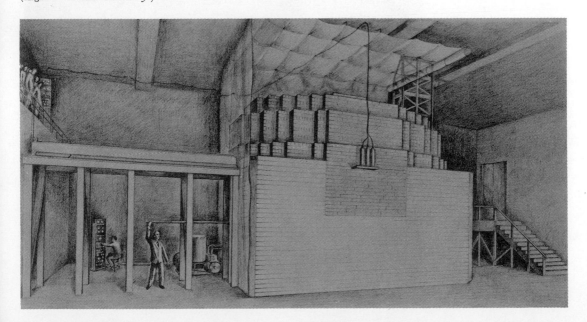

The basic component of a nuclear reactor is the moderator whose purpose is to decrease the average energy of the neutrons produced in fission from the initial value of approximately a million electron volts to the thermal energy of 0.04 electron volts. In the original reactor, the moderator was a cubical pile of graphite (carbon) bricks weighing a total of 385 tons. A matrix of holes in the graphite permitted inserting rods and/or cubes of natural uranium and natural uranium ore having a total weight of 45 tons. Another group of holes permitted the insertion of many cadmium rods into the graphite pile.

For the sake of discussion, let us assume that each fission releases exactly three neutrons (instead of an average of 2.3 per fission as stated earlier). Let us next assume that a stray neutron enters the graphite at high velocity. It makes hundreds of collisions with carbon nuclei and, losing 16 percent of its energy in each collision, soon becomes a thermal

FIG. 16-13

Construction details of the first nuclear reactor. This photograph shows the 19th layer (solid graphite) partially covering the 18th layer (graphite with uranium inserts).
(*Argonne National Laboratory.*)

neutron with an energy of approximately 0.04 eV. This neutron may have millions of additional collisions, and although it may wander out of the reactor or be absorbed in the cadmium, let us assume that the particular neutron we are considering finally enters one of the uranium rods. Here it encounters many $^{238}_{92}U$ atoms with which it does not react because $^{238}_{92}U$ does not capture thermal neutrons. However, it eventually may encounter a $^{235}_{92}U$ nucleus with which it reacts violently, producing fission. Suddenly, deep in the reactor, energy is released and three neutrons are created.

Although it is possible that these three neutrons will each produce a fission, it is far more likely that at least two will be used up some other way. However, if *on the average* these three neutrons produce one fission, a chain reaction can be said to be in progress and the reactor is described as "critical." If each fission produces slightly more than one fission *on the average*, the level of activity will grow indefinitely.

Should the reaction proceed too rapidly, the temperature in the interior may rise to the point that the metal parts may melt and the coolant explode. Elaborate systems, therefore, have been developed to control the reaction rate at a predetermined level. In addition, reactor engineers have developed intricate shields and fail-safe devices to further minimize the danger of such a catastrophe. The most important and immediate control of the reaction rate is provided by the cadmium rods, whose position and depth within the reactor determines the level of operation. If the activity is at too high a level, servomotors ("servant" motors) automatically push the cadmium rods deeper into the holes in the graphite where they absorb a larger fraction of the neutrons; if the activity is too low, the cadmium rods are withdrawn.

Much of the ease of controlling the rate of fission in a nuclear reactor must be credited to the existence of the delayed neutrons. Even though they make up less than 1 percent of the neutrons produced by fission, the fact that they are emitted seconds after the fission process has occurred makes the reactor just a little sluggish, thereby giving operators and mechanisms some time to respond.

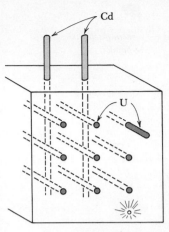

FIG. 16-14

A nuclear reactor. Rods of natural uranium and of cadmium are inserted into a large block of graphite. Neutrons emitted in the fission of U-235 (1) produce fission of other U-235 nuclei, (2) convert U-238 into Pu-239, and (3) are captured by the cadmium.

Breeder reactors

It has been stated that the capture of a neutron by a $^{238}_{92}U$ nucleus removes that neutron from the reactor. However, the capture of a neutron by $^{238}_{92}U$ has two interesting consequences: (1) at some energies, the neutron will produce fission in $^{238}_{92}U$ and release energy and (2) at certain energies, the neutron can initiate a nuclear reaction that produces $^{239}_{94}Pu$, another thermal-fissionable isotope. The reaction is

$$^{238}_{92}U + ^{1}_{0}n \Longrightarrow {}^{239}_{93}Np + {}^{0}_{-1}e \qquad (16\text{-}7a)$$

The neptunium isotope decays by emitting electrons according to

$$^{239}_{93}Np \Longrightarrow {}^{239}_{94}Pu + {}^{0}_{-1}e \qquad (16\text{-}7b)$$

The plutonium so formed has a half-life of 24,000 years.

The importance of this reaction is that it offers the possibility of a nuclear reactor which produces heat by fission while continually generating plutonium to replace the fuel being lost. Such "breeder" reactors require enormous care in the use of neutrons, but several models are now in operation. They offer a manyfold increase in available energy and may even have the potential of supplying energy adequate for the whole world. Certain very difficult technical problems, however, must be solved before any such cornucopia becomes a reality.

The atomic bomb

An "atomic" or fission bomb can be regarded as a nuclear reactor that has been deliberately pushed beyond the point of being "critical," each fission producing, on the average, say 1.1 additional fissions. The bomb differs from the reactor we have described in two important respects: (1) the fuel consists of uranium in which the component of $^{235}_{92}U$ has been enhanced and (2) no moderator is used but, instead, a substantial fraction of the fissions are produced by fast neutrons. Operating with fast neutrons, one generation of fission events follows another at intervals of approximately 10^{-9} sec. Therefore 1000 generations occur in 10^{-6} sec. Even with the relatively small "multiplication factor" of 1.1, a total of 10^{24} atoms* (approximately 0.86 lb of uranium) will split in 10^{-6} sec, releasing as much energy as 7700 tons of TNT. The bomb that fell in Hiroshima has been rated at "20,000 tons"; that is, it released as much energy as 20,000 tons of TNT. Presumably, approximately 1 kg of uranium was used by the reaction, but because a substantial fraction of the uranium surely was blown away without splitting, the actual amount of uranium present may have been many times greater.

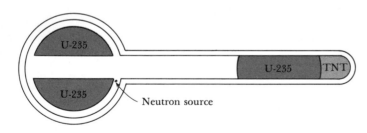

FIG. 16-15

An "armchair" model of an atomic bomb. The "multiplication factor" for nuclear fission suddenly changes from a value less than unity to a value greater than unity when the cylindrical section at the right is driven into the cylindrical hole in the sphere at the left.

16-6 THERMONUCLEAR FUSION REACTIONS

The energy of the Sun

One cannot lie on a beach on a clear day without marveling at the Sun and its astounding output of thermal energy. The phenomenon

---

* With a multiplication factor of 1.1, a total of 10^{43} neutrons would be available at the end of 1000 generations. However, as the uranium is used up (or blown away), the multiplication factor decreases.

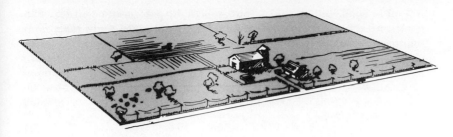

FIG. 16-16

Solar energy. The energy reaching a typical midwestern farm (1 mile square) is approximately 2.5×10^9 joules/sec, equal to the output of one of the largest industrial reactors.

is made the more remarkably by the fact that the Sun has poured out this wealth for eons, for 3 billion years at least, and manifests every prospect of continuing to do so for another 3 billion years. What can be the source of all this energy? How did it originate? Will it continue indefinitely? If not, what then?

Were one to possess the appropriate absorbers, thermometers, and accessory equipment, one would find that thermal energy reaches us at the rate of 1400 joules per m² per second. This quantity is known as the *solar constant S*.

Knowledge of the solar constant and of our distance from the Sun enables us to compute the total energy emitted each second. Since a sphere at our distance from the Sun (1.49×10^{11} m) has a surface area ($4\pi r^2$) of 27.9×10^{22} m², it follows that the total energy emitted by the Sun equals 3.92×10^{26} joules/sec. According to the mass-energy equation, the mass of the Sun, therefore, must decrease at the rate of 4.35×10^9 kg/sec ($= 4.7$ million tons per second!). However, there is no immediate cause for alarm. The Sun's mass is 1.97×10^{30} kg; at the above rate of loss of mass, the mass of the Sun will decrease by only 0.00001 percent in the next million years.

No chemical process could possibly account for all this energy. Even if the Sun were solid coal, burning efficiently on an endless supply of oxygen piped in from outer space, at this rate it would all be gone in 2000 years. Furthermore, we cannot explain it as a conversion of gravitational energy to thermal energy. Were this the source of the energy, the Earth would have been inside the fireball as recently as 20 million years ago, thus fixing an age for the Earth that is in conflict with every other measurement. We must look elsewhere for an explanation.

The answer, of course, is that the Sun is an enormous hydrogen fusion reactor, pouring out the energy that is released when hydrogen nuclei combine to form helium. Its constituent elements almost shout this process as the only possible explanation. At the present time, of the 1.97×10^{30} kg that represents the mass of the Sun, 74 percent by mass (1.47×10^{30} kg) is hydrogen, 24 percent (0.48×10^{30} kg) is helium, and the remaining 2 percent (0.04×10^{30} kg) is a mixture of heavier nuclei.

One can write many fusion reactions involving the hydrogen

isotopic nuclei, but the ones that appear to predominate in the Sun constitute a sequence of processes known as the *proton-proton chain*. The sequence is as follows:

$$_1^1H + _1^1H \Longrightarrow _1^2H + _{+1}^0e + _0^0\nu_e + Q_1 \qquad (16\text{-}8a)$$

$$_1^2H + _1^1H \Longrightarrow _2^3He + Q_2 \qquad (16\text{-}8b)$$

$$_2^3He + _2^3He \Longrightarrow _2^4He + 2_1^1H + Q_3 \qquad (16\text{-}8c)$$

This chain can be read as follows: Protons combine with protons to give deuterons, positrons, and neutrinos; deuterons combine with protons to produce helium-3 nuclei; helium-3 nuclei combine to form alpha particles and protons. At each stage, energy is released; a significant portion of this energy is given off as gamma rays because the nuclei ($_1^2H$, $_2^3He$, and $_2^4He$) may be formed in an excited state. Of course, steps (16-8a) and (16-8b) must occur twice to generate two $_2^3He$ nuclei for the last step.

If one is not concerned with the details of the process, it would be useful to notice that the overall effect of the proton-proton chain is to combine 6 protons to form an alpha particle, 2 positrons, two neutrinos, and 2 protons. With the recovery of 2 protons in each cycle, the net effect is

$$4_1^1H \Longrightarrow _2^4He + 2_{+1}^0e + 2_0^0\nu_e + Q \qquad (16\text{-}9)$$

Written in this form, the calculation of the energy released per cycle is moderately straightforward. It must be noted, however, that atomic masses as listed in tables represent the masses of whole atoms, nuclei plus their appropriate number of electrons. In this calculation we will subtract the appropriate number of electron masses from each atomic mass to secure the mass of the individual nucleus. (This extra

FIG. 16-17

The proton-proton chain. The nuclear reaction primarily responsible for the energy released by the Sun is the union of 6 protons to produce ultimately a helium nucleus, 2 protons, 2 positrons, and 2 neutrinos.

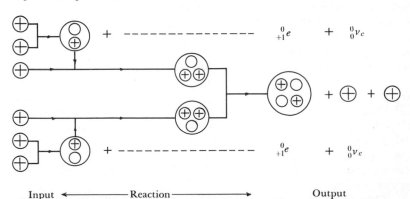

bookkeeping chore is made necessary because we do not have the same number of electrons attached to nuclei after the reaction as we had before the reaction.) In addition, a portion of the energy is emitted in the form of two neutrinos. Letting the total energy released per reaction cycle $(2{}_{0}^{0}\nu_e + Q)$ be represented by Q' we have

$Q' = $ [mass of 4 protons] $-$ [mass of alpha particle $+$ 2 positrons]

$= [4(1.007825 - 0.000549)]$

$\qquad\qquad -[(4.00260 - 2 \times 0.000549) + 2 \times 0.000549]$

$= 0.0265$ amu

$= 24.8$ MeV

$= 3.96 \times 10^{-12}$ joules

In order to release 3.92×10^{26} joules/sec, the chain of reactions described by Eqs. (16-8) must be carried out on the Sun at the rate of 10^{38} times each second.

According to present views of the origin of the universe, original matter consisted of nucleons—that is, of protons and/or neutrons. Considering the large percentage of the Sun's mass that remains as nucleons, the Sun must be regarded as a relatively young star. Older stars contain heavier elements, including carbon, and release energy in a thermonuclear process that involves the carbon present in its interior by a cycle of operations known as the carbon cycle.

Conditions for fusion to occur

Whether a given star generates energy by the proton-proton chain or by the carbon cycle, it must be realized that neither process can occur unless the nuclei collide at sufficiently high relative velocities that they can overcome the force of repulsion associated with their positive charges. The only way that this process can occur in a gas is for the gas to be at a very high temperature; in fact, the temperature must be in the millions of degrees!

In order to get an estimate of the temperature needed to cause fusion, let us apply some principles that have been introduced in earlier chapters to one of the specific reactions we have already mentioned, for example, to the fusion of protons. That is,

$${}_{1}^{1}\text{H} + {}_{1}^{1}\text{H} \Longrightarrow {}_{1}^{2}\text{H} + {}_{+1}^{0}e + {}_{0}^{0}\nu_e + Q \qquad (16\text{-}8a)$$

We will imagine the two protons in question to start at a considerable distance apart and to approach each other with the same velocity. They will, therefore, each have the same kinetic energy. We will assume that on collision, the two protons just touch each other—that is, that their centers are at a distance of $2r_p$ apart, where r_p is the radius of a proton $(= 0.8 \times 10^{-15}$ m). Letting KE represent the original kinetic energy of one of the protons and recalling the equation for the electrical potential

FIG. 16-18

Dynamics of a thermal reaction. In order to react with each other, two protons must get close enough so that nuclear forces can become significant—that is, at a distance comparable to the size of the nucleus. Because of the electrostatic field of the protons, this cannot happen unless the temperature of the proton gas is on the order of several million degrees.

energy of a charge in the electric field of another charge, we have

$$2(\text{KE}) = \frac{kq_1q_2}{2r_p} \tag{16-10}$$

Therefore $\text{KE} = \dfrac{kq_1q_2}{4r_p} = \dfrac{9 \times 10^9 \times (1.6 \times 10^{-19})^2}{4 \times 0.8 \times 10^{-15}}$

$$= 7.2 \times 10^{-14} \text{ joules}$$

In our discussion of the thermal energy of a gas, however, we learned that the average kinetic energy of a gas molecule is equal to $\frac{3}{2}(R/N)T$. Therefore, for a typical proton in a "proton gas" to fuse with another proton, the temperature of the gas must be related to the energy of the protons by the equation

$$\frac{3}{2}\frac{R}{N}T = 7.2 \times 10^{-14} \text{ joules} \tag{16-11}$$

Introducing $R = 8310$ and $N = 6.02 \times 10^{26}$ and solving for T, we find

$$T = 3.48 \times 10^9 \text{ °K}$$

According to the preceding calculation, a gas consisting of protons would have to acquire a temperature of several billion degrees in order for the average proton to fuse with another average proton. Since the temperature of the core of the Sun is "only" about 15 million degrees Kelvin, one must wonder how such a reaction can take place there. Two factors enable the reaction to proceed at lower temperatures:

1. Particles in a gas have a wide range of velocities; hence some protons in a proton gas at 10^7 °K will have the needed velocity.

2. A process known as "tunneling," the details of which go beyond the scope of this book, permits protons to fuse even if they lack the energy required by Newtonian principles.

Nuclear processes, of course, can take place only in the core of the

Sun because the surface is relatively cool, approximately 6000°K. Energy released near the core travels out from the center due to successive collisions of protons and through repeated absorption and reemission of gamma rays.

Thermonuclear bombs

Thermonuclear fusion reactions have achieved most of their attention and notoriety because of their use in hydrogen bombs. It is most probable that the bomb consists of a mixture of deuterium and tritium gas at high pressure in a thick-walled vessel. A uranium fission bomb is located near the center. The detonation of this bomb provides a temperature in the millions of degrees, which initiates the fusion of many combinations of the available isotopes. Unlike the fission bomb, whose maximum size is limited to a few kilograms of uranium, the fusion bomb is open-ended; it can be made as large as one chooses. Fusion bombs equivalent to 20 million tons of TNT (20 megaton bombs) have been built and successfully tested.

A nuclear reaction may occur if two nuclear particles are brought **SUMMARY** within range of nuclear forces. High-voltage machines provide kinetic energy to charged nuclear particles, sufficient to enable them to overcome the Coulomb force that each exerts on the other.

The energy absorbed or released in a nuclear reaction can be computed from the change in total mass: if there is a mass increase, the mass-energy equation [Eq. (14-12)] enables one to compute the minimum energy to cause the reaction to occur; if there is a mass decrease, the same procedure enables one to compute the energy released as kinetic energy of the fragments, as photons, and as neutrinos.

A low-energy neutron can initiate a nuclear reaction because,

FIG. 16-19

Schematic diagram of a hydrogen bomb. A conventional atomic bomb provides the fuel to bring the hydrogen to a temperature of several million degrees in order to initiate the fusion of protons or other light isotopes.

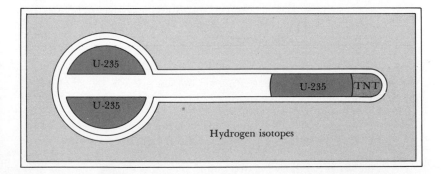

having no charge, it can enter a nucleus without overcoming a Coulomb force field.

Nuclear fission may be induced when a low-energy neutron encounters the nucleus of $^{235}_{92}U$, $^{233}_{92}U$, or $^{239}_{94}Pu$. Of these, only $^{235}_{92}U$ exists as a natural resource. A typical fission reaction is

$$^1_0n + ^{235}_{92}U \Longrightarrow ^{95}_{38}Sr + ^{138}_{54}Xe + 3^1_0n + Q \qquad (16\text{-}6a)$$

The fusion of the lighter isotopes may also release energy in a sustained nuclear reaction. Because of the Coulomb force of repulsion between nuclei, a sustained reaction is possible only if a quantity of "fuel" is brought to a temperature of millions of degrees, in which case the thermal motion of the nuclei is sufficient to cause fusion when they collide. The proton-proton cycle in the Sun is a complex series of fusion processes, which can be summarized by the reaction equation

$$4^1_1H \Longrightarrow ^4_2He + 2^{\;\;0}_{+1}e + 2^0_0\nu_e + Q \qquad (16\text{-}9)$$

**QUESTIONS
AND
PROBLEMS**

16-A1. A bowling ball travels with a constant velocity of 4 m/sec on a smooth, horizontal floor until it encounters a ramp. It rolls up this ramp, stops, and returns. How high was the ball above the floor at the instant it came to rest? By analogy to Rutherford's experiment, what could one learn about a ramp by rolling bowling balls toward it?

16-A2. If a nuclear process were to be discovered that involved the annihilation of protons, how much energy would be released each time a proton is annihilated?

16-A3. Starting with the definition of potential difference [Eq. (8.4)], show that an electron volt equals 1.6×10^{-19} joules.

16-A4. Distinguish between fission and fusion.

16-A5. List the various things that can happen to a neutron in a nuclear reactor. Describe each phenomenon briefly.

16-A6. What is a "chain reaction"?

16-A7. Assume that a drop of water is given an electric charge equal to 2×10^{-7} coul and that the drop splits into two equal parts, each having a diameter of 4 mm (= 0.004 m). (*a*) How much energy is converted into kinetic energy as the drops move apart? (*b*) Since the mass of each drop will be approximately 3.3×10^{-5} kg, what speed will each ultimately attain?

See Appendix 8 for atomic masses needed but not given in the statement of the following problems.

16-B1. What must be the energy of a proton that is brought to rest by the Coulomb field of a gold nucleus ($^{197}_{79}Au$) at a distance such that the centers of the two particles are 2.7×10^{-14} m apart? (Convert your answer in joules to electron volts.)

16-B2. (*a*) How much energy in atomic mass units (amu) should be emitted in the nuclear reaction

$$\frac{3}{2}\text{He} + \frac{4}{2}\text{He} = \frac{7}{4}\text{Be} + \gamma \text{ (gamma ray)}$$

given that the mass of $\frac{7}{4}\text{Be}$ equals 7.01693 amu?

(*b*) What would be the wavelength of the gamma rays, assuming that all the energy goes into one photon in each reaction?

16-B3. What is the least energy of an alpha particle that could conceivably cause the nuclear transmutation of $\frac{14}{7}\text{N}$ into $\frac{17}{8}\text{O}$ [Eq. (15-4)]?

16-B4. An electron passing near a nucleus emits an x-ray photon by the process known as Bremsstrahlung (Sec. 13-8). If the electron had an energy of 10 MeV initially and emerged with an energy of 9 MeV, what was the frequency and the wavelength of the photon that was produced?

When $\frac{226}{88}\text{Ra}$ disintegrates, it emits an alpha particle. The total kinetic energy of the alpha particle and the resultant $\frac{222}{86}\text{Rn}$ nucleus equals 4.87 MeV ($= 7.79 \times 10^{-13}$ joules). What is the atomic mass of radon-222, given that the mass of radium-226 equals 226.0254 amu?

16-B6. What is the temperature of a neutron gas in which a typical neutron has an energy of 1 MeV?

16-B7. What is the energy in electron volts of a relativistic neutron, defined as a neutron with a velocity of 3×10^7 m/sec?

16-B8. The typical fission event in the splitting of uranium-238 results in the release of 1.98×10^8 eV of energy. In our calculation based on the liquid-drop model of the nucleus, we arrived at an estimate of 2.53×10^8 eV. Since this is approximately 25 percent too large, it is apparent that our guess of the internuclear separation at the instant of fission was in error. Considering the actual value of the energy released in nuclear fission, how far apart were the nuclei at the instant of fission?

16-B9. The binding energy of a nucleus is defined as the difference between the mass-energy of the nucleus and its constituent neutrons and protons. (*a*) Starting with this definition, compute the binding energy of a carbon nucleus. (*b*) What is the numerical value of the binding energy per nucleon in carbon-12?

16-B10. In Sec. 16-5 we computed the energy released in fission by assuming that the fragments move violently apart because they are both positively charged. Why, then, is energy not released in the splitting, say, of an alpha particle? [*Note:* You can put your discussion on a quantitative basis by computing the binding energy per nucleon (defined in Prob. 16-B9) for $\frac{235}{92}\text{U}$, $\frac{138}{54}\text{Xe}$, $\frac{4}{2}\text{He}$, and $\frac{2}{1}\text{H}$. Let the mass of $\frac{138}{54}\text{Xe}$ be 137.91381 amu.]

References

Physics texts that discuss most of the topics presented in this text at approximately the same level

PHILLIP W. ALLEY and ROBERT L. SELLS, *Physics for Nonscientists* (Allyn and Bacon, 1971).

KENNETH R. ATKINS, *Physics: Once Over Lightly* (John Wiley and Sons, 1972).

ARTHUR BEISER, *Basic Concepts of Physics* (Addison-Wesley Publishing Company, 1961).

GEORGE GAMOW, *Matter, Earth, and Sky* (Prentice-Hall, 1965).

ELISHA R. HUGGINS, *Physics I* (Benjamin, 1968).

ROBERT KARPLUS, *Introductory Physics—A Model Approach* (Benjamin, 1969).

R. B. LINDSAY, *Basic Concepts of Physics* (Van Nostrand Reinhold Company, 1971).

ROBERT H. MARCH, *Physics for Poets* (Van Nostrand Reinhold Company, 1971).

PHYSICAL SCIENCE STUDY COMMITTEE, *College Physics* (Raytheon Education Company, 1968).

Physics texts that present a more explicit overview of the history of physics

ARNOLD B. ARONS, *Development of Concepts of Physics* (Addison-Wesley Publishing Company, 1965).

GEORGE GAMOW, *Biography of Physics* (Harper and Row, 1961).

GERALD HOLTON, *Introduction to Concepts and Theories in Physical Science* (Addison-Wesley Publishing Company, 1973).

GUY C. OMER, JR., H. L. KNOWLES, B. L. MUNDY, and W. H. YOHO, *Physical Science: Men and Concepts* (D. C. Heath and Company, 1962).

F. JAMES RUTHERFORD, GERALD HOLTON, and FLETCHER G. WATSON, *The Physics Project Course* (Holt, Rinehart and Winston, 1970).

**Texts that may present a more
comprehensive statement of
certain topics in modern physics**

LEON N. COOPER, *An Introduction to the Meaning and Structure of Physics* (Harper and Row, 1968).

KENNETH W. FORD, *The World of Elementary Particles* (Blaisdell Publishing Company, 1963).

M. GARDNER, *Relativity for the Million* (The Macmillan Company, 1962).

BANESH HOFFMAN, *Strange Story of the Quantum* (Dover, 1959).

DONALD J. HUGHES, *The Neutron Story* (Doubleday Anchor, 1963).

RALPH E. LAPP and HOWARD L. ANDREWS, *Nuclear Radiation Physics* (Prentice-Hall, 1972).

ALFRED ROMER, *The Restless Atom* (Doubleday and Company, Inc., 1960).

Appendices

APPENDIX 1 TABLES OF EQUIVALENTS

Metric units of mass and length and their British equivalents

MASS

$$1 \text{ kilogram} = 1000 \text{ grams} = 1,000,000 \text{ milligrams}$$

$$1 \text{ gram} = \frac{1}{1000} \text{ kilograms} = 1000 \text{ milligrams}$$

$$1 \text{ microgram} = \frac{1}{1,000,000} \text{ grams}$$

LENGTH

$$1 \text{ kilometer} = 1000 \text{ meters} = 100,000 \text{ centimeters}$$

$$1 \text{ meter} = 100 \text{ centimeters}$$

$$1 \text{ centimeter} = 10 \text{ millimeters} = \frac{1}{100} \text{ meter}$$

BRITISH EQUIVALENTS

1 foot = 30.5 centimeters = 0.305 meter
1 inch = 2.54 centimeters = 0.0254 meter
1 pound = 454 grams = 0.454 kilogram
1 meter = 39.37 inches = 3.28 feet
1 kilogram weighs 2.20 pounds

Miscellaneous derived metric units and their British equivalents

AREA

$1 \text{ cm}^2 = 100 \text{ mm}^2 = 0.0001 \text{ m}^2$
$1 \text{ ft}^2 = 144 \text{ in.}^2$
$1 \text{ in.}^2 = 6.45 \text{ cm}^2$

VOLUME

$1 \text{ m}^3 = 1,000,000 \text{ cm}^3$
$1 \text{ liter} = 1000 \text{ cm}^3$
$1 \text{ ft}^3 = 1728 \text{ in.}^3$

$1 \text{ gallon} = 231 \text{ in.}^3$
$1 \text{ ft}^2 = 28.32 \text{ liters}$
$1 \text{ liter} = 1.06 \text{ quarts}$

DENSITY

$$1 \frac{\text{gm}}{\text{cm}^3} = 62.4 \frac{\text{lb}}{\text{ft}^3} = 10^3 \frac{\text{kg}}{\text{m}^3}$$

ANGLE

1 revolution = 360 degrees = 2π radians
1 radian = 57.3 degrees

APPENDIX 2 EXPONENTIALS

1. Physics frequently deals with very small and very large quantities. In order to facilitate writing such quantities and handling them in equations, these quantities will be expressed in exponential form. For illustrative purposes, we list below several quantities written in exponential form.

$$12400 = 1.24 \times 10^4$$

$$0.000031 = 3.1 \times 10^{-5} = 31 \times 10^{-6}$$

$$510600 = 5.106 \times 10^5 = 0.5106 \times 10^6$$

2. The chief advantage of writing numbers in exponential form is found in multiplying or dividing small or large quantities. Let us recall a basic principle of algebra that governs such operations. When we multiply a^m and a^n, we find

$$a^m \times a^n = a^{m+n}$$

Similarly, in division of these quantities,

$$\frac{a^m}{a^n} = a^{m-n}$$

In all our applications of these rules, a will be equal to 10.

3. When applied to the multiplication and/or division of a few numbers chosen at random, we secure the results shown in the following examples:

$$10^4 \times 10^3 = 10^7$$

$$10^7 \times 10^{-5} = 10^2$$

$$2.1 \times 10^{-16} \times 3 \times 10^9 = 6.3 \times 10^{-7}$$

$$\frac{1.59 \times 10^9}{2.48 \times 10^3} = \frac{1.59}{2.48} \times 10^{9-3} = 0.642 \times 10^6$$

4. In extracting the square root of an exponential, the quantity should be written in the form of a number times 10 to an *even* power. The answer will then be given by a number times 10 to an integral power. For example,

$$\sqrt{144 \times 10^7} = \sqrt{14.4 \times 10^8} = 3.79 \times 10^4$$

APPENDIX 3   THE SOLAR SYSTEM

BODY	MASS EARTH = 1	SEMIMAJOR AXIS (m)	DIAMETER (m)	SIDEREAL PERIOD (days)	ACCELERATION OF GRAVITY (m/sec^2)
Sun	329390.	—	1.3906×10^9	—	274.4
Mercury	0.0549	0.058×10^{12}	0.514×10^7	87.97	3.92
Venus	0.8073	0.108×10^{12}	1.262×10^7	244.70	8.82
Earth	1.0000	0.149×10^{12}	1.2756×10^7	365.26	9.80
Mars	0.1065	0.228×10^{12}	0.686×10^7	686.98	3.92
Jupiter	314.5	0.778×10^{12}	14.360×10^7	4332.59	26.46
Saturn	94.07	1.426×10^{12}	12.060×10^7	10759.20	11.76
Uranus	14.40	2.868×10^{12}	5.34×10^7	30685.93	9.80
Neptune	16.72	4.494×10^{12}	4.97×10^7	60187.64	9.80
Pluto	<0.1	5.90×10^{12}	$\cdot 0.58 \times 10^7$	90885.	<5
Moon	0.01228	3.8×10^8	0.3476×10^7	27.32	1.67

Equatorial radius of the Earth $= 6.378 \times 10^6$ m

Mean distance to the Sun $= 1.495 \times 10^{11}$ m

Mean distance to the Moon $= 3.844 \times 10^8$ m

Mass of the Earth $= 5.983 \times 10^{24}$ kg

Mass of the Moon $= 7.347 \times 10^{22}$ kg

Mass of the Sun $= 1.970 \times 10^{30}$ kg

APPENDIX 4 PERIODIC TABLE OF THE ELEMENTS

PERIOD	I	II	III	IV	V	VI	VII	VIII			O
1	1H 1.00797										2He 4.0026
2	3Li 6.939	4Be 9.0122	5B 10.811	6C 12.01115	7N 14.0067	8O 15.9994	9F 18.9984				10Ne 20.183
3	11Na 22.9898	12Mg 24.312	13Al 26.9815	14Si 28.086	15P 30.9738	16S 32.064	17Cl 35.453				18A 39.948
4	19K 39.102	20Ca 40.08	21Sc 44.956	22Ti 47.90	23V 50.942	24Cr 51.996	25Mn 54.9380	26Fe 55.847	27Co 58.9332	28Ni 58.71	
4	29Cu 63.54	30Zn 65.37	31Ga 69.72	32Ge 72.59	33As 74.9216	34Se 78.96	35Br 79.909				36Kr 83.80
5	37Rb 85.47	38Sr 87.62	39Y 88.905	40Zr 91.22	41Nb 92.906	42Mo 95.94	43Tc [99]	44Ru 101.07	45Rh 102.905	46Pd 106.4	
5	47Ag 107.870	48Cd 112.40	49In 114.82	50Sn 118.69	51Sb 121.75	52Te 127.60	53I 126.9044				54Xe 131.30
6	55Cs 132.905	56Ba 137.34	57–71 Lanthanide series*	72Hf 178.49	73Ta 180.948	74W 183.85	75Re 186.2	76Os 190.2	77Ir 192.2	78Pt 195.09	
6	79Au 196.967	80Hg 200.59	81Tl 204.37	82Pb 207.19	83Bi 208.980	84Po [210]	85At [210]				86Rn [222]
7	87Fr [223]	88Ra [226.05]	89–Actinide series†								

* Lanthanide series:

57La 138.91	58Ce 140.12	59Pr 140.907	60Nd 144.24	61Pm [147]	62Sm 150.35	63Eu 151.96	64Gd 157.25	65Tb 158.924	66Dy 162.50	67Ho 164.930	68Er 167.26	69Tm 168.934	70Yb 173.04	71Lu 174.97

† Actinide series:

89Ac [227]	90Th 232.038	91Pa 231.0359	92U 238.03	93Np 237.0480	94Pu 239.0522	95Am [243]	96Cm [247]	97Bk [249]	98Cf [249]	99Es [253]	100Fm [255]	101Md [256]	102No [253]	103Lw [252]

Note: Atomic masses are based on $^{12}_{6}C$ (= 12.0000 amu).

APPENDIX 5 USEFUL PHYSICAL CONSTANTS

NAME	SYMBOL	NUMERICAL VALUE[*]
Speed of light	c	2.997925×10^8 m/sec
Gravitational constant	G	6.670×10^{-11} N m²/kg²
Volume of Kmol of gas	v_0	22.4136 m³/kmol
Universal gas constant	R	8314.34 joules/(kmol)(°K)
Boltzmann's constant	k	1.38054×10^{-23} joules/°K
Mechanical equivalent of heat	J	4184 joules/kcal
Avogadro's number	N_0	6.02252×10^{26} molecules/kmol
Faraday constant	F	9.64870×10^7 coul/kmol
Coulomb constant	K	8.9877×10^9 N m²/coul²
Planck's constant	h	6.62559×10^{-34} joule-sec
Atomic mass unit ($\frac{1}{12}$ mass of $^{12}_{6}$C)	amu	1.66043×10^{-27} kg
Energy equivalent of 1 amu		931.478 MeV
Electronic charge	e	1.60210×10^{-19} coul
Electron rest mass	m_0	$\begin{cases} 9.10908 \times 10^{-31} \text{ kg} \\ 5.48597 \times 10^{-4} \text{ amu} \end{cases}$
Energy equivalent of electron rest mass	$m_0 c^2$	0.511006 MeV
Proton rest mass	m_p	$\begin{cases} 1.67252 \times 10^{-27} \text{ kg} \\ 1.007277 \text{ amu} \end{cases}$
Neutron rest mass	m_n	$\begin{cases} 1.67482 \times 10^{-27} \text{ kg} \\ 1.008665 \text{ amu} \end{cases}$

[*] When solving problems at the ends of chapters, the numerical value of each of these physical constants usually can be rounded off to three significant figures. Thus the entries in this table can be read $c = 3 \times 10^8$ m/sec, $G = 6.67 \times 10^{-11}$ N m²/kg², $R = 8310$ joules/(kmol)(°K), etc.

APPENDIX 6 TEMPERATURE AS A MEASUREMENT OF THE AVERAGE KINETIC ENERGY OF GAS MOLECULES

1. Consider a cubical box L meters on a side containing one (1) molecule traveling with a velocity v along the x axis and periodically making collisions with opposite interior faces A and B. Since, by Newton's Second Law, the average force exerted on face A equals $m\,(\Delta v/\Delta t)$, and letting $\Delta v = 2v$ (since the velocity changes in a collision from v to $-v$) and the time between collisions $\Delta t = 2L/v$, we have for the average force exerted on face A,

$$\overline{F} = \frac{mv^2}{L} \qquad \text{(for one molecule)}$$

2. If the same box contains a kmol of gas (N molecules) and if we assume that they may be treated as if one-third are moving along each of the three coordinate axes, the force on any of the faces becomes

$$\overline{F} = = \frac{N}{3}\frac{mv^2}{L} \qquad \text{(for } N \text{ molecules)}$$

The pressure exerted by the gas is then

$$p = \frac{\overline{F}}{L^2} = \frac{N}{3}\frac{mv^2}{L^3} = \frac{2}{3}\frac{N}{L^3}\left(\frac{1}{2}mv^2\right)$$

3. From the general gas law, we know that for a kmol of gas

$$pL^3 = RT$$

Therefore
$$RT = \tfrac{2}{3}N(\overline{KE})$$

where $\overline{KE}$ equals the average kinetic energy of one molecule.

4. In conclusion, then,

$$\overline{KE} = \left(\frac{3}{2}\right)\frac{R}{N}\,T \qquad\qquad (7\text{-}3a)$$

or, alternately,
$$\overline{KE} = \frac{3}{2}kT \qquad\qquad (7\text{-}3b)$$

where $\quad k = \dfrac{R}{N} = \dfrac{8314.34}{6.02252 \times 10^{26}} = 1.38054 \times 10^{-23}$ joules/molecule °K

APPENDIX 7 CORRESPONDENCE BETWEEN GRAVITATIONAL AND ELECTRICAL FIELDS

DESCRIPTION	GRAVITATIONAL	ELECTRICAL
The force field	g (newtons/kg)	E (newtons/coul)
Force in a uniform field	mg newtons	QE newtons
Energy in a uniform field	mgh joules	QEd joules
Potential at a given location in a uniform field	gh joules/kg	$Ed \dfrac{\text{joules*}}{\text{coul}}$
Reference position	Any arbitrarily chosen but well-defined position	Any arbitrarily chosen but well-defined position
Force due to point mass or charge	$-\dfrac{Gmm'}{r^2}$ newtons	$+\dfrac{kQQ'}{r^2}$
Potential energy of two masses or two charges	$-\dfrac{Gmm'}{r}$ joules	$+\dfrac{kQQ'}{r}$ joules
Potential at r meters from a mass or a charge	$-\dfrac{Gm}{r} \dfrac{\text{joules}}{\text{kg}}$	$+\dfrac{kQ}{r} \dfrac{\text{joules*}}{\text{coul}}$
Reference position	$r = \infty$	$r = \infty$

* ≡ "volts."

MASSES OF CERTAIN PARTICLES
                 AND ISOTOPES

ATOMIC NUMBER	NAME	SYMBOL	MASS NUMBER	MASS* (amu)
−1	Electron	e	0	0.0005486
1	Proton	p	1	1.007277
0	Neutron	n	1	1.008665
1	Hydrogen	^1_1H	1	1.007825
1	Deuterium	^2_1H	2	2.01410
1	Tritium	^3_1H	3	3.01605
2	Helium-3	^3_2He	3	3.01603
2	Helium-4	^4_2He	4	4.00260
2	Helium-5	^5_2He	5	5.01230
3	Lithium-6	^6_3Li	6	6.01512
3	Lithium-7	^7_3Li	7	7.01600
6	Carbon-12	$^{12}_6\text{C}$	12	12.00000
6	Carbon-13	$^{13}_6\text{C}$	13	13.00335
6	Carbon-14	$^{14}_6\text{C}$	14	14.00324
7	Nitrogen-13	$^{13}_7\text{N}$	13	13.00574
7	Nitrogen-14	$^{14}_7\text{N}$	14	14.00307
7	Nitrogen-15	$^{15}_7\text{N}$	15	15.00011
7	Nitrogen-16	$^{16}_7\text{N}$	16	16.00610
8	Oxygen-15	$^{15}_8\text{O}$	15	15.00308
8	Oxygen-16	$^{16}_8\text{O}$	16	15.99492
8	Oxygen-17	$^{17}_8\text{O}$	17	16.99913
8	Oxygen-18	$^{18}_8\text{O}$	18	17.99916
17	Chlorine-35	$^{35}_{17}\text{Cl}$	35	34.96885
17	Chlorine-37	$^{37}_{17}\text{Cl}$	37	36.96590
45	Rhodium-103	$^{103}_{45}\text{Rh}$	103	102.9055
47	Silver-107	$^{107}_{47}\text{Ag}$	107	106.9051
47	Silver-109	$^{109}_{47}\text{Ag}$	109	108.9048
54	Xenon-134	$^{134}_{54}\text{Xe}$	134	133.9054
74	Tungsten-184	$^{184}_{74}\text{W}$	184	183.9510
86	Radon-222	$^{222}_{86}\text{Rn}$	222	222.0175
88	Radium-226	$^{226}_{88}\text{Ra}$	226	226.0254
90	Thorium-234	$^{234}_{90}\text{Th}$	234	234.0436
92	Uranium-234	$^{234}_{92}\text{U}$	234	234.0409
92	Uranium-235	$^{235}_{92}\text{U}$	235	235.0439
92	Uranium-238	$^{238}_{92}\text{U}$	238	238.0508

* These masses are all relative to the mass of the carbon-12 isotope, which is assigned a mass of precisely 12.00000. Isotopic masses include the mass of the electrons associated with the normal atom. Ref: *Nuclear Physics* **67**, 1 (1965).

Solutions
to
Problems

1. (*a*) 91.5 m; (*b*) 3360 kg.

3. (*a*) 2120 ft; (*b*) 6.99 m; (*c*) 5.40 ton; (*d*) 6.8 mi; (*e*) 0.35 kg.

5. Kg m/sec.

7. (*a*) 3.147×10^2; (*b*) 12.9×10^{-6}; (*c*) 9.3164×10^{10};
(*d*) 2.196×10^{10}; (*e*) 3.48×10^{-8}; (*f*) 5.490×10^1.

9. (*a*) 74.0×10^{-6}; (*b*) 6.77×10^6.

11. (*a*) 2.30×10^{-30} kg; (*b*) 1.67×10^{-27} kg for both.

Chapter 1

1-A3. 46.3 mi at 27.2° west of north.

1-A5. (*a*) 9.84×10^8 ft/sec; (*b*) 186,000 mi/sec.

1-B1. 9.6 m west of starting point.

1-B3. 8.1 m at 23° west of south.

1-B5. (*a*) 98 m/sec; (*b*) 49 m/sec; (*c*) 490 m.

1-B9. 1020 m.

1-B11. (*a*) 5 m/sec; (*b*) 2.5 m/sec^2.

2-A3. 9.8 m/sec^2.

2-B1. 3450 N.

2-B3. 687.0 N down; 684.6 N up; Net = 2.4 N down.

2-B5. Check point: At $r = 26{,}380$ m, F = 47 N.

3-A5. (*a*) 500 joule; (*b*) none.

3-B1. (*a*) 1600 N; (*b*) 1890 N (net force); (*c*) 2.51×10^6 joules; (*d*) none.

3-B3. (*a*) 3.6×10^{-17} joule; (*b*) 3.6×10^{-17} joule; (*c*) 8.9×10^8 m/sec.

3-B5. (*b*) 60×10^6 joule; (*c*) 59.5×10^6 joule.

3-B7. (*a*) 28.2×10^6 joule; (*b*) 2.38×10^3 m/sec.

3-B9. 194 m/sec.

4-A1. Same.

4-B1. 1.73×10^6 m.

4-B3. 7.27×10^{22} kg.

4-B7. (*a*) 20 m^2; (*b*) 72° change in direction.

5-B1. (*a*) 1.257 m/sec; 3.95 m/sec^2;   (*b*) 1.257 m/sec; 3.95 m/sec^2;
    (*c*) v_{max} at center; a_{max} at ends.
5-B3. (*a*) 10 N; (*b*) 6.67 m/sec^2; (*c*) none.
5-B5. (*a*) 0.6 joule; (*b*) 0.447 m/sec; (*c*) yes.
5-B7. 221 m.
5-B9. 11.5° (1st nodal line).

6-A3. (*a*) 2120 lb/ft^2; (*b*) 2120 lb of force; (*c*) equal force on other side.
6-B1. 1.393 × 10^5 N/m^2
6-B3. (*a*) 0.506 joule/°K; (*b*) 2.03 × 10^5 N/m^2.
6-B5. 6.28 × 10^4 joule.
6-B7. 4.8°C.

7-A3. 7500 cm^3 of hydrogen; 2500 cm^3 of nitrogen.
7-A5. (*a*) 4.0026 kg; 16.04303 kg; 26.0179 kg; 44.0100 kg; (*b*) 0.669
    × 10^{-26} kg; 2.66 × 10^{-26} kg; 4.32 × 10^{-26} kg; 7.31 × 10^{-26} kg.
7-B5. (*a*) 2.24 × 10^{16} m^3; (*b*) 0.00033 m or 0.033 cm.
7-B7. (*a*) 0.581 × 10^{-20} joule; (*b*) 399 m/sec; 642 m/sec.

8-A1. 2.30 × 10^{-8} N.
8-A3. 55,800 N/coul.
8-A7. 8 × 10^{-7} joule/sec.

8-B1. (*a*) 90,000 N/coul; (*b*) 1.44 × 10^{-17} joule/proton; (*c*) 90 volt.
8-B3. (*a*) 2.4 × 10^{-16} joule; (*b*) 0; (*c*) 2.4 × 10^{-16} joule.
8-B5. (*a*) 56.2 × 10^{19} atoms; (*b*) 28.1 × 10^{19} molecules;
    (*c*) 6.29 × 10^{26} molec/kmol.
8-B7. 9.06 × 10^{-31} kg.

9-A3. 1.25 wb/m^2.
9-A5. 0.00688 wb.
9-B1. (*a*) 1.92 × 10^{-13} N; (*b*) 11.5 × 10^{13} m/sec^2; (*c*) 0.0782 m.
9-B3. (*a*) 0.192 × 10^{-19} N; (*b*) 0.384 × 10^{-20} joule;
    (*c*) 0.024 volt; (*d*) no change.
9-B5. (*a*) 0.09 volt; (*b*) charges are displaced.
9-B7. (*a*) left ⟶ right; (*b*) right ⟶ left; (*c*) right ⟶ left.

10-A7. $Vq = \frac{1}{2}mv^2$.
10-B1. (*a*) 400 volt; (*b*) 12.8 × 10^{-16} N; 7.66 × 10^{11} m/sec^2.
10-B3. (*a*) 4.38 × 10^5 m/sec; (*b*) 0.25 m
10-B5. (*a*) 7.11 × 10^{-5} m; (*b*) 44.7 × 10^{-12} sec.
10-B7. (*a*) 4 × 10^5 m/sec; (*b*) no; (*c*) no.
10-B9. 4.03021 amu (model 1);   4.03188 amu (model 2);   4.00150 amu
    (helium nucleus).

11-A3. (*a*) 5.08×10^{14} vib/sec; (*b*) 3×10^{10} vib/sec; (*c*) 10^{18} vib/sec.
11-A5. (*a*) 7.5×10^{14} vib/sec; (*b*) same; (*c*) 0.30×10^{-6} m.
11-B1. 34.7 rev/sec.
11-B3. 1*st* max at 23.5° to right or left of central reinforcement.

Chapter 12

12-A1. Zn, Cu, gold.
12-A5. 1.87×10^{7} m/sec; 3.89×10^{-11} m.
12-B1. (*a*) 0.256×10^{-6} m; (*b*) 2.19×10^{-19} joule.
12-B3. (*a*) 3.77×10^{9} photons/sec; (*b*) 6.03×10^{-10} ampere.
12-B5. 941 volts.

13-A1. approximately 2×10^{8} atoms.
13-A3. (*a*) 33 m; (*b*) 132 m.
13-B1. 1.14×10^{-10} m.
13-B3. 0.656×10^{-6} m; 0.486×10^{-6} m; 0.434×10^{-6} m.
13-B5. (*a*) 2.19×10^{6} m/sec; (*b*) 3.32×10^{-10} m.
13-B7. 21,800 volts.

14-A3. 10.6 m.
14-A5. 1.15 m.
14-A7. 0.995 c.
14-B5. 14.2×10^{-6} sec.
14-B7. 3×10^{8} m/sec.

15-A1. (*a*) 1 throw; (*b*) 7 throws.
15-B1. $^{228}_{88}$Ra, $^{228}_{89}$Ac, $^{228}_{90}$Th, $^{224}_{88}$Ra, $^{220}_{86}$Rn, $^{216}_{84}$Po, $^{212}_{82}$Pb, $^{212}_{83}$Bi, $^{212}_{84}$Po, $^{208}_{82}$Pb.
15-B3. 85 m.
15-B5. (*a*) $^{0}_{-1}$e; (*b*) $^{2}_{1}$H; (*c*) $^{6}_{3}$Li.

16-A1. 0.816 m.
16-A7. (*a*) 0.0225 joule; (*b*) 26.1 m/sec.
16-B1. 6.74×10^{-13} joule or 4.21 MeV.
16-B3. 1.94×10^{-13} joule or 1.21 MeV.
16-B5. 222.0176 amu.
16-B7. 4.71×10^{6} eV.
16-B9. (*a*) 0.09894 amu or 92.1 MeV; (*b*) 0.00825 amu or 7.7 MeV.

Index

Energy (*cont.*)
from fission of nucleus, 319–320, 324, 327
from fusion, 328
gained by a charge, 177
gravitational potential, 73–74
kinetic and potential, 56–64, 284
molecular, 141
nuclear, 56, 288, 308–329
of photoelectrons, 231–233
relativistic kinetic, 284
state, 256
of the sun, 324–327
thermal, 56, 65, 66, 104–116, 235
thermal equivalent of mechanical, 116, 235
units in nuclear physics, 313–314
Escape kinetic energy, 85
Ether, 220, 267–269
eV, 313, 314
"Exact" science, 2, 55
Exothermic, 295
Explorer 34, 78
Explosion, 70
Exponentials, 8, 335

F

Fahrenheit temperature scale, 105, 107, 113
"Fallout" from atomic bombs, 320,
Faraday charge, 154–156, 162–164
Faraday constant, 155, 163
Faraday dark space, 158
Faraday's Law of Electromagnetic Induction, 178
Fermi, Enrico, 317, 321
Fictitious forces, 43–44, 50
Field:
electric, 185–187
gravitational, 48
magnetic, 168–174, 176, 179–181
First Law of Motion, 34–36, 44–45
First nuclear reactor, 321
Fission, nuclear, 304, 317–324, 330
bomb, 324
fragments, 319
Fissionable isotopes, 317
Fixed points in thermometry, 107
"Fixed" stars, 266
Force, 39–40
centripetal or central, 42–44
on charges and currents, 174–175
conservative, 64–65
Coulomb's law, 146
effective, 54–55
fictitious, 43–44
frictional, 44–45

lines of magnetic, 171
magnetic interaction, 172–175
nuclear and electric compared, 308–311, 329
unbalanced (net), 39–40
Force constant, 61
"Force-field" interpretation of *g*, 48–49, 50
Frames of reference, 35–37, 44, 266–267
Freely falling object, 22, 24
Frequency, 91, 98, 101, 188, 210, 221, 234
Friction, 34, 44–45, 64, 116
Fringe, 267
Frisch, Otto, 317
Fusion reaction, 327–329, 330

G

Galilean transformation of velocity, 280–281
Gamma rays, 223, 290–291, 305
Gas thermometer temperature scale, 110–112
Gases, 124–125
Boyle's law, 110
general gas law, 114, 135–137
kinetic theory of, 138–141, 339
Geiger, H., 251, 252
Geiger-Muller counter, 203–204
General gas constant, 135–137, 338
General gas law, 114, 135–137
Generating electric energy, 177
Geocentric solar model, 78
Geometric simple harmonic motion, 88–90, 100
Germer, L. H., 243
Gold-leaf electroscope, 145
Gravitation:
analogy with electricity, 166, 340
constant *G*, 47–48, 73, 235, 338
field intensity, 48, 50, 61–64, 340
Law of Universal Gravitation, 45–48, 50, 73–74, 79–81
lines of force, 49
potential, 64
potential energy, 62–64, 73–74
relationship of *G* to *g*, 81
Gravity, acceleration of, 24, 81, 336

H

"Half-life," 292–295, 305
Hahn, Otto, 317

Heat, mechanical equivalent of, 115–117
Heated cathode, 260
Heliocentric solar model, 78
Helium nuclei:
    alpha particles, 194, 290
    in fusion reaction, 324–327
    in nuclear reactions, 194, 290–293, 295–300, 309–314
    structure of, 258, 290
Hertz, 103
Hess, Victor F., 300
High-energy physics, 313–314
High-voltage machines, 194–199, 261, 296, 309–311, 329
Hydrogen:
    as bearer of unit charge, 155
    Bohr model of atom, 254–258, 264
    bomb, 312, 329
    fusion reactor, 325
    mass of, 134, 337, 341
    in periodic table, 258, 337
    spectrum of, 249–250, 256–257
    structure of hydrogen molecule, 130–135
Hydrogenlike elements, 259

I

Ice point, 112–113, 117, 118
Induced current, 176
Induced radioactivity, 311
Induced voltage, 175–179, 182
Ineffective component, 16, 55
Inelastic collision, 66, 70
Inertia, 35–38, 49, 82
Inertial frames, 35–37, 44, 49, 266–267
Inertial mass, 37–38, 50
Inertial system, 266
Instantaneous velocity and acceleration, 20
Intensity:
    of an electric field, 148–149, 165
    of a gravitational field, 61–63
    of a magnetic field, 171–175
Interactions, elastic and inelastic, 65-67, 298
Interference of waves, 98–101, 214–217
Interferometer, 267–268
Intermittent voltage, 156 n
Inverse square law, 29, 47–49, 61–64, 79–80, 146–147, 153–154
Ion gun, 185, 193, 197, 199
Ionization chamber, 200–201, 204
Ions, 155, 192–193
Ions, q/M of positive, 193–194
Isotopes, 127, 288–289, 317, 341
    discovery of, 311–312
    fissionable, 317

masses of the, 193–194, 341
naturally radioactive, 291–292
radioactive, 312

J

Joliot, Frederic, 296, 311
Joule, J. P., 54
Joules, 54, 56, 115–116, 313–314
Jupiter, 78, 336

K

K x-ray series, 262
Kelvin temperature, 113
Kelvin temperature scale, 112–114, 117
Kepler, Johann, 75
Kepler constant, 77, 78, 84
Kepler's laws, 46, 75–80, 83
Kilogram, 4, 38, 334
Kilogram calorie, 116–117
Kilogram molecular mass (kmol), 135, 141
Kinetic energy, 56–58, 70, 284
    escape, 85
    of gas molecules, 138–141, 339
    in theory of relativity, 284
Kinetic interpretation of temperature, 138–141, 339
Kinetic theory of gases, 138, 141, 339
kmol, 135–136, 141

L

Lamp, strobe, 22
Lavoisier, A. L., 125, 247
Lawrence, E. O., 196, 311
Laws:
    Boyle's, 109–110, 114, 117
    Combining Gas Volumes, 129–130, 140
    Conservation of Angular Momentum, 82
    Conservation of Energy, 115–117
    Conservation of Momentum, 67–70
    Definite Proportions, 129–130, 140
    Electromagnetic Induction (Faraday), 178
    Inertia, 35–37, 49
    Kepler's, 46, 75–80, 83
    Newton's laws of motion, 34–35, 40, 42, 49, 65
    Universal Gravitation, 45–48, 73, 79

Moving particle, wave nature of, 242–245
Multiplication factor in a nuclear reactor, 324
Muon, 286
Musical sounds, 86

N

Natural philosophy, 1
Natural radioactivity, 289–293
Negative carriers, 175
Negative charge, 144, 164
Negative force, 61, 87
Negative potential energy, 62–63
Neptune, 77, 336
Neptunium, 293, 323
Net force, 39
Neutrino, 291, 320, 329
Neutron:
    as a "building stone," 289
    capture, 318
    "delayed," 319, 323
    discovery of, 296–300, 306
    disintegration of, 315–316
    in fission reaction, 319
    mass of, 300, 341
    relativistic, 331
    slow or "thermal," 316, 322–323
Newton, 40–41
Newton, Isaac, 11
Newton-meters (joules), 54
Newton's Law of Universal Gravitation, 50, 79–81
Newton's laws of motion, 34-35, 40, 42, 49, 65
Nitrogen point, 118
Nodal lines, 99–100, 101, 217
Noise, 86
Nonconservative forces and systems, 64–65, 116–117
Noninertial frame, 44
Nonrelativistic particles, 195
North-seeking poles, 168
Nuclear:
    atom, 251–252
    bomb, 308, 319
    density, 304–305
    disintegration, probability in, 294
    energy, 56, 288, 308–329
    fission, 317–324, 330
    forces, 308–311, 329
    fragments in fission, 318–319
    mass, 288, 341
    model, 288–289, 303–305
    radius, 252–253, 289, 304, 306
    reactions, 311–316, 329
    reactor, 318–324
    transmutation, 296–300, 309, 311
Nuclei, liquid-drop properties of, 304–305

Nucleons, 308, 320, 327
Nucleus, hydrogen, 258 . *See also* Proton
Nucleus, size of the, 252–253, 304, 306
Numbers, pure, 7

O

Object, freely falling, 22, 24
Ocean waves, 93, 210 n
Oersted, H. C., 179
Oersted's discovery, 179–180
Oil-drop experiment, 160–162
Oliphant, M. L. E., 312
Operational definitions, 3, 104
"Orderliness of chance," 295

P

Pair production, 302–303 ,306
Paradox of the double-slit experiment, 239–242
"Parent" isotope, 292–293
Particle:
    accelerated, 195–196
    accelerators, 194, 204
    alpha, 290, 305, 309, 312
    behavior of, 207–209
    beta, 291–305
    detection of a moving charged, 200–203
    elimination, 288–289
    momentum of, 65, 242
    probability of finding, 244
    reflection and refraction of, 209
    relativistic, 195, 285
    wave nature of, 242–245
Particle nature of electrification, 154–156
Paschen Series, 250
Period:
    from cyclotron principle, 188
    of planets, 77, 336
    relation to frequency, 188
    in simple harmonic motion, 90
Periodic table, 137–138, 258–260, 337
Phase, 91, 102
Phases of matter, 124–125
Philosophy, natural, 1
Photoelectric effect, 229–235
Photon, 233–239, 244
    discovery, 229–235
    energy of, 234
    gamma ray, 291, 312
    mass of, 235
    momentum of, 235–237
    probability of finding, 241–242
    x rays, 261–264

Physical constants:
    masses of certain particles, 341
    metric (MKS) units, 334
    periodic table, 337
    solar system, 336
    useful selection of, 338
Physical second, 4 n
Physics, 1 (defined), 2, 13
Pictures of motion, 18–29, 37
Pitch, 86
Planck's constant:
    energy of photon, 234
    measurement of, 229–235
    momentum of photon, 237
    wavelength of particle, 242
Planets:
    Earth, 77-78, 81–82, 171, 293, 336
    other, 77–78, 336
Plum-pudding atomic model, 250–251
Pluto, 77, 336
Plutonium, 323
Polarized light, 221
Poles, N and S, 168
Position vs. time graph, 23, 24
Positive carriers, 175
Positive charges, 144, 164
Positive rays:
    discovery of, 192–193
    mass spectograph, 193
    ratio of charge to mass, 193–194
Positron (positive electron), 299–303, 306
Postulates of relativity, 269–270
Potential:
    defined, 152, 164
    gravitational, 64
    inverse square field, 153–154, 164, 340
    uniform field, 152–153, 340
Potential difference:
    defined, 151, 165
    gravitational, 64
    inverse square field, 154, 340
    uniform field, 152, 340
Potential energy:
    defined, 58–64
    gravitational, 62, 64
    negative, 62
    reference position, 59–62
    related to area, 60–63
    of a spring, 60–61, 70
Pound, 4
Pressure:
    defined, 109
    of gas, 110–114, 136–137, 339
    of light, 236–237
Primary cosmic rays, 300
Primary inertial frame, 35–36, 267
Primary standards, 3 n, 4, 4 n
Primitive-particle model for light, 207–208, 214, 229
Probability:
    of finding a particle, 244
    of finding a photon, 241–242

in nuclear disintegration, 294
in throwing dice, 293
Procedural definitions, 3, 104
Propagation of a pulse, 91
Propagation of waves, 91
Proper length, mass, and time, 271–274, 276, 283–284
Proton, 165, 182, 254, 300, 309, 341
Proton Compton effect, 298
Proton gun, 193, 199
Proton-proton chain, 326–327, 330
Ptolemy of Alexandria, 75, 77
Pure numbers, 7
pV as a thermometer property, 110–111

Q

q/M of positive ions, 193–194
Quanta, 244
Quantities, basic and derived, 3–7
Quantities, scalar and vector, 3–16, 30
Q-v-B Rule:
    applications, 176, 177, 187, 301
    statement of, 173

R

Radian, 7
Radio wave, 92, 223
Radioactive decay, 293–295
Radioactive isotopes, 291, 305, 312
Radioactivity, 288–306
    induced, 311
    natural, 289–293
Radius:
    of an atom, 248, 288
    of atomic nuclei, 253, 288, 304, 306, 309
    of Earth, 81
    of the planets, 336
    of a proton, 309
Ratio of charge to mass:
    of electrons, 158, 163, 182, 191–192
    of positive ions, 193–194
Reaction, 42
Reactor, nuclear, 318–324
Reference frame, 35–37, 44, 266–267
"Reference level" of potential energy, 59–62
Reference system, 36
References, 332–333
Reflection:
    diffuse, 213
    grating, 217
    of light, 213-214
    of a particle, 208–209
    specular, 213
    of a wave, 210

Thomson model of atom, 251
Thorium series, 293, 307
Time, 4, 271–276
    dilation, 274–276, 285
    measuring, 271–274
Trajectory in an electric field, 185–187
Trajectory in a magnetic field, 187–189
Transformation of velocities, 280–282
Transformations, Lorentz, 278–280
Transition, atomic, 256, 264
Transmission grating, 217
Transmutation, nuclear, 296–300, 306,
    309, 311
Transuranic elements, 317
Transverse length, 272
Transverse light clock, 274–278
Transverse waves, 93–94, 101, 221
Triple point, 113, 113 n
Tritium, 306, 312, 341
Trough of a wave, 91
Tunneling, 328
Tycho Brahe, 75–76

U

Ultraviolet light, 223, 231
Unbalanced force, 39–40
Uniform circular motion, 25–28, 33
Uniform electric field, 150, 185–187
Uniform magnetic field, 187–189
Uniformly accelerated motion, 24, 33,
    57–58
Unitless quantities, 7
Units, 334
    conversion of units, 5, 334
    MKS system of units, 4, 38
    Units Rule, 6–7, 21, 40
Unity, 5
Universal gravitation, 45–48, 50, 79–81
Universal gravitational constant, 47
Uranium-235, 317–321, 341
Uranium-238, 290, 314, 341
Uranium series, 292
Uranus, 78, 336

V

Vacuum, best obtainable, 142
Vacuum, velocity of light in, 211–212,
    269–270
Van de Graaff, R. J., 195
Van de Graaff accelerator, 195, 310,
    312
Vector quantities, 13–16

Velocity, 16–17, 20, 30
    of cathode particles, 190
    dependence of mass on, 282–283
    graph, vs. time, 19, 24
    of light, 211–213, 235
    selector, 190, 205
    speed and velocity compared, 17
    transformation of, 280–282
    of waves, 91–92, 101, 210, 221
Venus, 77, 336
Vibration, 22–23, 25, 86–90
Visible light, 231
Voltage:
    induced, 175–179, 182
    stopping, 232
Volume, 7

W

Walton, E. T. S., 311
Water waves, 93–94
Wave fronts, 95, 98
Wave generator, 94–95, 221, 223
Wave motion, 91–94
Wave nature of light, 207–225
Wave nature of a moving particle,
    242–245
Wavelength, 91–92, 97, 98, 209–210,
    221
    de Broglie, 244
    of light, 215–220
    of a particle, 242, 244–245
Waves:
    behavior of, 96–99, 209–211
    electromagnetic, 222–224, 242, 260,
        267
    frequency vs. wavelength, 92
    longitudinal and transverse, 93–94,
        100–101, 221
    matter, 244
    polarized, 221
    propagation of, 91–94
    reflection and refraction, 210
    variety of, 93
Waves and particles, 229–246
Wave theory of light, 100, 220
Webers, 178
Webers per square meter, 173, 181
Weight, 41, 50
Wilson, C. T. R., 160
Work, 53–55, 70
Work, thermal equivalent of, 116

X

X rays, 223, 260–263